T0192543

Essentials of Inorganic Chemistry

Essentials of Inorganic Chemistry

For Students of Pharmacy, Pharmaceutical Sciences and Medicinal Chemistry

KATJA A. STROHFELDT

School of Pharmacy, University of Reading, UK

WILEY

This edition first published 2015

Registered office
John Wiley & Sons Ltd, The Atrium, Southern Gate, Chichester, West Sussex, PO19 8SQ, United Kingdom

For details of our global editorial offices, for customer services and for information about how to apply for permission to reuse the copyright material in this book please see our website at www.wiley.com.

Library of Congress Cataloging-in-Publication Data

Strohfeldt, A. Katja.
Essentials of inorganic chemistry : for students of pharmacy, pharmaceutical sciences and medicinal chemistry / Dr Katja A. Strohfeldt
pages cm
Includes index.
ISBN 978-0-470-66558-9 (pbk.)
1. Chemistry, Inorganic–Textbooks. I. Title.
QD151.3.S77 2015
546 – dc23

2014023113

A catalogue record for this book is available from the British Library.

ISBN: 9780470665589

Cover image: Test tubes and medicine. Photo by Ugurhan.

Typeset in 10/12pt TimesLTStd by Laserwords Private Limited, Chennai, India

1 2015

To my dear Mum, who suddenly passed away before this book was finished, and my lovely husband Dave, who is the rock in my life and without whose support this book would have never been finished.

Contents

Preface

The aim of this book is to interest students from pharmacy, pharmaceutical sciences and related subjects to the area of inorganic chemistry. There are strong links between pharmacy/pharmaceutical sciences and inorganic chemistry as metal-based drugs are used in a variety of pharmaceutical applications ranging from anticancer drugs to antimicrobial eye drops.

The idea of this introductory-level book is to teach basic inorganic chemistry, including general chemical principles, organometallic chemistry and radiochemistry, by using pharmacy-relevant examples. Each chapter in this book is dedicated to one main group of elements or transition-metal group, and typically starts with a general introduction to the chemistry of this group followed by a range of pharmaceutical applications. Chemical principles are introduced with relevant pharmaceutical examples rather than as stand-alone concepts.

Chapter 1 gives an introduction to medicinal inorganic chemistry and provides an overview of the basic inorganic principles. The electronic structures of atoms and different bond formations are also discussed.

Chapter 2 is dedicated to alkali metals. Within this chapter, the basic chemistry of group 1 elements is discussed, together with the clinical use of selected examples. The reader is introduced to the clinical use of lithium salts in the treatment of bipolar disorder together with its historical development. In addition, the central role of sodium and potassium ions in many physiological functions is discussed within this chapter. Furthermore, the reader is introduced to a variety of chemical concepts, such as oxidation states, reduction and oxidation reactions, osmosis and others.

The chemistry of alkaline-earth metals and their clinical applications are the topic of Chapter 3. The potential biological role, clinical use and toxicity of a variety of examples are covered in this chapter. This includes issues relating to excessive beryllium uptake and the central physiological role magnesium and calcium play in the human body as well as the clinical use of barium salts and their potential toxicity.

After an introduction to the general chemistry of group 13 elements, the clinical uses of multivalent boron, aluminium and gallium are discussed in Chapter 4. The concept of metalloids is introduced, together with the general chemical behaviour of group 14 elements.

Chapter 5 concentrates on the general chemistry of group 14 elements and the clinical application of silicon- and germanium-based compounds. Silicon-based compounds are under discussion as novel drug alternatives to their carbon-based analogues. Germanium-based compounds have a very varied reputation for clinical use, ranging from food supplementation to proposed anticancer properties.

The biological role of phosphate and its clinical use together with potential drug interactions are discussed in Chapter 6. Furthermore, this chapter focusses on the long-standing research history of arsenic-based drugs. During the development of the most famous arsenic-based drug, Salvarsan, Ehrlich created the term the *Magic Bullet* – a drug that targets only the invader and not the host. This is seen as the start of chemotherapy.

Chapter 7 gives an overview of the area of transition-metal-based drugs with cisplatin being the most widely used example. In addition, developments in the area of iron and ruthenium-based compounds for clinical use are also discussed. Other topics include the clinical use of coinage metals and the biological role of zinc. The reader is introduced to a variety of concepts in connection to d-block metals including crystal field theory.

The concept of organometallic chemistry with a focus on d-block metals is introduced in Chapter 8. Clinical developments in the area of ferrocenes, titanocenes and vanadocenes are used as examples for current and future research.

In Chapter 9, the reader is introduced to f-block metals and their clinical applications. The topics discussed include the use of lanthanum carbonate as a phosphate binder, the use of gadolinium in MRI contrast agents and the potential use of cerium salts in wound healing.

Chapter 10 is dedicated to the concept of radioactivity. Topics such as radiopharmacy and its use in therapy and diagnostics are discussed. Clinical examples include the use of radioactive metals in therapy, for example, ^{131}I and ^{89}Sr, and in imaging, such as ^{99m}Tc, ^{67}Ga and ^{201}Tl. The final chapter of the book introduces the reader to the concept of chelation and its clinical application in the treatment of heavy-metal poisoning.

This book certainly does not aim to cover every clinical or preclinical example in the area of metal-based drugs. The chosen examples are carefully selected according to their relevance to the pedagogical approach used in this book. The idea is to introduce the reader to the main concepts of inorganic chemistry and reiterate those with pharmacy-relevant examples. For those who wish to study this area in more depth, there are excellent books available which are given under 'Further Reading' at the end of each chapter. I recommend any interested reader to have a look at these.

About the Companion Website

This book is accompanied by a companion website:

www.wiley.com/go/strohfeldt/essentials

The website includes:

- Answers to chapter exercises
- PowerPoint files of all figures from the book

The Periodic Table

0.98 — Pauling electronegativity 3 — Atomic number Li — Element 6.941 — Atomic weight (^{12}C)

1 2.20 **H** 1.008	2 **He** 4.003

Group: 3 – 12 ← d transition elements →

Group 1	Group 2	3	4	5	6	7	8	9	10	11	12	Group 13	Group 14	Group 15	Group 16	Group 17	Group 18
3 0.98 **Li** 6.941	4 1.57 **Be** 9.012											5 2.04 **B** 10.811	6 2.55 **C** 12.011	7 3.04 **N** 14.007	8 3.44 **O** 15.999	9 3.98 **F** 18.998	10 **Ne** 20.179
11 0.93 **Na** 22.990	12 1.31 **Mg** 24.305											13 1.61 **Al** 26.98	14 1.90 **Si** 28.086	15 2.19 **P** 30.974	16 2.58 **S** 32.054	17 3.16 **Cl** 35.453	18 **Ar** 39.948
19 0.82 **K** 39.102	20 1.00 **Ca** 40.08	21 **Sc** 44.956	22 **Ti** 47.90	23 **V** 50.941	24 **Cr** 51.996	25 **Mn** 54.938	26 **Fe** 55.847	27 **Co** 58.933	28 **Ni** 58.71	29 **Cu** 63.546	30 **Zn** 65.37	31 1.81 **Ga** 69.72	32 2.01 **Ge** 72.59	33 2.18 **As** 74.922	34 2.55 **Se** 78.96	35 2.96 **Br** 79.909	36 **Kr** 83.80
37 0.82 **Rb** 85.47	38 0.95 **Sr** 87.62	39 **Y** 88.906	40 **Zr** 91.22	41 **Nb** 92.906	42 **Mo** 95.94	43 **Tc** (99)	44 **Ru** 101.07	45 **Rh** 102.91	46 **Pd** 106.4	47 **Ag** 107.87	48 **Cd** 112.40	49 1.78 **In** 114.82	50 1.96 **Sn** 118.69	51 2.05 **Sb** 121.75	52 2.10 **Te** 127.60	53 2.66 **I** 126.90	54 **Xe** 131.30
55 0.79 **Cs** 132.91	56 0.89 **Ba** 137.34	57 **La** 138.91	72 **Hf** 178.49	73 **Ta** 180.95	74 **W** 183.85	75 **Re** 186.2	76 **Os** 190.2	77 **Ir** 192.22	78 **Pt** 195.09	79 **Au** 196.97	80 **Hg** 200.59	81 2.04 **Ti** 204.37	82 2.32 **Pb** 207.19	83 2.02 **Bi** 208.98	84 **Po** (210)	85 **At** (210)	86 **Rn** (222)
87 **Fr** (223)	88 **Ra** 226.025	89 **Ac** 227.0	104 **Rf** (261)	105 **Db** (262)	106 **Sg** (263)	107 **Bh**	108 **Hs**	109 **Mt**	110 **Uun**	111 **Uuu**	112 **Unb**						

58 **Ce** 140.12	59 **Pr** 140.91	60 **Nd** 144.24	61 **Pm** (147)	62 **Sm** 150.35	63 **Eu** 151.96	64 **Gd** 157.25	65 **Tb** 158.92	66 **Dy** 162.50	67 **Ho** 164.93	68 **Er** 167.26	69 **Tm** 168.93	70 **Yb** 173.04	71 **Lu** 174.97
90 **Th** 232.04	91 **Pa** (231)	92 **U** 238.03	93 **Np** (237)	94 **Pu** (242)	95 **Am** (243)	96 **Cm** (247)	97 **Bk** (247)	98 **Cf** (249)	99 **Es** (254)	100 **Fm** (253)	101 **Md** (253)	102 **No** (256)	103 **Lw** (260)

1
Introduction

Many metal ions play a vital role in living organisms. Metal ions are also involved in a variety of processes within the human body, such as the oxygen transport or the formation of the framework for our bones. Haemoglobin is an iron-containing metalloprotein which carries oxygen from the lungs to the various tissues around the human body. Calcium (Ca) ions are a vital component of our bones. Elements such as copper (Cu), zinc (Zn) and manganese (Mn) are essential for a variety of catalytic processes (Figure 1.1).

Nevertheless, metals are very often perceived as toxic elements. Very often, the toxicity of a metal in a biological environment depends on the concentration present in the living organism. Some metal ions are essential for life, but concentrations too high can be highly toxic whilst too low concentrations can lead to deficiency resulting in disturbed biological processes [2]. The so-called Bertrand diagram visualises the relationship between the physiological response and the metal concentration. There are concentration ranges that allow the optimum physiological response, whilst concentrations above and below this range are detrimental to life. The form of this diagram can vary widely depending on the metal, and there are metals with no optimum concentration range [3]. Nevertheless, living organisms, including the human body, have also found very sophisticated solutions to mask the toxicity of those metals (Figure 1.2).

Researchers have questioned whether metal ions can and should be introduced into the human body artificially and, if so, what the consequences are. Indeed, the use of metals and metal complexes for clinical applications gives access to a wide range of new treatment options.

1.1 Medicinal inorganic chemistry

Medicinal inorganic chemistry can be broadly defined as the area of research concerned with metal ions and metal complexes and their clinical applications. Medicinal inorganic chemistry is a relatively new research area grown from the discovery of the anticancer agent cisplatin. Indeed, the therapeutic value of metal ions has been known for hundreds and thousands of years. Metals such as arsenic have been used in clinical studies more than 100 years ago, whilst silver, gold and iron have been involved in 'magic cures' and other therapeutic applications for more than 5000 years.

Nowadays, the area of metal-based drugs spans a wide range of clinical applications including the use of transition metals as anticancer agents, a variety of diagnostic agents such as gadolinium or technetium,

Essentials of Inorganic Chemistry: For Students of Pharmacy, Pharmaceutical Sciences and Medicinal Chemistry,
First Edition. Katja A. Strohfeldt.

Companion website: www.wiley.com/go/strohfeldt/essentials

H																	He
Li	Be											B	C	N	O	F	Ne
Na	Mg											Al	Si	P	S	Cl	Ar
K	Ca	Sc	Ti	V	Cr	Mn	Fe	Co	Ni	Cu	Zn	Ga	Ge	As	Se	Br	Kr
Rb	Sr	Y	Zr	Nb	Mo	Tc	Ru	Rh	Pd	Ag	Cd	In	Sn	Sb	Te	I	Xe
Cs	Ba	La-Lu	Hf	Ta	W	Re	Os	Ir	Pt	Au	Hg	Tl	Pb	Bi	Po	At	Rn
Fr	Ra	Ac-Lr	Rf	Db	Sg	Bh	Hs	Mt	Ds	Rg	Uub						

Figure 1.1 Periodic table of elements showing metals (grey), semimetals (light grey) and nonmetals (white). Elements believed to be essential for bacteria, plants and animals are highlighted [1] (Reproduced with permission from [1]. Copyright © 2013, Royal Society of Chemistry.)

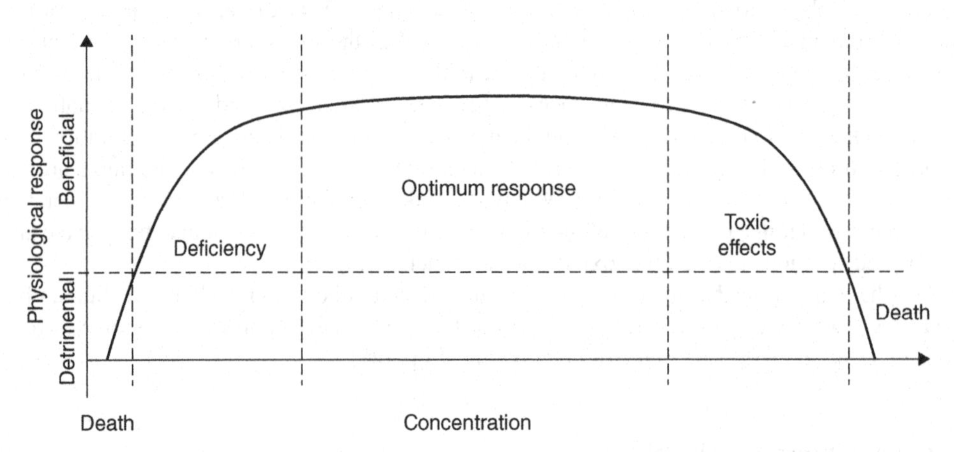

Figure 1.2 Bertrand diagram showing the relationship between the physiological response and metal concentration [4] (Reproduced with permission from [4]. Copyright © 1994, John Wiley & Sons, Ltd.)

lanthanum salts for the treatment of high phosphate levels and the use of gold compounds in the treatment of rheumatoid arthritis. In general, research areas include the development of metal-based therapeutic agents, the interaction of metals and proteins, metal chelation and general functions of metals in living systems [5].

1.1.1 Why use metal-based drugs?

Metal complexes exhibit unique properties, which, on one hand, allow metal ions to interact with biomolecules in a unique way and, on the other, allow scientists to safely administer even toxic metal ions to the human body. Coordination and redox behaviour, magnetic moments and radioactivity are the main unique properties displayed by metal centres together with the high aqueous solubility of their cations. The ability to be involved in reduction and oxidation reactions has led to the use of metal complexes in photodynamic therapy (PDT). In particular, transition metals are able to coordinate to electron-rich biomolecules such as DNA. This can lead to the deformation of DNA and ultimately to cell death. Therefore, transition metals are under scrutiny as potential anticancer agents. Metals that display a magnetic moment can be used as imaging reagents in magnetic resonance imaging (MRI). Many metals have radioactive isotopes, which can be used as so-called radiopharmaceuticals for therapy and imaging.

There is a huge array of clinical applications for most elements found in the periodic table of elements. This book tries to give an idea of the core concepts and elements routinely used for therapy or imaging.

1.2 Basic inorganic principles

It is important to understand the basic inorganic principles in order to evaluate the full potential of inorganic compounds in clinical applications. In the following sections, aspects such as atomic structures, chemical bonds and the set-up of the periodic table will be discussed.

1.2.1 Electronic structures of atoms

1.2.1.1 What is an atom

An atom is defined as the smallest unit that retains the properties of an element. The most famous definition has been published by Dalton in his *Atomic Theory* [6]:

> *All matter is composed of atoms and these cannot be made or destroyed. All* ***atoms*** *of the same element are identical and different elements have different types of atoms.* ***Chemical reactions*** *occur when atoms are rearranged* [7].

After Dalton's time, research showed that atoms actually can be broken into smaller particles, and with the help of nuclear processes it is even possible to transform atoms. Nevertheless, these processes are not necessarily considered as chemical processes. Probably, a better definition is that atoms are units that cannot be created, destroyed or transformed into other atoms in a chemical reaction [8].

Atoms consist of three fundamental types of particles: protons, electrons and neutrons. Neutrons and protons have approximately the same mass and, in contrast to this, the mass of an electron is negligible. A proton carries a positive charge, a neutron has no charge and an electron is negatively charged. An atom contains equal numbers of protons and electrons and therefore, overall, an atom has no charge. The nucleus of an atom contains protons and neutrons only, and therefore is positively charged. The electrons occupy the region of space around the nucleus. Therefore, most of the mass is concentrated within the nucleus.

Figure 1.3 shows the typical shorthand writing method for elements, which can also be found in most periodic tables of elements. Z (atomic number) represents the number of protons and also electrons, as an element has no charge. The letter A stands for the mass number, which represents the number of protons and neutrons in the nucleus. The number of neutrons can be determined by calculating the difference between the mass number (A) and the atomic number (Z).

Within an element, the atomic number (Z), that is, the number of protons and electrons, is always the same, but the number of neutrons and therefore the mass number (A) can vary. These possible versions of an element are called *isotopes*. Further discussion on radioisotopes and radioactivity can be found in Chapter 10.

$$^{A}_{Z}E$$

E = element symbol

A = number of protons + number of neutrons = mass number
Z = number of protons = number of electrons = atomic number

Figure 1.3 *Shorthand writing of element symbol*

$^{1}_{1}H$ Protium $^{2}_{1}H$ Deuterium $^{3}_{1}H$ Tritium

Figure 1.4 *Isotopes of hydrogen*

Atoms of the same element can have different numbers of neutrons; the different possible versions of each element are called ***isotopes****. The numbers of protons and electrons are the same for each isotope, as they define the element and its chemical behaviour.*

For example, the most common isotope of hydrogen called *protium* has no neutrons at all. There are two other hydrogen isotopes: deuterium, with one neutron, and tritium, with two neutrons (Figure 1.4).

1.2.1.2 Bohr model of atoms

In 1913, Niels Bohr published his atomic model stating that electrons can only circle the nucleus on fixed orbits in which the electron has a fixed angular momentum. Each of these orbits has a certain radius (i.e. distance from the nucleus), which is proportional to its energy. Electrons therefore can only change between the fixed energy levels (quantisation of energy), which can be seen as light emission. These fixed energy levels are defined as the principal quantum number n, which is the only quantum number introduced by the Bohr model of the atom. Note that, as the value of n increases, the electron is further away from the nucleus. The further away the electron is from the nucleus, the less tightly bound the electron is to the nucleus (Figure 1.5).

1.2.1.3 Wave mechanics

In 1924, Louis de Brogli argued that all moving particles, especially electrons, show a certain degree of wave-like behaviour. Therefore, he proposed the idea of wave-like nature of electrons, which became known as the *phenomenon of the wave–particle duality* [9].

Schrödinger published in 1926 the famous wave equation named after him. Electrons are described as wave functions rather than defined particles. Using this approach, it was possible to explain the unanswered questions from Bohr's model of the atom. Nevertheless, if an electron has a wave-like consistency, there are important and possibly difficult-to-understand consequences; it is not possible to determine the exact momentum and the exact position at the same moment in time. This is known as *Heisenberg's Uncertainty Principle.* In order to circumvent this problem, the probability of finding the electron in a given volume of space is used.

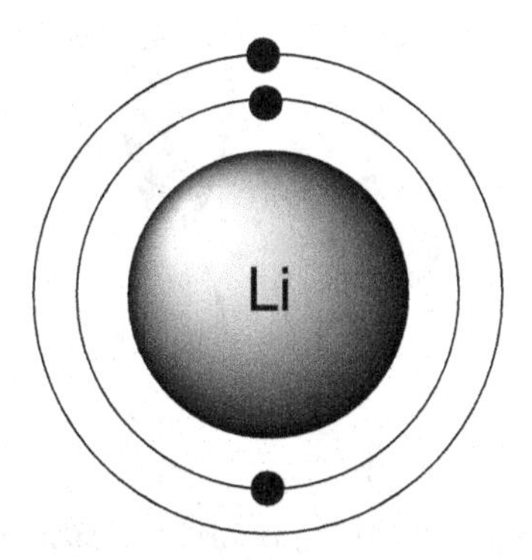

Figure 1.5 Bohr model of the atom

The Schrödinger wave equation delivers information about the wave function, and it can be solved either exactly or approximately. Only hydrogen-like atoms or ions, that is, the ones containing a nucleus and only one electron, can be exactly solved with the Schrödinger wave equation. For all other atoms or ions, the equation can be solved only approximately.

Solving the Schrödinger wave function gives us information about (i) the region or volume of space where the electron is most likely to be found, that is, where the probability of finding the electron is highest. This volume of space is called an *atomic orbital* (AO), which is defined by a wave function. (ii) *Energy values* associated with particular wave functions can be obtained by solving the Schrödinger wave equation. (iii) It can be shown that there is a *quantisation of energy levels*, similar to the observation described by Bohr.

1.2.1.4 Atomic orbitals

Each AO is defined by three so-called quantum numbers (n, l, m_l):

The principal quantum number n has already been introduced with the Bohr model of atoms. It can take values of $1 \leq n \leq \infty$, and is the result of the radial part of the wave function being solved.

> *Each atomic orbital is defined by a set of* ***three quantum numbers****: the principal quantum number (n), the orbital quantum number (l) and the magnetic quantum number (m_l).*

The quantum numbers l and m_l are obtained when the angular part of the wave function is solved. The quantum number l represents the shape of the AO. It is called the *orbital quantum number* as it represents the orbital angular momentum of the electron. It can have values of $l = 0, 1, 2, \ldots, (n-1)$, which correspond to the orbital labels s, p, d and f (see Figure 1.6).

The magnetic quantum number m_l provides information about the orientation (directionality) of the AO and can take values between $+l$ and $-l$. This means that there is only one direction for an s-orbital, as $l = 0$, and therefore m_l also is equal to 0. For a p-orbital, $l = 1$ and therefore m_l is -1, 0 or 1, which means it can occupy three orientations. In this case, they are classified as the p_x, p_y and p_z orbitals (see Figure 1.7). In the case of a d orbital ($l = 2$), the quantum number $m_l = -2, -1, 0, 1$ or 2. Therefore, there are five d orbitals with different orientations (see Figure 1.8).

The state of **each individual electron** can be described by an additional fourth quantum number, the so-called spin quantum number s (value of either $+1/2$ or $-1/2$). Each orbital can be filled with one or two

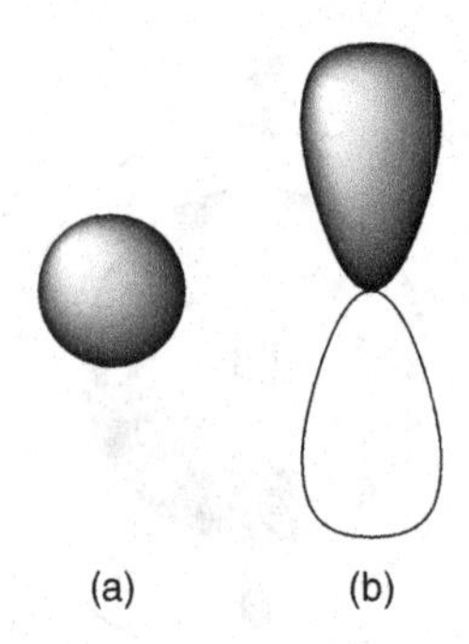

Figure 1.6 *Boundary surfaces of (a) s-orbital and (b) p-orbital*

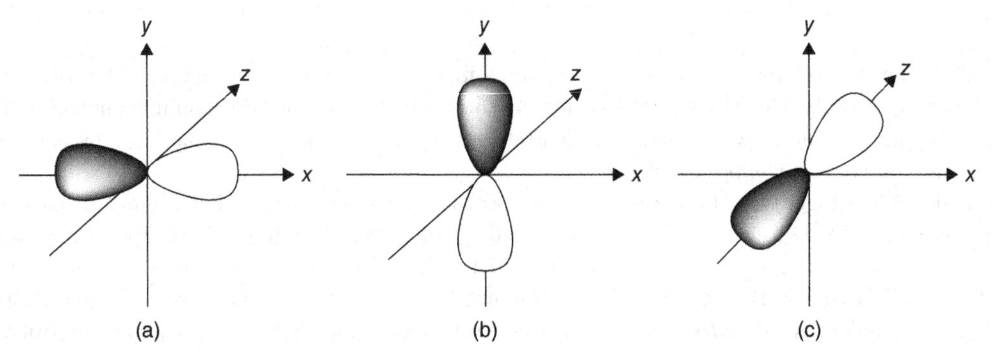

Figure 1.7 *Orientation of (a) p_x-orbital, (b) p_y-orbital and (c) p_z-orbital*

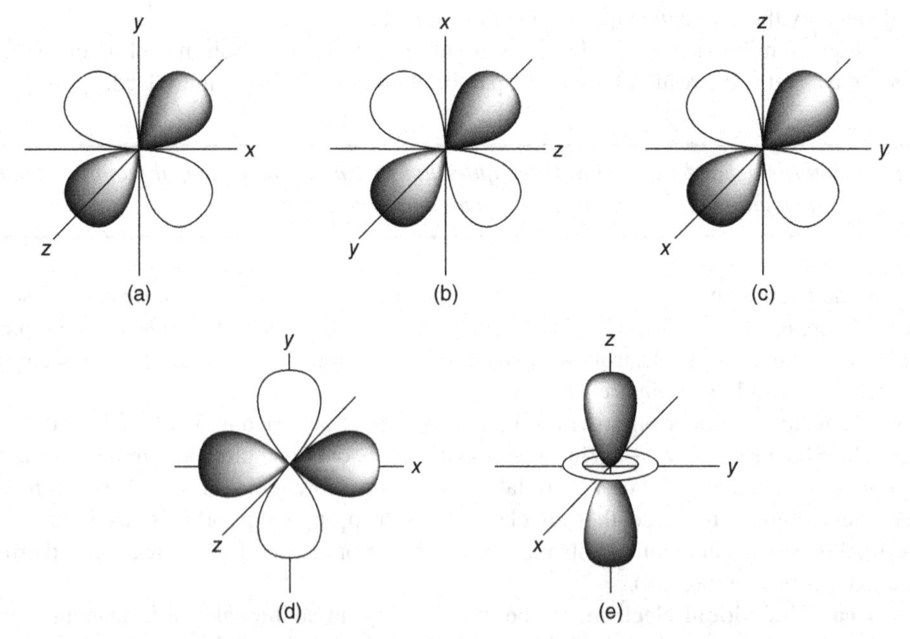

Figure 1.8 *Boundary surfaces of five d orbitals: (a) d_{xy}, (b) d_{xz}, (c) d_{yz}, (d) $d_{x^2-y^2}$ and (e) d_{z^2} [10] (Reproduced with permission from [10]. Copyright © 2009, John Wiley & Sons, Ltd.)*

electrons. Once an orbital is filled with two electrons, they will occupy opposite spin directions in order to fulfil the *Pauli Exclusion Principle* [8].

> *The* ***Pauli Exclusion Principle*** *states that no two electrons in the same atom can have the same values for their four quantum numbers.*

1.2.1.5 Electron configuration and Aufbau principle

Each element in the periodic table is characterised by a set of electrons, and their configuration can be described with the help of the quantum numbers that have been introduced in Section 1.2.1.4.

The electronic configuration is mostly used to describe the orbitals of an atom in its ground state and shows how these are distributed between the different orbitals. Nevertheless, this model can also be used to show valence electrons or ions.

> *A* ***valence electron*** *is defined as an electron that is part of an atom and can participate in the formation of a chemical bond. In main group elements, the valence electron is positioned in the outermost shell.*

The principle of electron configuration of an atom was already established with the Bohr model of atoms, and therefore very often the terms 'shell' and 'subshell' are used. In this nomenclature, the electron shell describes a set of electrons that occupy the same principal quantum number n. The respective electron shell n can be filled with $2n^2$ electrons. This means the first electron shell can be filled with a maximum of two electrons, as $n = 1$. Note that, according to the *Pauli Exclusion Principle*, these two electrons have an opposite spin direction (see Section 1.2.1.4). The second electron shell can accommodate up to eight electrons with $n = 2$, and so on. The subshells are defined by the quantum number l ($l = n - 1$), which as previously described correspond to the orbital labels s, p, d and f. The number of electrons that can be placed in each subshell can be determined by the following equation: $2(2l + 1)$. This allows two electrons to be placed in the s subshell ($l = 0$), six electrons in the p subshell ($l = 1$) and 10 electrons in the d subshell ($l = 2$). This information can be translated into the occupation of the corresponding orbitals. There are three p orbitals with differing orientations as defined by the quantum number m_l. Each of those can accommodate two electrons (which will occupy opposite spin directions as explained above). This means the three p orbitals can be filled with six electrons in total (Figure 1.9).

The so-called *Aufbau principle* (German for 'building-up principle') helps us to determine the electron configuration of an atom. It describes the hypothetical process of filling the orbitals of an atom with the given number of electrons. The first orbitals filled are the ones with the lowest energy levels before going onto the next higher energy level. This means the 1s orbital is filled before the 2s orbital. According to the so-called *Hund's Rule*, orbitals of the same energy level (such as p or d orbitals) are filled with one electron first before the electrons are paired within the same orbital. Pairing electrons requires additional energy, as the spin of the second electron has to be reversed in order obey the *Pauli Exclusion Principle*. The *Madelung Energy Ordering Rule* helps us to determine which orbitals are filled first. Orbitals are arranged by increasing energy, which means that order of occupation of the relevant orbitals is visualised by the arrow. Note that there are exceptions to this rule, which can be seen especially for electron configurations where d and/or f orbitals are occupied (Figure 1.10).

Scientists use the following standard notation to indicate the electronic configuration of an atom: Basically, the sequence of the occupied orbitals is written in order, with the number of electrons occupying each orbital as superscript number. In the simplest case hydrogen (H), which has only one electron occupying the 1s orbital,

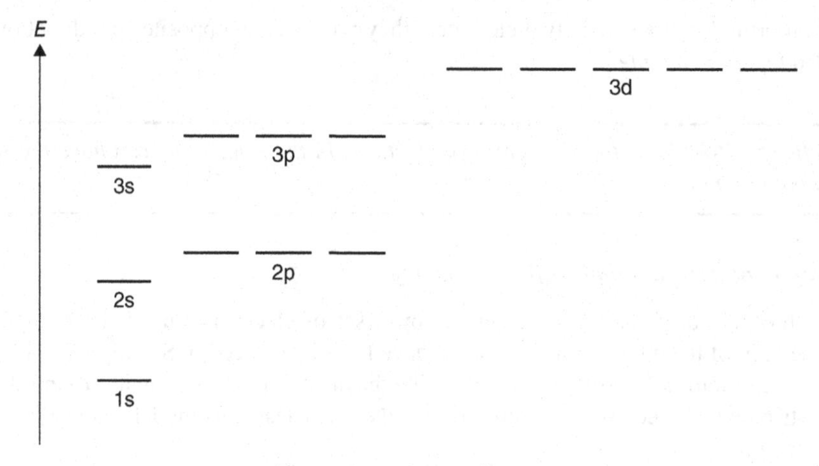

Figure 1.9 Energy diagram

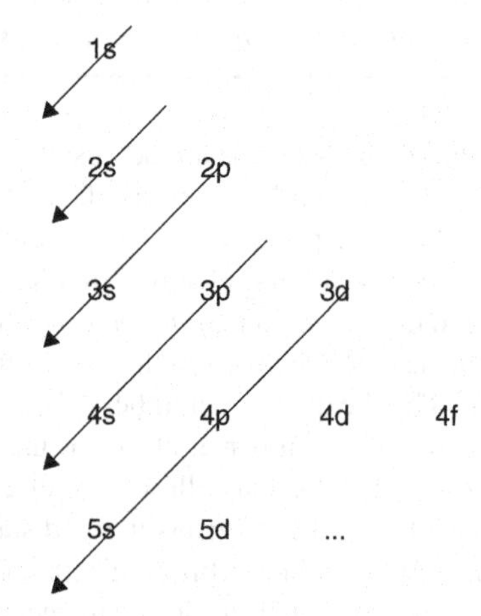

Figure 1.10 Madelung energy ordering rule

the electronic configuration is written as $1s^1$. Helium (He) has two electrons in the 1s orbital and therefore its electronic configuration can be denoted as $1s^2$. Lithium (Li) has three electrons in total, and its electron configuration is $1s^22s^1$. A more complicated example is nitrogen (N), which has seven electrons, and its electronic configuration can be written as $1s^12s^22p^3$. When more electrons are involved, the electronic configurations can get increasingly complicated. Therefore, lengthy notations can be shortened if some of the subshells are identical to those of noble gases. Sulfur (S) has the electronic configuration $\underline{1s^22s^22p^6}3s^23p^4$. The first three subshells (underlined) are identical to the electronic configuration of the noble gas neon ($1s^22s^22p^6$), which has all subshells completely filled, as this is characteristic for noble gases. The shortened notation for the electronic configuration of sulfur can therefore be written as [Ne]$3s^23p^4$, where [Ne] is the short form for the electron configuration of neon.

***Noble gas configuration** is the term used for the description of the electronic configuration of noble gases.*

This notation is also useful to identify the valence electrons of an atom, which are located in the outer shells. These electrons typically determine the chemical behaviour, as elements strive to achieve the noble gas configuration by gaining or removing electrons from these shells.

1.2.2 Bonds

1.2.2.1 Introduction

Historically, the formation of ionic species was seen as a result of the transfer of electrons between atoms. The result is the formation of an anion (negatively charged partner) and a cation (positively charged partner), which form a strong bond based on electrostatic attraction. The bonding situation in covalently bonded molecules was described as a sharing of valence electrons. The valence electrons involved belong to neither of the atoms involved completely. Nowadays, chemical bonding is mainly explained by the application of wave mechanics and described either by the valence bond (VB) or molecular orbital (MO) theory.

> *A* ***chemical bond*** *is defined as an attraction between atoms, which leads to the formation of chemical substances containing two or more atoms. The bond is a result of the electrostatic attraction between opposite charges, such as electrons or nuclei or dipole attraction.*

The forces behind this attraction are electrostatic forces of varying degrees, ranging from weak dipole interactions to strong attraction between opposite charges. Strong bonds are ionic bonds and covalent bonds, whilst weak chemical bonding can be observed where dipole interaction and hydrogen bonding is seen.

Lewis structures are used to simply describe how valence electrons are arranged in molecules and how they are involved in chemical bonds. Basically, dots are used to visualise the number of valence electrons, whereas the elemental symbol represents the nuclei. As a basic rule, electrons should be ruled. Paired electrons are sometimes also represented by a line, which can be interpreted as a single covalent bond. An element with a single electron represents a radical. Electron pairs not contributing to any bonds are called *lone pairs*. The Lewis structure of water can be found in Figure 1.11. Oxygen (O) has six valence electrons and hydrogen (H) has one valence electron each, adding up to a total of eight electrons around the O centre. Two electrons each are used for the formation of a covalent single bond between the O and the H nucleus, whilst O keeps two electron pairs as lone pairs.

Double and triple bonds can also be symbolised using Lewis structures. Using the Lewis structure nitrogen (N) as an example (see Figure 1.12), it can be seen that three pairs of electrons form the N—N bond, whilst each N centre keeps one electron pair as lone pair. Again, eight electrons in total directly surround each N centre – two from the lone pair and six from the triple bond.

Figure 1.11 *Lewis structure of water*

:N≡N:

Figure 1.12 *Lewis structure of nitrogen*

Each atom (at least of atomic numbers <20) tends to form molecules in this way so that it has eight electrons in its valence shell. This gives them the same electronic configuration as noble gases. This rule of thumb is often used for main group elements and is called the *Octet Rule*. Lewis structures are often used to visualise the valence electrons. Note that valence electrons used in bonds are counted twice, once for each bond partner.

The VSEPR (valence shell electron pair repulsion) rules are a set of rules used to predict the shape of a molecule. The basic principle is that valence electrons around the centre atom repel each other and therefore will form an arrangement in which they are situated furthest from each other. Also, lone pairs are included in this electrostatic repulsion. Double and triple bonds are seen the same as single bonds. The resulting geometry

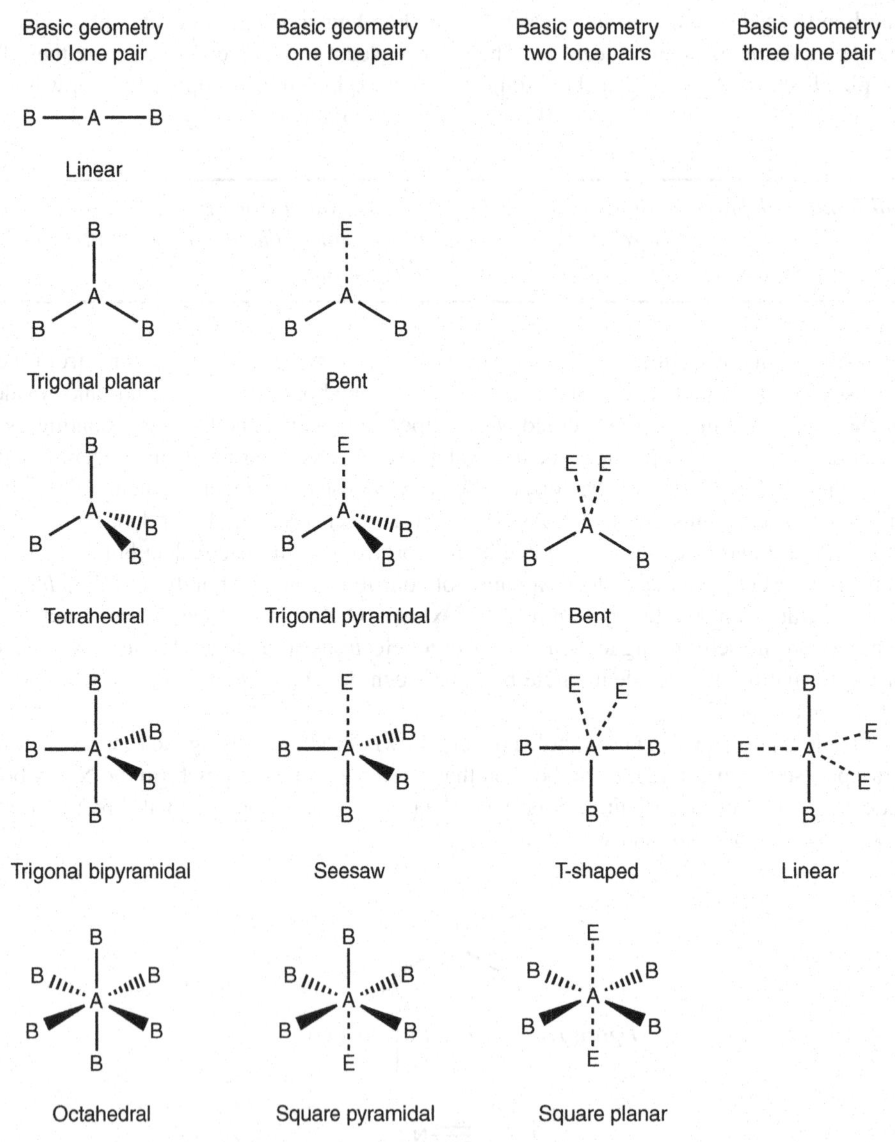

Figure 1.13 *Examples of geometries depending on number of ligands (L) and lone pairs (E) around the central atom (A)*

depends on the number of binding and nonbinding electron pairs, which can be summarised as the steric number. Figure 1.13 gives a selection of geometries found in typical main group molecules. Note that E denotes lone pairs, whereas B represents a covalent bond atom or group. Lone pairs occupy slightly more space than binding electron pairs, which explains the smaller angle in water of 104.9° rather than the standard tetrahedral angle of 108.9°.

1.2.2.2 Covalent bond

The simplest way to describe a covalent bond is based on the picture of sharing electron pairs between two atoms. Each atom would contribute electrons to this bond. In a dative covalent bond, one bond partner would donate all the electrons needed to form the bond. This type of binding is very important for biological actions of various elements.

*A **covalent bond** is defined as a chemical bond that is based on the sharing of electrons. Often, this leads to full outer shells for the binding partner to obtain the noble gas configuration.*

Within *homonuclear species* (chemical bond occurs between two atoms of the same element), the binding electron pair is evenly distributed between the two partners. In a *heteronuclear species* (chemical bond occurs between two atoms of different elements), the electrons are more attracted/polarised to one partner than the other, depending on the so-called electronegativity (EN).

***Electronegativity** describes the tendency of an atom to attract electrons or electron density towards itself.*

EN depends on the atomic number and the distance of the valence electrons from the nucleus. There are different scales for calculating the relative values, but the Pauling scale is the most commonly used one. Visualising EN trends on the periodic table shows an increase of EN within the row and a decrease within the group (Figure 1.14).

Using a more advanced approach, chemical bonding is nowadays mainly explained by the application of wave mechanics and described either by the VB or MO theory. The VB theory describes the formation of a covalent bond as the overlapping of two half-filled valence AOs from each binding partner, which contains one electron each. The simplest example is probably the molecule H_2. Hydrogen has only one valence electron ($1s^1$), and therefore two hydrogen 1s orbitals filled with one electron ($1s^1$), can overlap and form a chemical bond (Figure 1.15).

In a HF molecule, the 1s orbital of H and the $2p_z$ orbital of F, each filled with one unpaired electron, overlap and form a covalent bond (Figure 1.16).

MO theory approaches chemical binding from a more advanced point of view, where MOs are formed covering the whole molecule. The principal idea is that the AOs (as discussed in Section 1.2.1.4) of both binding partners are combined and form binding and nonbinding MOs. Those MO are filled with electrons contributed from both binding partners (Figure 1.17).

1.2.2.3 Ionic bond

Ionic bonds are strong bonds based on the transfer of electrons between the atoms and the resulting electrostatic attraction between the negatively and positively charged bond partners (Figure 1.18).

> *The term **ion** is used for atoms or molecules in which the total number of electrons is different from the number of protons and therefore carries a positive or negative charge. An **anion** is an atom or molecule that has a negative charge (has more electrons than protons). In contrast to this, a **cation** is defined as an atom or molecule with a positive charge, that is, it has more protons than electrons.*

H *2.1*																	He
Li *1.0*	**Be** *1.5*											**B** *2.0*	**C** *2.5*	**N** *3.0*	**O** *3.5*	**F** *4.0*	Ne
Na *0.9*	**Mg** *1.2*											**Al** *1.5*	**Si** *1.8*	**P** *2.1*	**S** *2.5*	**Cl** *3.0*	Ar
K *0.8*	**Ca** *1.0*	Sc	Ti	V	Cr	Mn	Fe	Co	Ni	Cu	Zn	**Ga** *1.6*	**Ge** *1.8*	**As** *2.0*	**Se** *2.4*	**Br** *2.8*	Kr
Rb *0.8*	**Sr** *1.0*	Y	Zr	Nb	Mo	Tc	Ru	Rh	Pd	Ag	Cd	**In** *1.7*	**Sn** *1.8*	**Sb** *1.9*	**Te** *2.1*	**I** *2.5*	Xe
Cs *0.7*	**Ba** *0.9*	La-Lu	Hf	Ta	W	Re	Os	Ir	Pt	Au	Hg	**Tl** *1.8*	**Pb** *1.8*	**Bi** *1.9*	**Po** *2.0*	**At** *2.2*	Rn
Fr *0.7*	**Ra** *0.9*	Ac-Lr	Rf	Db	Sg	Bh	Hs	Mt	Ds	Rg	Uub						

Figure 1.14 *Pauling's electronegativity values (italic) for the most common oxidation state of each element [8] (Reproduced with permission from [8]. Copyright © 1996, John Wiley & Sons, Ltd.)*

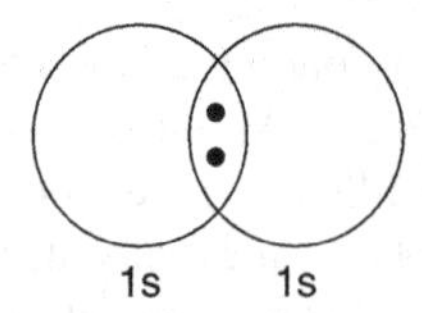

Figure 1.15 *VB model of H_2*

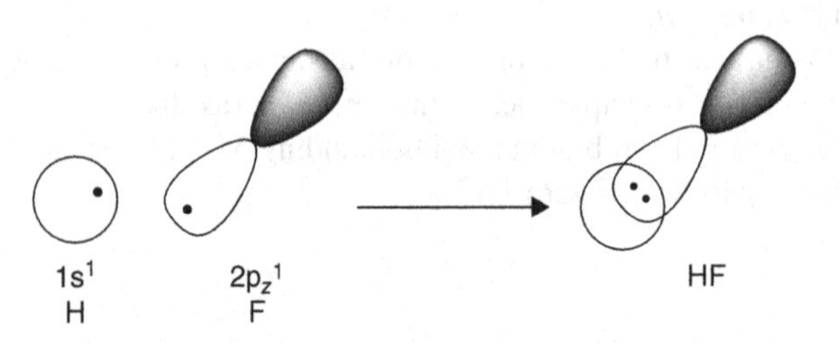

Figure 1.16 *VB model of HF*

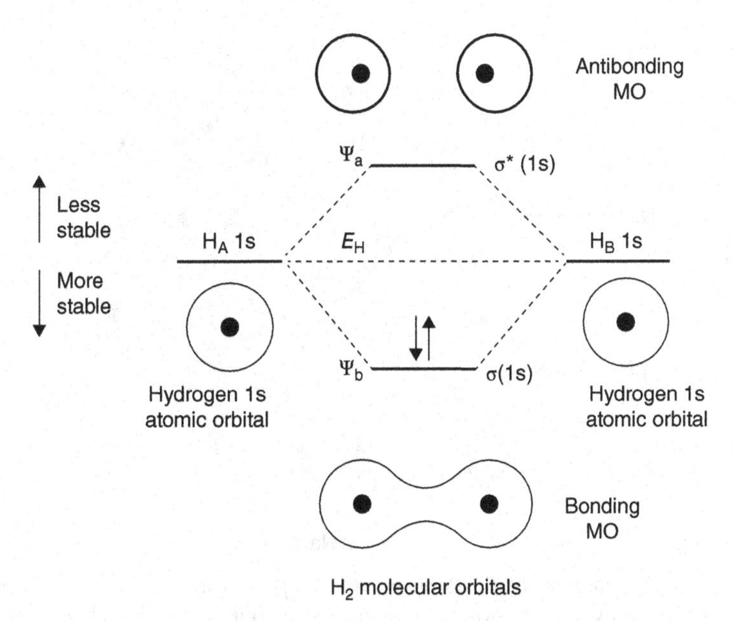

Figure 1.17 *Binding and nonbinding MOs in H_2 [10] (Reproduced with permission from [10]. Copyright © 2009, John Wiley & Sons, Ltd.)*

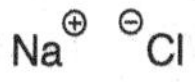

$Na^{\oplus}\ {}^{\ominus}Cl$

Figure 1.18 *NaCl as an example for an ionic species*

Ionic bonds are strong but short-range bonds and they have no defined direction in space, as they are 'only' the result of electrostatic attraction. Therefore, these attractions are not only limited to one directional partner each but also with each ion around them. As a result, a whole network of ions will be formed with anions and cations occupying specified spaces. The result is called a *salt*, which typically have high melting points (Figure 1.19).

1.2.2.4 Metallic bond

A metallic bond is most commonly described as a type of chemical bond where the metal atom donates its valence electrons to a 'pool' of electrons that surrounds the network of metal atoms. Electrons are not anymore identified with one particular atom but are seen as delocalised over a wide range. This is a very strong type of chemical bond. Electrical and thermal conductivity as well as malleability of metals can be explained using this model.

1.2.2.5 Intermolecular forces

After discussing intramolecular forces such as covalent bonding, it is also important to be aware of the interactions between molecules, the so-called intermolecular forces. These forces are much weaker than any of the types of binding discussed previously.

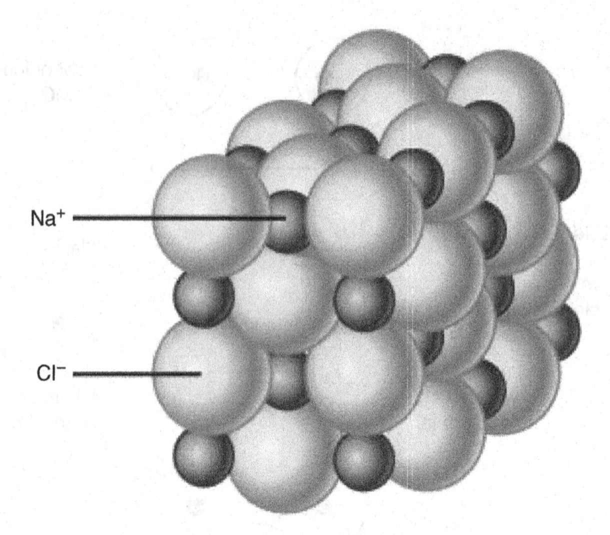

Figure 1.19 *Example of a salt structure (NaCl) Reproduced with permission from Gerard J. Tortora and Bryan H. Derrickson,* Principles of Anatomy and Physiology: Organization, Support and Movement, and Control Systems of the Human Body, 13th Edition International Student Version. *Copyright © 2011, John Wiley & Sons, Ltd*

$$\overset{\delta^+}{\mathrm{H}}-\overset{\delta^-}{\mathrm{Cl}}\cdots\overset{\delta^+}{\mathrm{H}}-\overset{\delta^-}{\mathrm{Cl}}$$

Figure 1.20 *Dipole–dipole interaction*

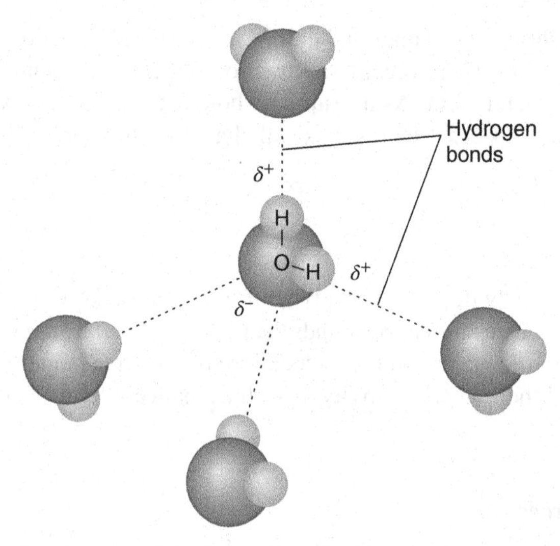

Figure 1.21 *Hydrogen bonding in water [11] (Reproduced with permission from [11]. Copyright © 2009, John Wiley & Sons, Ltd.)*

van der Waals forces are the weakest forces occurring between molecules. They can be found between molecules that do not have a permanent dipole. Electrons are not static but move around, and therefore moments occur when the electron distribution is not even, leading to the formation of temporary dipoles. These temporary dipoles cause a polarisation in the neighbouring molecules. As a result, molecules are (weakly) attracted to each other.

Dipole–dipole interaction is an electrostatic interaction of permanent dipoles. In heteronuclear molecules, a polarisation of the bond is caused by the difference in electronegativity between the two atoms forming the covalent bond. This leaves a partially positive charge (δ^+) on the less electronegative partner and, formally, a partially negative charge (δ^-) on the more electronegative partner. These molecules are attracted to each other as they can align in such way that opposite charges are next to each other, leading to intramolecular forces between the molecules (Figure 1.20).

Hydrogen bonding is the strongest of the intermolecular forces discussed here and seen as the particularly strong electrostatic interactions occurring between molecules of the type H—X. X is an electronegative atom such as F, O or N. O and N have also the advantage of possessing lone-pair orbitals. Hydrogen bonding can be seen as a strong and specialised form of dipole–dipole interaction and is the reason for the high boiling point of water (Figure 1.21).

1.3 Exercises

1.3.1 Write the electron configuration for the following elements or ions:

(a) F
(b) Cl
(c) Cl^-
(d) Na
(e) Na^+
(f) Si

1.3.2 Draw the energy diagram for the following elements or ions:

(a) Cl
(b) Cl^-
(c) Mg
(d) Mg^{2+}

1.3.3 Draw the Lewis structure for the following molecules:

(a) NH_3
(b) H_2O_2
(c) F_2
(d) O_2
(e) SiH_4
(f) BF_3
(g) N_2

1.3.4 Calculate the ΔEN for each bond within the following molecules:

(a) HCl
(b) HBr
(c) NH_3
(d) NaCl
(e) SiH_4

1.3.5 Use the VSEPR model to predict the shape of the following molecules.

(a) NH_3
(b) SiH_4
(c) PF_5
(d) SF_6

References

1. N. P. E. Barry, P. J. Sadler, Exploration of the medical periodic table: towards new targets. *Chem. Commun.* **2013**, *49*, 5106–5131.
2. K. H. Thompson, C. Orvig, Boon and bane of metal ions in medicine. *Science* **2003**, *300*, 936–939.
3. C. Orvig, M. J. Abrams, Medicinal inorganic chemistry: Introduction. *Chem. Rev.* **1999**, *99*, 2201–2203.
4. W. Kaim, B. Schwederski, *Bioinorganic chemistry: inorganic elements in the chemistry of life: an introduction and guide*, Wiley, Chichester, **1994**.
5. (a) Z. J. Guo, P. J. Sadler, Metals in medicine. *Angew. Chem. Int. Ed.* **1999**, *38*, 1513–1531; (b) A. M. Barrios, S. M. Cohen, M. H. Lim, *Chem. Commun.* **2013**, *49*, 5910–5911.
6. J. Dalton, *A new system of chemical philosophy*, Printed by S. Russell for R. Bickerstaff, Manchester, **1808**.
7. J. Dalton, *Atom theory: citation from lecture given at the royal institution in 1803*.
8. J. D. Lee, *Concise inorganic chemistry*, 5th ed., Chapman & Hall, London, **1996**.
9. R. M. Eisberg, R. Resnick, *Quantum physics of atoms, molecules, solids, nuclei, and particles*, Wiley, New York; Chichester, **1985**.
10. J. C. Dabrowiak, *Metals in medicine*, Wiley-Blackwell, Oxford, **2009**.
11. G. J. Tortora, B. Derrickson, *Principles of anatomy and physiology*, 12th ed., international student/Gerard J. Tortora, Bryan Derrickson. ed., Wiley [Chichester: John Wiley, distributor], Hoboken, N.J., **2009**.

Further Reading

E. Alessio, *Bioinorganic medicinal chemistry*, Wiley-VCH, Weinheim, **2011**.

H.-B. Kraatz, N. Metzler-Nolte, *Concepts and models in bioinorganic chemistry*, Wiley-VCH [Chichester: John Wiley, distributor], Weinheim, **2006**.

G. A. McKay, M. R. Walters, J. L. Reid, *Lecture notes. Clinical pharmacology and therapeutics*, 8th ed., Wiley-Blackwell, Chichester, **2010**.

R. M. Roat-Malone, *Bioinorganic chemistry: a short course*, Wiley, Hoboken, N.J. [Great Britain], **2002**.

E. R. Tiekink, M. Gielen, *Metallotherapeutic drugs and metal-based diagnostic agents: the use of metals in medicine*, Wiley, Chichester, **2005**.

2
Alkali Metals

Members of group 1 of the periodic table (first vertical column) with exception of hydrogen are called *alkali metals*. Under the term *alkali metals*, the following elements are included: lithium (Li), sodium (Na), potassium (K), rubidium (Rb), caesium (Cs) and francium (Fr). Generally, francium is not included in further discussions, as only artificial isotopes are known with ^{223}Fr having the longest half-life $T_{1/2} = 21.8$ min (Figure 2.1) [1].

In terms of a clinical use, sodium and potassium are essential ions for the human body and any imbalance in them has to be corrected. Lithium is medically used to treat bipolar disorder (BD), and the application of lithium salts is further discussed within this chapter.

2.1 Alkali metal ions

This group of elements belongs to the so-called s-block metals as they only have one electron in their outer shell, which is of s type. The chemistry of the metals is characterised by the loss of this s electron to form a monocationic ion M^+, which results from the relatively low ionisation energy of this electron (Table 2.1).

> *The term* ***ionisation energy*** *(IE) is defined as the energy that is required to remove the outer electron of an atom or molecule. The tendency to lose the outer electron is directly correlated to the ionisation energy – the lower the ionisation energy, the easier the removal of the electron.*
>
> *Within the group of alkali metals, the ionisation energy for the removal of the outer electron decreases as a result of the increasing distance of this electron from the nucleus.*

The loss of the outer s electron within the group of alkali metals results in the formation of the M^+ ion as mentioned. Consequently, most of the compounds of group 1 elements tend to be ionic in nature and form salts. In all pharmaceutical applications, only the salts of alkali metals are used, as most of the pure metals react violently with water.

Essentials of Inorganic Chemistry: For Students of Pharmacy, Pharmaceutical Sciences and Medicinal Chemistry,
First Edition. Katja A. Strohfeldt.

Companion website: www.wiley.com/go/strohfeldt/essentials

H																	He
Li	Be											B	C	N	O	F	Ne
Na	Mg											Al	Si	P	S	Cl	Ar
K	Ca	Sc	Ti	V	Cr	Mn	Fe	Co	Ni	Cu	Zn	Ga	Ge	As	Se	Br	Kr
Rb	Sr	Y	Zr	Nb	Mo	Tc	Ru	Rh	Pd	Ag	Cd	In	Sn	Sb	Te	I	Xe
Cs	Ba	La-Lu	Hf	Ta	W	Re	Os	Ir	Pt	Au	Hg	Tl	Pb	Bi	Po	At	Rn
Fr	Ra	Ac-Lr	Rf	Db	Sg	Bh	Hs	Mt	Ds	Rg	Uub						

Figure 2.1 *Periodic table of elements; group 1 metals are highlighted*

Table 2.1 *First and second ionisation energies (kJ/mol) for group 1 metals*

	First	Second
Li	520	7296
Na	496	4563
K	419	3069
Rb	403	2650
Cs	367	2420

Source: J. D. Lee, *Concise inorganic chemistry*, 5th ed., Chapman & Hall, London, **1996**.

2.1.1 Extraction of alkali metals: an introduction to redox chemistry

The main sources for Na and K are rock salts. Both elements do not naturally occur in their elemental state. In contrast, Li, Rb and Cs have small natural abundances, but also occur as rock salts. As previously mentioned, only artificial isotopes of Fr are known.

Sodium is manufactured by the so-called Downs' process, which is the electrolysis of molten NaCl and represents the major production process for sodium metal and is also a minor source of industrial chlorine. The process is based on a redox reaction (see Section 2.1.2) where the reduction of liquid Na^+ to liquid Na takes place at the cathode and the oxidation of liquid chloride (Cl^-) to chlorine (Cl_2) gas at the anode:

$$\text{Reduction at the cathode}: \quad Na^+_{(l)} + e^- \rightarrow Na_{(l)}$$

$$\text{Oxidation at the anode}: \quad 2Cl^-_{(l)} \rightarrow Cl_{2(g)} + 2e^-$$

$$\text{Overall redox reaction}: \quad 2Na^+_{(l)} + 2Cl^-_{(l)} \rightarrow 2Na_{(l)} + Cl_{2(g)}$$

Molten NaCl is used as the electrolyte medium within the electrolytic Downs' process and $CaCl_2$ is added in order to decrease the operating temperature. The melting point of NaCl is 800 °C, whereas the addition of $CaCl_2$ lowers it to around 600 °C. The design of the electrolysis cell is crucial in order to prevent oxidation and hydrolysis of freshly produced sodium and recombination to NaCl. Chlorine (Cl_2) is produced at the

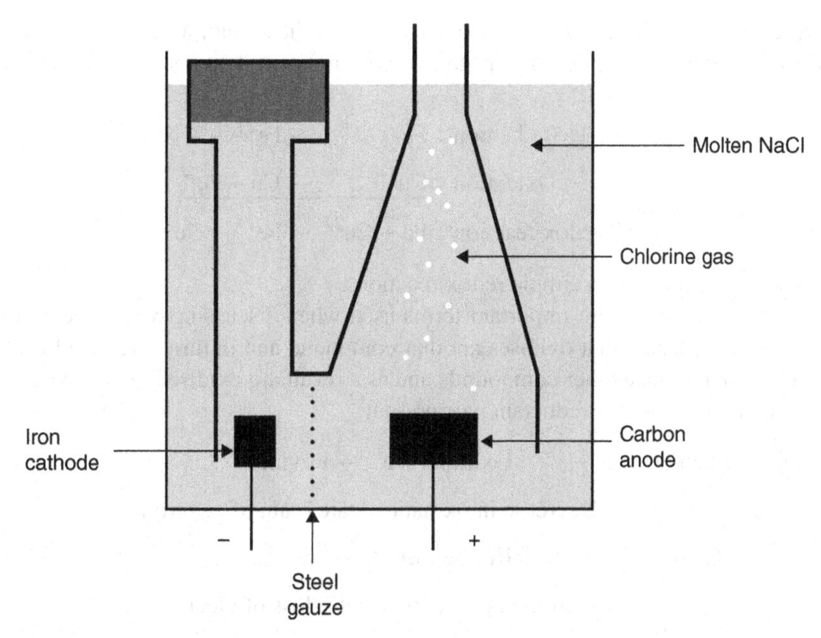

Figure 2.2 *Schematic diagram of Down's cell*

positive graphite anode and captured in form of gas, whereas Na^+ is reduced to liquid sodium at the negative iron cathode and collected in its liquid form.

The anode is defined as the electrode at which the oxidation takes place, which, means in this example, that the anode is positively charged. Conversely, the cathode is defined as the place where the reduction takes place and therefore it is negatively charged in electrolysis processes. Note that the same definition for electrodes applies in a galvanic cell (e.g. in batteries), but as a result of the electron flow the anode has a negative charge whereas the cathode is positively charged (Figure 2.2).

Lithium can be isolated in a similar electrolytic process using lithium chloride (LiCl). Spodumene ($LiAlSi_2O_6$) is the main source for LiCl, which is reacted to LiOH by heating with CaO and subsequent conversion to LiCl prior to the electrolysis. Potassium can also be obtained via electrolysis of KCl, but there are more efficient methods, for example, the reaction to a Na-K alloy which can be subsequently distilled.

2.1.2 Excursus: reduction–oxidation reactions

The term redox (short for reduction–oxidation) reaction describes all chemical reactions in which atoms change their oxidation number. Redox reactions have many applications, which can range from industrial processes (see Downs' process) to biological systems, for example, oxidation of glucose in the human body.

A redox reaction contains the two 'half-reactions', namely reduction and oxidation, which are always one set; this means that there is never a reduction reaction without an oxidation.

Redox reactions *are a family of reactions that consist of oxidation and reduction reactions. These are always a matched set, that is, there is no oxidation reaction without a reduction reaction. They are called half-reactions, as two half-reactions are needed to form a redox reaction. Oxidation describes the loss of electrons (e^-), whereas the term reduction is used for the gain of electrons (e^-).*

- *Reduction*: Describes the *gain* of electrons or the *decrease* in the oxidation state of an atom or molecule.
- *Oxidation*: Describes the *loss* of electrons or the *increase* in the oxidation state of an atom or molecule.

$$\text{Reduction}: \text{Fe} \rightarrow \text{Fe}^{2+} + 2\text{e}^-$$

$$\text{Oxidation}: \text{Cu}^{2+} \rightarrow \text{Cu} + 2\text{e}^-$$

$$\text{Overall redox reaction}: \text{Fe} + \text{Cu}^{2+} \rightarrow \text{Fe}^{2+} + \text{Cu} \quad (2.1)$$

Equation 2.1 gives an example for a simple redox reaction.

The terms *oxidant* and *reductant* are important terms used when discussing redox equations. An oxidant is defined as the atom or molecule that oxidises another compound and in turn is reduced itself. Reductants are molecules or atoms that reduce other compounds and as a result are oxidised themselves. Within a redox equation, there is always the oxidant–reductant pair present.

$$\text{Reduction}: \quad \textbf{Oxidant} + \text{e}^- \rightarrow \text{Product}$$

Decrease in oxidation state (gain of electrons)

$$\text{Oxidation}: \quad \textbf{Reductant} \rightarrow \text{Product} + \text{e}^-$$

Increase in oxidation state (loss of electrons) (2.2)

Equation 2.2 shows the involvement of an oxidant and a reductant in redox reactions.

2.1.2.1 The oxidation state

It is actually not precisely correct to describe oxidation/reduction reactions as loss/gain of electrons. Both reactions are better referred to the changes in oxidation state rather than to the actual transfer of electrons; there are reactions classified as redox reaction even though no transfer of electrons takes place.

*The **oxidation state** of an atom describes how many electrons it has lost whilst interacting with another atom compared to its original state (in its elemental form). The oxidation state is a hypothetical charge that an atom would have if all bonds within the molecule are seen as 100% ionic. When a bond is formed, it is possible for an atom to gain or lose electrons depending on its electronegativity. If an atom is electropositive, it is more likely to take electrons away from another atom, and vice versa.*

The oxidation number relates to the number of electrons that has been allocated to each atom; that is, if an atom has gained an electron, its oxidation state is reduced by 1 and therefore this atom has the oxidation state of –I.

IUPAC has defined the oxidation state as follows: A measure of the degree of oxidation of an atom in a substance. It is defined as the charge an atom might be imagined to have when electrons are counted according to an agreed-upon set of rules:

1. The oxidation state of a free element (uncombined element) is zero.
2. For a simple (monatomic) ion, the oxidation state is equal to the net charge on the ion.
3. Hydrogen has an oxidation state of 1 and oxygen has an oxidation state of −2 when they are present in most compounds (exceptions to this are that hydrogen has an oxidation state of −1 in hydrides of active metals, e.g. LiH, and oxygen has an oxidation state of −1 in peroxides, e.g. H_2O_2).

+I −II	+III −II
H_2O	NO_2^-
+I +VII −II	+I +IV −II
$KMnO_4$	H_2SO_3
+VI −II	−IV +I
$Cr_2O_7^{2-}$	CH_4
+III −II	+V −II
Fe_2O_3	BrO_3^-

Figure 2.3 *Example of oxidation states*

4. The algebraic sum of the oxidation states of all atoms in a neutral molecule must be zero, whilst in ions the algebraic sum of the oxidation states of the constituent atoms must be equal to the charge on the ion. For example, the oxidation states of sulfur in H_2S, S_8 (elementary sulfur), SO_2, SO_3 and H_2SO_4 are, respectively: −2, 0, +4, +6 and +6. 'The higher the oxidation state of a given atom, the greater its degree of oxidation; the lower the oxidation state, the greater its degree of reduction (Figure 2.3)' [2].

2.1.2.2 How to establish a redox equation?

In order to describe the overall redox reaction, the establishing and balancing of the half-reaction is essential. In aqueous reactions, generally electrons, H^+, OH^- and H_2O can be used to compensate for changes, whereas it has to be kept in mind whether the reaction takes place under acidic or basic conditions. There are five steps to follow in order to successfully establish a redox equation:

Example

Determine the redox equation for the reaction of MnO_4^- to Mn^{2+} and Ag to Ag^+ under acidic conditions.

1. Determine the oxidation state of each element involved (oxidation number is stated in Roman numbers):

$$\overset{+VII\ -II}{MnO_4^-} \rightarrow \overset{+II}{Mn^{2+}}$$

$$\overset{0}{Ag} \rightarrow \overset{+I}{Ag^+}$$

2. Add electrons to the equation; the number of electrons must reflect the changes in the oxidation state:

$$\overset{+VII\ -II}{MnO_4^-} + 5e^- \rightarrow \overset{II}{Mn^{2+}}$$

$$\overset{0}{Ag} \rightarrow \overset{+I}{Ag^+} + e^-$$

3. Determine which reaction is oxidation or reduction:

$$\text{Red.}: \overset{+VII\ -II}{MnO_4^-} + 5e^- \rightarrow \overset{II}{Mn^{2+}}$$

$$\text{Ox.}: \overset{0}{Ag} \rightarrow \overset{+I}{Ag^+} + e^-$$

4. Balance reduction and oxidation reaction by using either H^+/H_2O (acidic conditions) or OH^-/H_2O (basic conditions), depending on the reaction conditions, which can be acidic or basic (in this example: acidic conditions):

$$\text{Red.}: \overset{+VII\ -II}{MnO_4^-} + 5e^- + 8H^+ \rightarrow \overset{II}{Mn^{2+}} + 4H_2O$$

$$\text{Ox.}: \overset{0}{Ag} \rightarrow \overset{+I}{Ag^+} + e^-$$

5. Formulate the redox equation keeping in mind that the number of electrons has to be equal in the reduction and oxidation reactions. If necessary, the oxidation reaction has to be multiplied by the number of electrons in the reduction step, or vice versa.

$$\text{Red.}: \overset{+VII\ -II}{MnO_4^-} + 5e^- + 8H^+ \rightarrow \overset{II}{Mn^{2+}} + 4H_2O \quad / * 1$$

$$\text{Ox.}: \overset{0}{Ag} \rightarrow \overset{+I}{Ag^+} + e^- \quad / * 5$$

$$\text{Redox}: MnO_4^- + 5Ag + 8H^+ \rightarrow Mn^{2+} + 5Ag^+ + 4H_2O$$

Within the above example, the reaction conditions are stated as being acidic. A similar example using basic conditions is stated below and hydroxyl ions and water (OH^-/H_2O) are used in order to balance the half-reactions.

Example

Determine the redox equation for the reaction of MnO_4^- *to* MnO_2 *and* Fe^{2+} *to* Fe^{3+} *under basic conditions.*

1. Determine the oxidation state of each element involved:

$$\overset{+VII\ -II}{MnO_4^-} \rightarrow \overset{+IV}{MnO_2}$$

$$\overset{-I}{Br^-} \rightarrow \overset{+V}{BrO_3^-}$$

2. Add electrons to the equation; the number of electrons must reflect the changes in the oxidation state:

$$\overset{+VII\ -II}{MnO_4^-} + 3e^- \rightarrow \overset{+IV}{MnO_2}$$

$$\overset{-I}{Br^-} \rightarrow \overset{+V}{BrO_3^-} + 6e^-$$

3. Determine the redox and the oxidation reaction:

$$\text{Red.}: \overset{VII\ -II}{MnO_4^-} + 3e^- \rightarrow \overset{+IV}{MnO_2}$$

$$\text{Ox.}: \overset{-I}{Br^-} \rightarrow \overset{+V}{BrO_3^-} + 6e^-$$

4. Balance reduction and oxidation reaction by using OH^-/H_2O:

$$\text{Red.}: \overset{VII\ -II}{MnO_4^-} + 3e^- + 2H_2O \rightarrow \overset{+IV}{MnO_2} + 4OH^-$$

$$\text{Ox.}: \overset{-I}{Br^-} + 6OH^- \rightarrow \overset{+V}{BrO_3^-} + 6e^- + 3H_2O$$

5. Formulate the redox equation:

$$\begin{array}{lll} \text{Red.}: & \overset{+VII\ -II}{MnO_4^-} + 3e^- + 2H_2O & \rightarrow \overset{+IV}{MnO_2} + 4OH^- \ / * 2 \\ \text{Ox.}: & \overset{-I}{Br^-} + 6OH^- & \rightarrow \overset{+V}{BrO_3^-} + 6e^- + 3H_2O \ / * 1 \\ \hline \text{Redox}: & 2MnO_4^- + Br^- + H_2O & \rightarrow 2MnO_2 + BrO_3^- + 2OH^- \end{array}$$

2.1.2.3 *How to calculate the redox potential*

The spontaneous reaction of $Zn_{(s)}$ metal with $Cu^{2+}{}_{(aq)}$ from a Cu(II) solution is the prime example for a redox reaction.

$$Zn_{(s)} + Cu^{2+}{}_{(aq)} \rightarrow Zn^{2+}{}_{(aq)} + Cu_{(s)}$$

In order to calculate the overall redox potential, it is important to formulate the individual half-equations as reductions first. The standard reduction potential E^0_{red} for each reaction can be taken from standard tables:

$$Cu^{2+}{}_{(aq)} + 2e^- \rightarrow Cu_{(s)} \quad E^0_{red} = +0.340\,V$$

$$Zn^{2+}{}_{(aq)} + 2e^- \rightarrow Zn_{(s)} \quad E^0_{red} = -0.763\,V$$

The half-equation with the more positive value is the reduction, whilst the other reaction is the oxidation. In this example, Cu^{2+} will be reduced to Cu whilst $Zn_{(s)}$ metal will be oxidised to Zn^{2+}. The standard potential E^0 of an oxidation half-equation is the negative value of E^0_{red}.

$$\text{Reduction}: \; Cu^{2+}_{(aq)} + 2e^- \rightarrow Cu_{(s)} \quad E^0_{red} = +0.340\,V$$

$$\text{Oxidation}: \; Zn_{(s)} \rightarrow Zn^{2+}_{(aq)} + 2e^- \quad E^0_{ox} = -(-0.763\,V)$$

In order to calculate the voltage produced by an electrochemical cell (E^0_{cell}), the potentials of all half-equations are added up.

$$E^0_{cell} = E^0_{red} + E^0_{ox}$$

$$\text{For the example above}: \; E^0_{cell} = 0.340\,V + [-(-0.763\,V)] = 1.103\,V$$

If you deal with more complicated redox reactions and half-equations have to be multiplied because of the different numbers of electrons in each half-equation, the standard potential E^0 will not be affected by these coefficients. Furthermore, it is important to note that the above calculations can be used only if both reaction partners are present in the same concentration. Concentration can have an effect, and the overall potential is then calculated by using the so-called Nernst equation (see Section 2.4.2).

Table 2.2 *Table of standard reduction potential under acidic conditions at 25°C [1]*

	E^0 (V)
$Li^+ + e^- \rightarrow Li$	−3.05
$K^+ + e^- \rightarrow K$	−2.93
$Ba^{2+} + 2e^- \rightarrow Ba$	−2.90
$Na^+ + e^- \rightarrow Na$	−2.71
$Mg^{2+} + 2e^- \rightarrow Mg$	−2.37
$Al^{3+} + 3e^- \rightarrow Al$	−1.66
$Mn^{2+} + 2e^- \rightarrow Mn$	−1.18
$Ga^{3+} + 3e^- \rightarrow Ga$	−0.53
$Tl^+ + e^- \rightarrow Tl$	−0.34
$Fe^{2+} + 2e^- \rightarrow Fe$	−0.44
$Cr^{3+} + 3e^- \rightarrow Cr$	−0.74
$Cr^{3+} + e^- \rightarrow Cr^{2+}$	−0.41
$Ni^{2+} + 2e^- \rightarrow Ni$	−0.25
$Cu^{2+} + e^- \rightarrow Cu^+$	+0.15
$Cu^{2+} + 2e^- \rightarrow Cu$	+0.35
$Cu^+ + e^- \rightarrow Cu$	+0.50
$I_2 + 2e^- \rightarrow 2I^-$	+0.54
$Fe^{3+} + e^- \rightarrow Fe^{2+}$	+0.77
$Br_2 + 2e^- \rightarrow 2Br^-$	+1.07
$MnO_2 + 2e^- \rightarrow Mn^{2+}$	+1.23
$Cl_2 + 2e^- \rightarrow 2Cl^-$	+1.36
$MnO_4^- + 5e^- \rightarrow Mn^{2+}$	+1.54
$F_2 + 2e^- \rightarrow 2F^-$	+2.65

2.1.3 Chemical behaviour of alkali metals

Most alkali metals have a silvery white appearance with the exception of caesium which is golden yellow. They are all soft metals and typically can be cut with a knife. The softness of the metal increases within the group; caesium is the softest of the alkali metals.

Alkali metals are generally very reactive and oxidise in the air. The reactivity increases within the group, with lithium having the lowest reactivity and caesium the highest. Therefore, all alkali metals except lithium have to be stored in mineral oil. Lithium as an exception is normally stored under inert gas such as argon. Nevertheless, lithium, sodium and potassium can be handled in air for a short time, whereas rubidium and caesium have to be handled in an inert gas atmosphere.

All alkali metals react violently with water with the formation of the metal hydroxide and hydrogen. Again, lithium is the least reactive alkali metal and reacts 'only' quickly with water, whereas potassium, rubidium and caesium are more reactive and react violently with water.

$$2Li_{(s)} + 2H_2O \rightarrow 2LiOH + H_{2(g)} \uparrow \tag{2.3}$$

In terms of their pharmaceutical applications, alkali metals are not directly useable mainly because of their reaction behaviour in aqueous media. Nevertheless, alkali metal halides, some oxides, carbonates, citrates and other salts are of medicinal interest. NaCl and KCl solutions are important as oral rehydration salts, and KCl can also be used to treat potassium depletion.

In general, alkali metal halides can be prepared by the direct combination of the elements, that is, the reaction of an alkali metal with halogens. Alkali metal halides are very soluble in water, which is important for a potential pharmaceutical application, and partly soluble in organic solvents.

$$2M + X_2 \rightarrow 2MX \tag{2.4}$$

Alkali metal oxides can be synthesised by heating alkali metals in an excess of air. Thereby, the oxide, peroxide or superoxide formation can be observed depending on the metal.

$$\begin{aligned} 4Li + O_2 &\rightarrow 2Li_2O \quad \text{oxide formation} \\ 2Na + O_2 &\rightarrow Na_2O_2 \quad \text{peroxide formation} \\ K + O_2 &\rightarrow KO_2 \quad \text{superoxide formation} \end{aligned} \tag{2.5}$$

Alkali metal carbonates and bicarbonates have wide-ranging pharmaceutical applications. Lithium bicarbonate or citrate is used in the treatment of BD, whereas potassium bicarbonate or citrate is used in over-the-counter drugs as active pharmaceutical ingredients (APIs) against urinary-tract infections (increasing the pH of the urine) in the United Kingdom. Their solubility is highly dependent on the metal and varies from sparingly soluble (e.g. Li_2CO_3) whereas others are very soluble (Figure 2.4).

Sodium carbonate is produced by the so-called Solvay process – one of the most important industrial processes, which was developed in the 1860s by Ernest Solvay. Na_2CO_3 has extensive uses ranging from glass production to its application as a water softener. The starting materials, which are NaCl and $CaCO_3$ (lime stone), are inexpensive and easily available.

$$2NaCl + CaCO_3 \rightarrow Na_2CO_3 + CaCl_2 \tag{2.6}$$

$CaCO_3$ is heated to around 1000 °C, when it is converted to CO_2 and CaO (quicklime).

$$CaCO_3 \rightarrow CO_2 + CaO \tag{2.7}$$

CO_2 is then passed through an aqueous solution of NaCl and NH_3 (ammonia). NH_3 buffers the solution at a basic pH and $NaHCO_3$ (sodium bicarbonate) precipitates out of this solution. $NaHCO_3$ is less water soluble

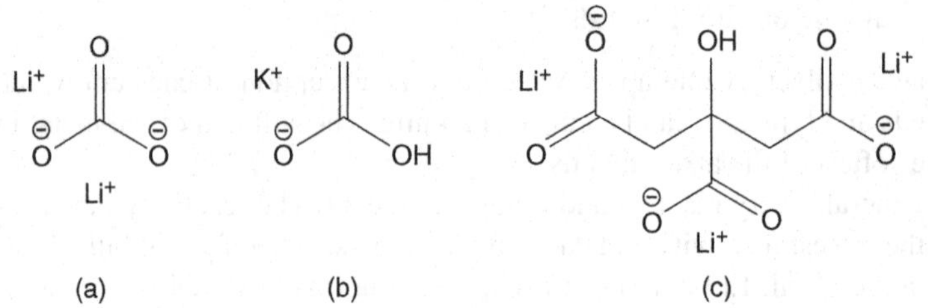

(a) (b) (c)

Figure 2.4 *Chemical structures of (a) lithium carbonate, (b) potassium bicarbonate and (c) lithium citrate*

than NaCl at a basic pH. Without the addition of NH_3, the solution would be acidic, as HCl is produced as a by-product.

$$NaCl + CO_2 + NH_3 + H_2O \rightarrow NaHCO_3 + NH_4Cl \tag{2.8}$$

$NaHCO_3$ is filtered, and the remaining solution of NH_4Cl reacts with the CaO from step 1. The produced $CaCl_2$ is usually used as road salts, and NH_3 is recycled back to the initial reaction of NaCl (reaction step 2).

$$2NH_4Cl + CaO \rightarrow 2NH_3 + CaCl_2 + H_2O \tag{2.9}$$

In a final step, $NaHCO_3$ is converted to Na_2CO_3 (sodium carbonate) by calcination (heating to 160–230 °C) with the loss of water.

$$2NaHCO_3 \rightarrow Na_2CO_3 + H_2O + CO_2 \tag{2.10}$$

Again, the CO_2 can be recycled and reused within the process. This means that the Solvay process consumes only a very small amount of NH_3 and the only ingredients are NaCl and $CaCO_3$ (Figure 2.5).

In terms of pharmaceutical applications, the various salts of sodium, lithium and potassium have the widest use and are discussed in detail in the following sections. These applications range from the use of sodium and potassium salts in dehydration solutions in order to restore replenished mineral balances to the treatment of BD with simple lithium salts.

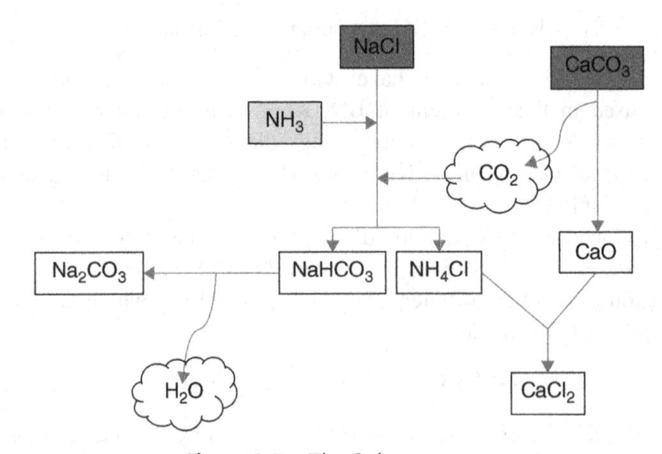

Figure 2.5 *The Solvay process*

2.2 Advantages and disadvantages using lithium-based drugs

The name 'lithium' stems from the Greek word 'lithos' which means stone. Lithium salts are well known for their use in batteries, metal alloys and glass manufacture. Nevertheless, Li also has a clinical application in the treatment of manic depression or BD. BD affects 1–2% of the population and severely reduces the quality of life for the patients and also increases the likelihood of patients committing suicide. Research has shown that lithium salts are very successful in the treatment of BD, and a broad research in this area has been stimulated [3]. The main question addressed within this research is how the lithium ion (Li^+) can be modified in order to improve its activity to patent their findings, which would result in income for the manufacturer. Unfortunately, the simple ion is the active ingredient and this makes it difficult to patent and secure any intellectual property.

Li is a member of group 1 alkali metals and is the lightest and smallest solid element. Li has an atomic number of 3 and contains a single valence electron, which determines its redox chemistry and reactivity. In nature, Li is found as ores, for example, spodumene [$LiAl(SiO_3)_2$], or at low concentrations as salts, for example, in rivers. The metal itself is soft and white in appearance; it has the lowest reactivity within the group of alkali metals. The Li^+ ion has the smallest ionic radius of all known metals.

2.2.1 Isotopes of lithium and their medicinal application

Lithium occurs as mixture of two stable, naturally occurring isotopes (see Section 1.2.1.1), namely ^6Li with an occurrence of 7.59% and the major isotope ^7Li (92.41%). The nucleus of ^6Li contains three protons and three neutrons, whilst ^7Li contains three protons and four neutrons.

^6Li and ^7Li are both NMR (nuclear magnetic resonance) active nuclei, which means their presence can be monitored via NMR technology. Using this analytical tool, it is possible to differentiate between intra- and extracellular Li^+ concentrations and therefore, the uptake of Li^+ into different cells can be monitored. The use of ^6Li results in sharper NMR spectra because of its properties, but it also has a lower intensity. ^6Li salts can also be used to monitor the distribution of lithium in the tissue. A further application of ^6Li is the production of tritium atoms ($^3{}_1H$) and their use in atomic reactors. In this process, ^6Li is bombarded with neutrons, which results in the production of tritium atoms and radioactive α-particles ($^4{}_2He$):

$$^6{}_3Li + ^1{}_0n \rightarrow ^3{}_1H + ^4{}_2He \tag{2.11}$$

Equation 2.11 shows the activation of ^6Li.

2.2.2 Historical developments in lithium-based drugs

The first medical use of Li^+ was described in 1859 for the treatment of rheumatic conditions and gout. The theory at that time was based on the ability of lithium to dissolve nitrogen-containing compounds such as uric acid. Their build-up in the body was believed to cause many illnesses such as rheumatic conditions and gout problems. In 1880, Li^+ was first reported as being used in the treatment of BD, and in 1885 lithium carbonate (Li_2CO_3) and lithium citrate [$Li_3C_3H_5O(COO)_3$] were included in the British Pharmacopoeia. It also became clear that there is a direct link between NaCl intake and heart diseases as well as hypertension. Therefore, LiCl was prescribed as replacement for NaCl in the diet of affected patients [3b].

The urea hypothesis and the connection of NaCl to heart diseases stimulated the use of lithium salts in common food. The prime example is the soft drink 7Up©, which has been marketed in 1929 under the label

Bib-Label Lithiated Lemon-Lime Soda. 7Up contained lithium citrate and was also marketed as a hang-over cure. The actual Li^+ was subsequently removed in 1950 [3b].

John Cade's experiments on guinea pigs in 1949 initiated the discovery of Li^+ and its sedative and mood-control properties. Uric acid was known to have mood-controlling properties and Cade used lithium urate as a control solution. To his surprise, he discovered that lithium urate had tranquillising properties, and after further experiments he concluded that this was caused by the lithium ion [4].

Nevertheless, there were drawbacks, especially when the FDA banned Li^+ salts following the death of four US patients. These patients had an average intake of 14 g of lithium chloride (LiCl) per day in order to replace NaCl. Another stumbling stone in the way of success of Li^+ was the discovery of chlorpromazine, the first antipsychotic drug, which is still used for the treatment of BD. In the early 1970s, Li^+ was re-approved by FDA and is now used in 50% of the treatment of BD [5].

2.2.3 The biology of lithium and its medicinal application

Lithium salts are used in the prophylaxis and treatment of mania, and in the prophylaxis of BD and recurrent depression. Lithium therapy is taken orally, usually as lithium carbonate (Li_2CO_3) or lithium citrate [$Li_3C_3H_5O(COO)_3$], with a total dose of up to 30 mmol/day. Li_2CO_3 is the preferred lithium salt used, as it causes the least irritation to the stomach. The treatment has to be closely monitored, and Li^+ blood concentrations are measured 12 h after administration to achieve a serum lithium concentration of 0.4–1 mmol/l. The therapeutic index (concentration window from efficacy to toxicity) for Li^+ is very narrow, and plasma Li^+ concentrations above 2 mM require emergency treatment for poisoning (Figure 2.6) [6].

The administered Li^+ is distributed uniformly in the body tissues and in the blood plasma, with the external cell Li^+ concentration being below 2 mM. Experiments studying the lithium distribution in rats after administration of a high dose of $^6Li^+$ showed that there was no exceptional accumulation of $^6Li^+$ in the brain. $^6Li^+$ distributes fairly uniformly in the body, with bones and endocrine glands showing higher concentrations (Figure 2.7) [7].

Li^+ ions are not soluble in lipids and therefore do not cross plasma membranes. The transport into cells occurs via exchange mechanism by lithium–sodium counter-transport, anion exchange (so-called $Li^+/CO_3{}^{2-}$ co-transport) and other unrelated transport molecules. The specific mode of action of the simple Li^+ ion is currently unknown, but it is clear that a displacement of Mg^{2+} by Li^+ is involved. Therefore, an alteration of the Mg^{2+} balance in the blood and the urine can be observed in patients treated with Li^+ [3b]. This displacement is actually not surprising because the properties of Li^+ and Mg^{2+} are similar, which can be explained by the concept of *diagonal relationship* (see Section 2.2.4).

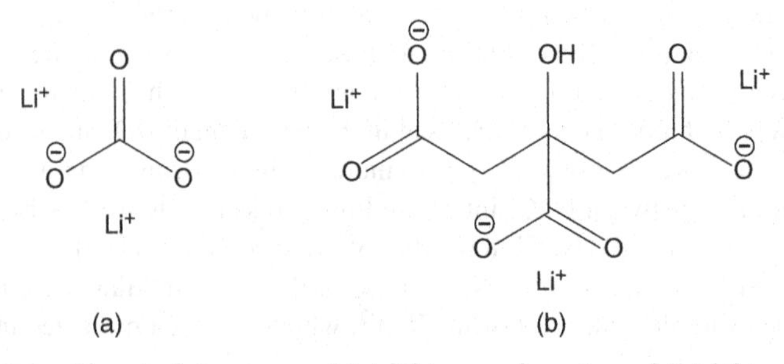

Figure 2.6 *Chemical structures of (a) lithium carbonate and (b) lithium citrate*

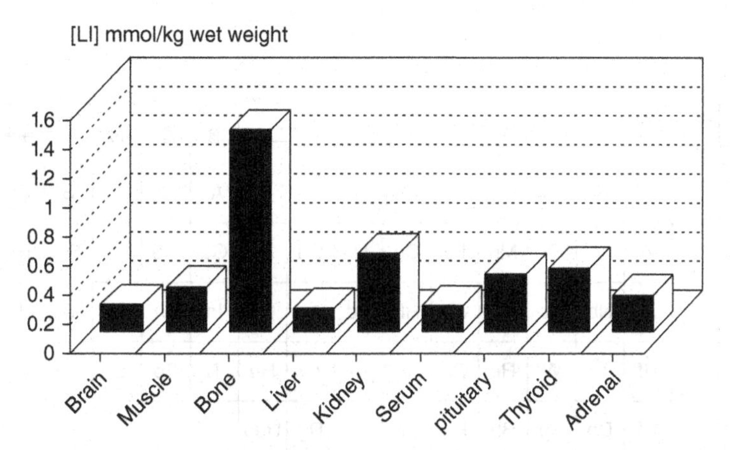

Figure 2.7 *Distribution of $^{6}Li^{+}$ in various tissues of rats after chronic administration [7] (Reproduced with permission from [7]. Copyright © 1999, Royal Society of Chemistry.)*

2.2.4 Excursus: diagonal relationship and periodicity

Within the periodic table, the element pairs Li/Mg, Be/Al, B/Si and others form a so-called *diagonal relationship* to each other. With this concept, similarities in biological activity can be explained as the element pairs have similar properties.

*Diagonally adjacent elements of the second and third periods have similar properties – this concept is called **diagonal relationship**. These pairs (Li/Mg, Be/Al, B/Si, etc.) have similarities in ion size, atomic radius, reactive behaviour and other properties.*

Diagonal relationship is a result of opposite effects, which crossing and descending within the periodic table has. The size of an atom decreases within the same period (from left to right). The reason is that a positive charge is added to the nucleus together with an extra electron orbital. Increasing the nucleus charge means that the electron orbitals are pulled closer to the nucleus. The atomic radius increases when descending within the same group, which is due to the fact that extra orbitals with electrons are added (Figure 2.8) [1].

Trends can be seen for the electronegativity and the ionisation energy; both increase when moving within the same period from left to right and decrease within the same group. Within a small atom, the electrons are located close to the nucleus and held tightly, which leads to a high ionisation energy. Therefore, the ionisation energy decreases within the group and increases within the period (from left to right) – showing the opposite trend compared to the atomic radius. A similar explanation can also be used for summarising the trends seen for the electronegativity: small atoms tend to attract electrons more strongly than larger ones. This means fluorine is the most electronegative element within the Periodic Tables of Elements (Figure 2.9).

This structure of trends is summarised under the term *periodicity of the elements*. This means that within the periodic table, all elements follow the above-mentioned trends in a repetitive way. This allows predicting trends for atomic and ionic radii, electronegativity and ionisation energy (Figure 2.10).

In relation to the diagonal relationship, it becomes clear by studying the described trends that these effects cancel each other when descending within the group and crossing by one element within the PSE. Therefore,

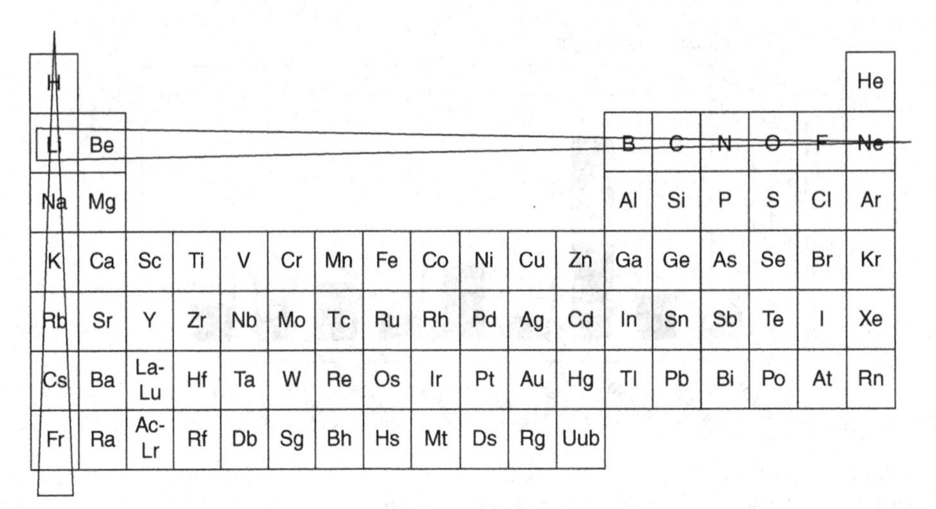

Figure 2.8 *Atomic radii*

H																	He
Li	Be											B	C	N	O	F	Ne
Na	Mg											Al	Si	P	S	Cl	Ar
K	Ca	Sc	Ti	V	Cr	Mn	Fe	Co	Ni	Cu	Zn	Ga	Ge	As	Se	Br	Kr
Rb	Sr	Y	Zr	Nb	Mo	Tc	Ru	Rh	Pd	Ag	Cd	In	Sn	Sb	Te	I	Xe
Cs	Ba	La-Lu	Hf	Ta	W	Re	Os	Ir	Pt	Au	Hg	Tl	Pb	Bi	Po	At	Rn
Fr	Ra	Ac-Lr	Rf	Db	Sg	Bh	Hs	Mt	Ds	Rg	Uub						

Electronegativity

Figure 2.9 *Electronegativity*

elements diagonally positioned within the periodic table have similar properties, such as similar atomic size, electronegativity and ionisation energy (Figure 2.11).

The concept of the diagonal relationship is crucial for the biological activity of lithium drugs, which is mainly due to the properties of the Li^+ ion being similar to the Mg^{2+} ion. In comparison, the size of the Li^+ ion is similar to that of Mg^{2+} and therefore they compete for the same binding sites in proteins. Nevertheless, lithium has relatively specific effects, and so only proteins with a low affinity for Mg^{2+} are targeted. Li^+ and Mg^{2+} salts have similar solubility, for example, CO_3^{2-}, PO_4^{3-}, F^- salts have a low water solubility, and halide and alkyl salts are soluble in organic solvents. Li^+ and Mg^{2+} compounds are generally hydrated, for example, $LiCl{\cdot}3H_2O$ and $MgCl_2{\cdot}6H_2O$. Similarities in ionic size, solubility, electronegativity and solubility result in similar biological activity and therefore pharmaceutical application [3b].

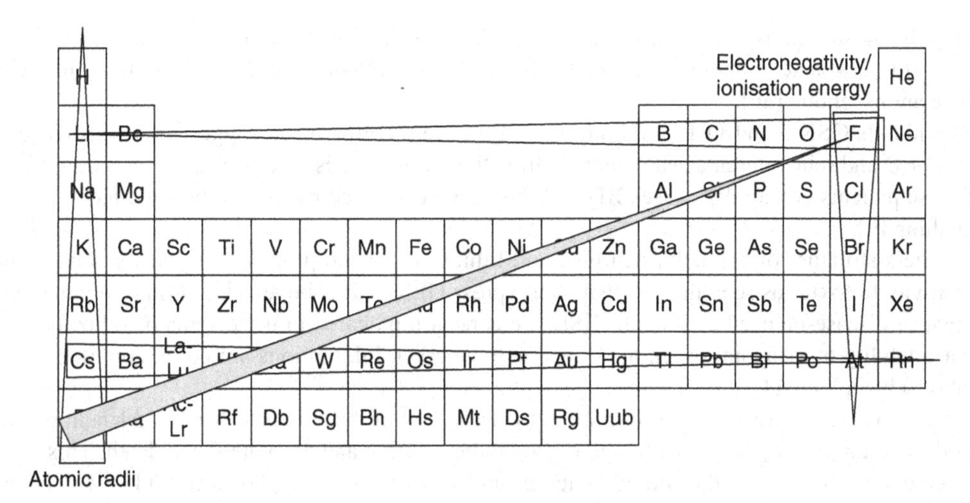

Figure 2.10 Periodicity showing the 'metallic character' trend (highlighted in grey) within the periodic table

H																	He
Li	Be											B	C	N	O	F	Ne
Na	Mg											Al	Si	P	S	Cl	Ar
K	Ca	Sc	Ti	V	Cr	Mn	Fe	Co	Ni	Cu	Zn	Ga	Ge	As	Se	Br	Kr
Rb	Sr	Y	Zr	Nb	Mo	Tc	Ru	Rh	Pd	Ag	Cd	In	Sn	Sb	Te	I	Xe
Cs	Ba	La-Lu	Hf	Ta	W	Re	Os	Ir	Pt	Au	Hg	Tl	Pb	Bi	Po	At	Rn
Fr	Ra	Ac-Lr	Rf	Db	Sg	Bh	Hs	Mt	Ds	Rg	Uub						

Figure 2.11 Diagonal relationship

2.2.5 What are the pharmacological targets of lithium?

The precise mechanism of action of lithium ions as mood stabilisers is unknown. Current research shows that there are two main targets in the cell – the enzymes glycogen synthase kinase-3 (GSK-3) and the phosphomonoesterases family (PMEs) [3b].

GSK-3 is a serine/threonine protein kinase, which is known to play an important role in many biological processes. GSK-3 is an enzyme that mediates the addition of phosphate molecules onto the hydroxyl groups of certain serine and threonine amino acids, in particular cellular substrates. Li^+ inhibits GSK-3 enzymes via competition for Mg^{2+} binding [8].

Phosphoric monoester hydrolases (PMEs) are enzymes that catalyse the hydrolysis of O—P bonds by nucleophilic attack of phosphorous by cysteine residues or coordinated metal ions. Inositol monophosphatase (InsP) is the best known member of this family. Li^+ and Mg^{2+} both have a high affinity to bind to phosphate groups. Li^+ inhibits the enzymatic function of InsP and prevents phosphate release from the active site. Generally, inositol phosphatases are Mg^{2+}-dependent, but Li^+ binds to one of the catalytic Mg^{2+} sites. Binding Li^+ to phosphate-containing messenger molecules could perturb the transcellular communication and thus be antipsychotic [3b].

Lithium inhibits GSK-3 and InsP, and both pathways have therefore been suggested to be involved in the treatment of BD and schizophrenia. The theory behind this hypothesis is that overactive InsP signalling in the brain of these patients potentially causes BD and this may be reduced by the inhibitory effect of lithium on such signalling [9].

GSK-3 alters structure of cerebella neurons, and lithium is a neuroprotective agent that can reduce the hypersensitivity to toxins as seen in cells that overexpress GSK-3. It is believed that lithium potentially can protect against disease-induced cell death. GSK-3 has been implicated in the origins of schizophrenia, but with the availability of many antipsychotic drugs on the market, lithium ions are not in common use for the treatment of schizophrenia [10].

There are also several direct roles of lithium in the treatment of Alzheimer's disease. Alzheimer's disease is a neurodegenerative brain disorder causing neuronal dysfunction and ultimately cell death. This leads ultimately to dementia, affecting in the United States around 10% of people aged over 65 and 48% aged over 85. Onset occurs with the accumulation of extracellular senile plaques composed of amyloid-β peptides and with the accumulation of intercellular neurofibrillary tangles. GSK-3 is necessary for the accumulation of tangles, and subsequently GSK-3 inhibition reduces the production of peptides – this is where the Li^+ interaction comes into play [3b].

2.2.6 Adverse effects and toxicity

Lithium has a very narrow therapeutic window, which makes the monitoring of blood levels essential during the treatment (see Section 2.2.3). Blood plasma concentrations of more than 1.5 mmol/l Li^+ may cause toxic effects, usually tremors in the fingers, renal impairment and convulsion. Also, memory problems are a very common side effect, and complaints about slowed mental ability and forgetfulness are commonly reported. Extreme doses of lithium can cause nausea and diarrhea, and doses above 2 mmol/l require emergency treatment, as these levels may be fatal. Weight gain and decreased thyroid levels are also commonly reported problems. The blood serum level of Li^+ has to be monitored 12 h after administration, and health care professionals also have to be aware that the mood-stabilising effects of lithium take a couple of days to take effect [6].

Lithium salts have severe adverse effects on the renal system. Lithium therapy can damage the internal structures of the kidneys, which are the tubular structure of the nephrons, and can lead to diabetes insipidus. One out of five patients experience polyuria – excess production of urine, which is the number one symptom of diabetes insipidus. Therefore, the kidney function has to be closely monitored for patients undergoing long-term therapy, with a full test of the kidney function every 6–12 months for stabilised regimes [3a, 6].

Great care has to be taken when lithium is given with nonsteroidal anti-inflammatory drugs (NSAIDs) because up to 60% increase of Li^+ concentration in blood has been observed. NSAIDs reduce the clearance of lithium through kidneys, and as a result lithium poisoning is possible. Furthermore, the concurrent use of diuretics, which can result in sodium depletion, can make lithium toxicity worse and can be hazardous [3a].

2.3 Sodium: an essential ion in the human body

Sodium has atomic number 11 and has the symbol Na, derived from the Latin name 'natrium'. Sodium ions (Na^+) are soluble in water and therefore present in large quantities in the oceans. Na^+ is also part of minerals and an essential element for all animal life.

The main biological roles of sodium ions are the maintenance of body fluids in humans and the functioning of neurons and transmission of nerve impulses. Na^+ is an important electrolyte and a vital component of the extracellular fluid. Therefore, one of its roles is to maintain the fluid in the human body via osmoregulation, a passive transport mechanism (Section 2.3.1). Na^+ ions also play a crucial role in the contraction of muscles and in the mode of action of several enzymes. In the human body, Na^+ is often used to actively build up an electrostatic potential across membranes, with potassium ions (K^+) being the counter-ion (Section 2.3.2). The build-up of an electrostatic potential across cell membranes is important to allow the transmission of nerve impulses.

2.3.1 Osmosis

Osmosis *is defined as the physical process of diffusion of a solvent (water) through a semi-permeable membrane towards an area of high solute (salt) concentration. This means that solvent (water) follows the osmotic gradient by moving across the semi-permeable membrane from one solution where there is a lower salt concentration towards a second solution with a high salt concentration in order to dilute this and to equalise the concentrations.*

Sodium is an essential mineral for the human body and crucial for the regulation of the body fluid via its osmosis activity. Sodium ions account for over 90% of all ions in the plasma and in the interstitial fluid, which are involved in osmosis processes. Furthermore, it is the most abundant cation in the extracellular fluid, and therefore the Na^+ content controls the extracellular volume. In particular, the kidneys play an important role in regulating the fluid level of the body as well as the filtration, secretion and re-absorption of Na^+ in the nephrons, the functional unit of the kidney. Na^+ ions are used in the human body to establish osmotic gradients, which in turn is crucial to control the water balance. Furthermore, decreases in blood pressure and in Na^+ concentrations are sensed by the kidneys, and hormones (e.g. renin, antidiuretic hormones (ADHs), atrial natriuretic peptide) are released that control the blood pressure, osmotic balances and water-retaining mechanisms.

In general, if a medium is

- *hypertonic*, that means the solution has a higher concentration of solutes than the surrounding area. This area will lose water through osmosis;
- *isotonic*, that means the solution has the same concentration of solutes as the surrounding area. No movement of water will occur;
- *hypotonic*, that means the solution has a lower concentration of solutes than the surrounding area. This area will gain water through osmosis (Figure 2.12).

A net movement of the solvent (water) occurs from the hypotonic solution to the solution with the higher concentration in order to reduce the difference in concentrations. The osmotic pressure is defined as the pressure that is required to establish equilibrium with no movement of solvents. It is important to mention that osmotic pressure depends on the number of ions or molecules in the solution, not the identity of those. The unit often being found to describe the osmotic pressure is the osmole (osmol or osm), which is a non-SI unit that defines the numbers of moles of a compound that contributes to the osmotic pressure of a solution.

In general, osmotic processes are important for many biological processes. Plants use osmosis to transport water and solutes through their systems and the osmotic gradient to establish the turgor within cells. The human body uses osmosis for many processes, the excretion of urine being one of the most prominent one.

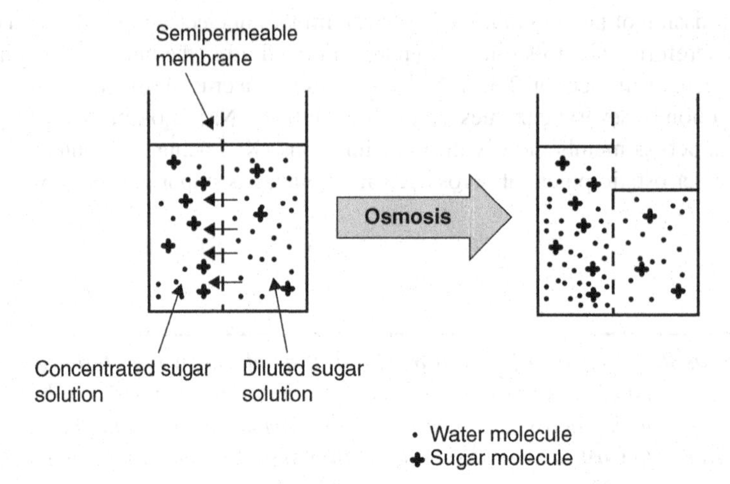

Figure 2.12 Schematic representation of osmosis

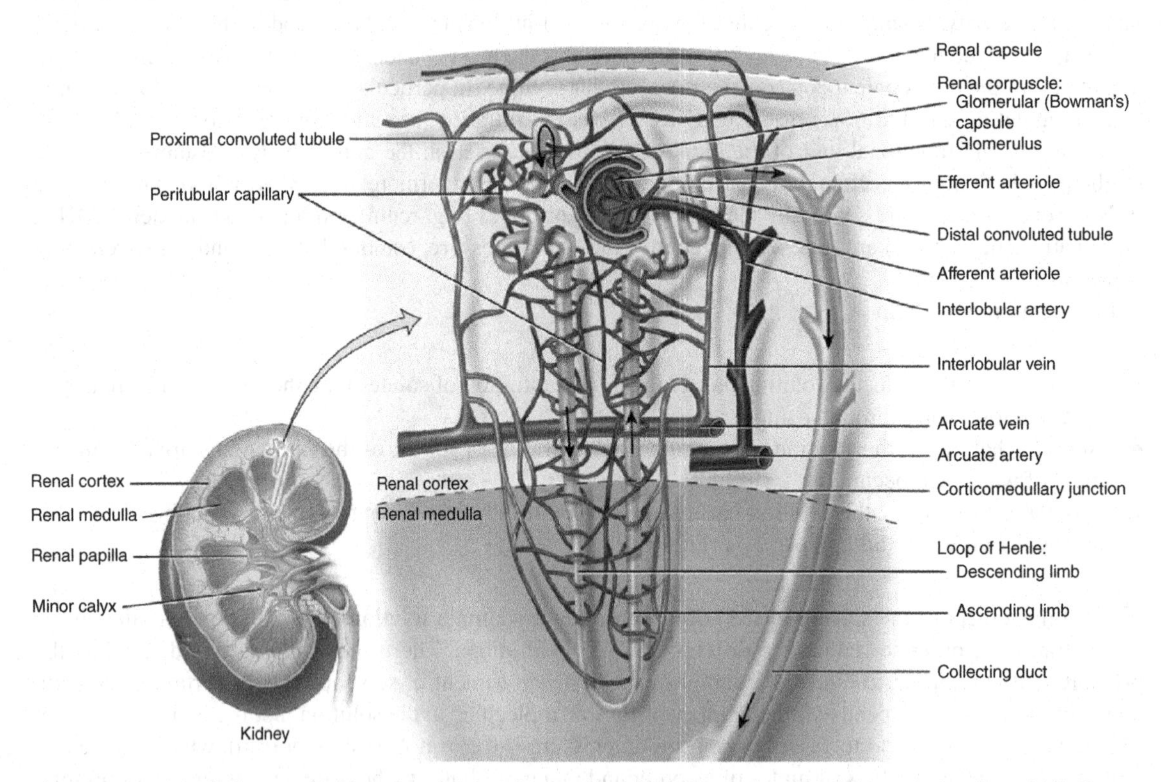

Figure 2.13 The kidney and its functional unit – the nephron [11] (Reproduced with permission from [11]. Copyright © 2009, John Wiley & Sons, Ltd.)

Urine production takes place within the kidney, more specifically at the nephrons which are the functional units of the kidneys. Approximately 150–180 of plasma is filtered every day through the glomerulus, which is a part of the nephron, in order to produce the urine. The nephron also consists of the proximal tubule, the Loop of Henle and the distal tubule, which leads to the collecting duct and ultimately to the ureter (Figure 2.13) [11].

Filtration takes places at the glomerulus, whereas the remaining parts of the nephron are responsible for the secretion and re-absorption of ions in order to regulate imbalances and manage the urine volume before the urine is stored in the bladder. This secretion and re-absorption can occur via an active or a passive transport across the nephron membrane. Na^+ is usually actively transported across via Na^+ pumps in order to establish the correct Na^+ concentration in the blood plasma, which is responsible for maintaining the correct osmotic pressure. Via this process, an osmotic gradient is established within the kidney parenchyma, which is used to conserve water. The ascending limb of the Loop of Henle is impermeable to water but permeable to Na^+. As a result, an osmotic gradient is established. The descending limb of the Loop of Henle is permeable to water and, as a result of the osmotic gradient, water moves to the interstitial fluid and urine is concentrated. The collecting ducts can be permeable to water if the body sends out a signal that water has to be conserved. Again, water will passively follow the osmotic gradient and urine will be concentrated even more (Figure 2.14).

2.3.2 Active transport of sodium ions

As previously mentioned, the active transport of sodium ions is crucial for the functioning of, for example, neurons and the subsequent transmission of a nerve impulse. This can be achieved by the active build-up of a concentration gradient along the cell membrane using Na^+/K^+ pumps as the active unit. This active

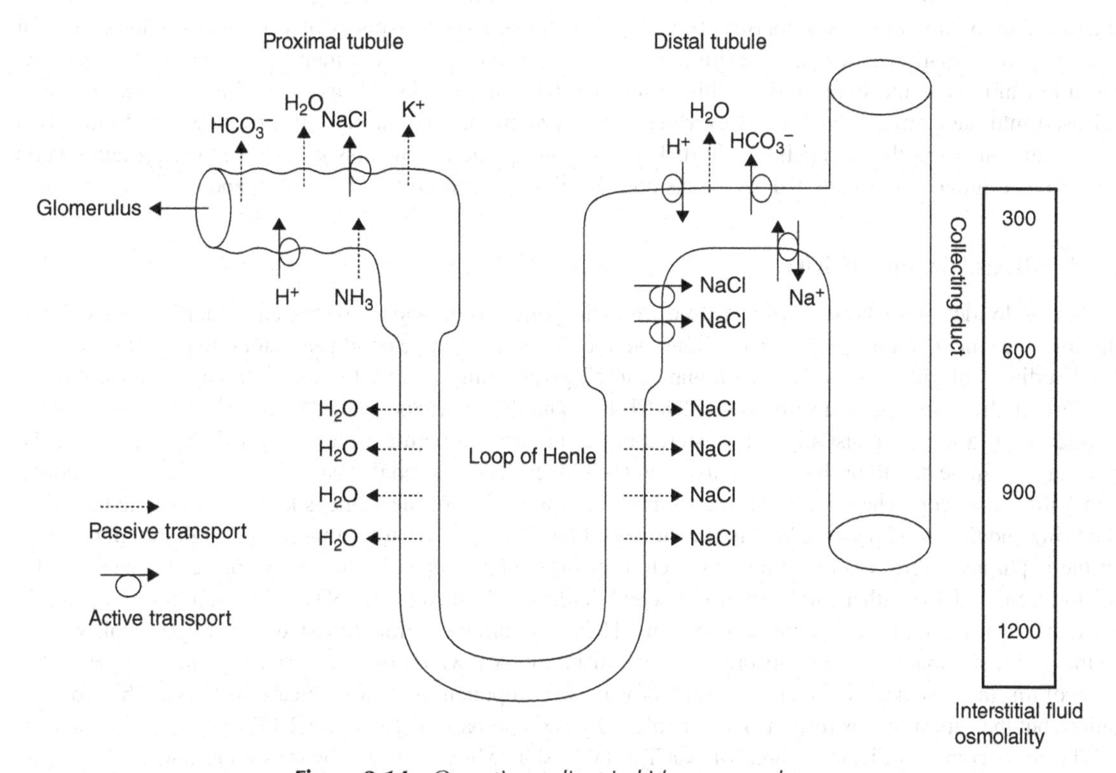

Figure 2.14 Osmotic gradient in kidney parenchyma

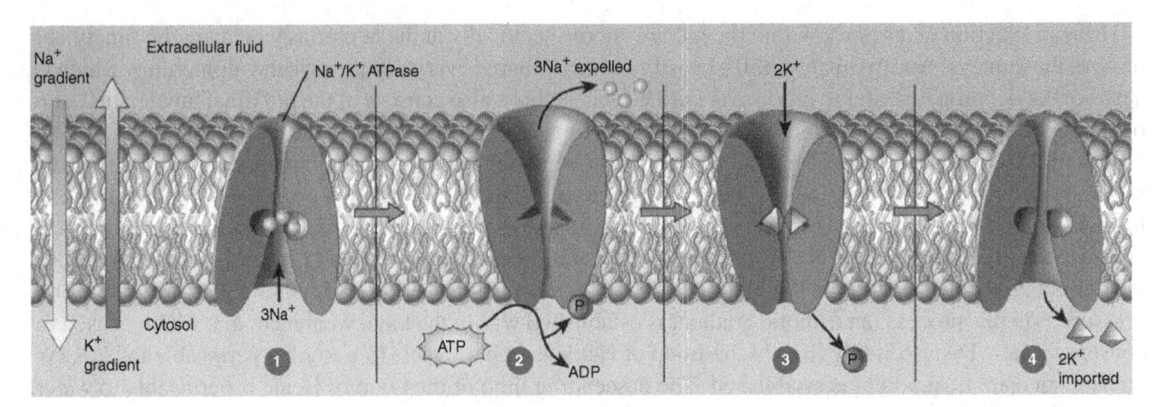

Figure 2.15 *Mode of action of the Na+/K+-ATPase [11] (Reproduced with permission from [11]. Copyright © 2009, John Wiley & Sons, Ltd.)*

transport is responsible for the cells containing relatively high concentrations of potassium ions and low concentrations of sodium ions. The resulting electrostatic potential that is built up along the cell membrane is called *action potential* and is subsequently responsible for the transmission of nerve impulses (see Section 2.4.1) (Figure 2.15).

The Na^+/K^+ pumps facilitate an active transport process which is based on the conformational changes of the cross-membrane protein and driven by the breakdown of ATP. In the initial step, three Na^+ ions bind the cross-membrane protein on the cytosolic side. This causes the protein to change its confirmation and makes it accessible to ATP. In its new confirmation, the protein becomes phosphorylated by ATP, which results in a second conformational change. The three Na^+ ions are located across the membrane, and the protein now has a low affinity to the sodium ions. This means that the sodium ions are dissociated from the protein and released into the extracellular fluid. Nevertheless, the protein has now a high affinity to K^+ and binds two potassium ions from the extracellular fluid. The bond phosphate is now dissociated, and the protein reverts back to its original confirmation. This means both K^+ ions are exposed to the cytosol and can be released.

2.3.3 Drugs, diet and toxicity

Sodium chloride solutions are normally used when the patient is diagnosed with sodium depletion and dehydration. Treatment is mostly administered intravenously, but in chronic conditions (mild to moderate sodium loss) sodium chloride or sodium bicarbonate can be given orally. Oral rehydration therapies usually use a mixture of alkali metal-based salts such as NaCl, KCl and their citrates (Figure 2.16) [6].

Sodium bicarbonate is usually administered orally in order to regulate the serum pH. Imbalances of the plasma pH can be due to problems occurring in the kidneys such as renal tubular acidosis. This is a medical condition that occurs where the body accumulates acid as a result of the kidneys failing to regulate the pH of the urine and the blood plasma. Within the kidneys, blood is filtered before it passes through the tubular part of the nephrons where re-absorption or secretion of important salts and others takes place. In renal tubular acidosis, the kidneys either fail to filter or secrete acid ions (H^+) from the plasma (secretion takes place in the distal tubule), or to recover bicarbonate ions (HCO_3^-) from the filtrate (passive re-absorption takes place in the proximal tubule, active re-absorption at the distal tubule), which is necessary to balance the pH. In the view of this mode of action, the pharmaceutically active component of sodium bicarbonate is the bicarbonate anion, but the cation Na^+ is responsible for solubility and compatibility (Figure 2.17) [3a].

The most common dietary source of NaCl is table salt, which is used for seasoning and pickling (the high NaCl content inhibits the bacterial and fungal growth as a result of the osmotic gradient). The daily recommended NaCl intake varies depending on the country and the age group. Within the United Kingdom,

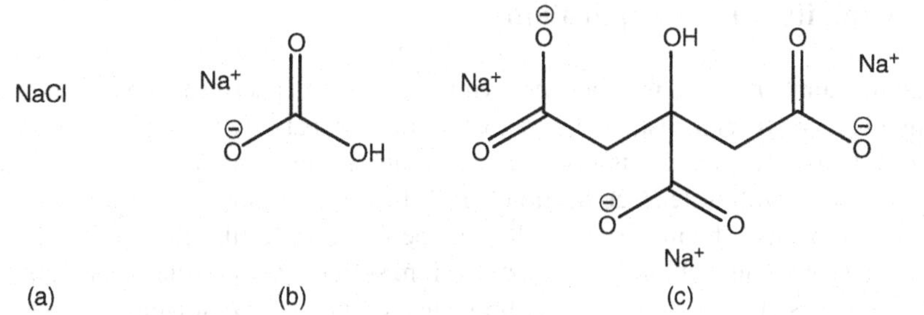

Figure 2.16 Chemical structures of (a) sodium chloride, (b) sodium bicarbonate and (c) sodium citrate

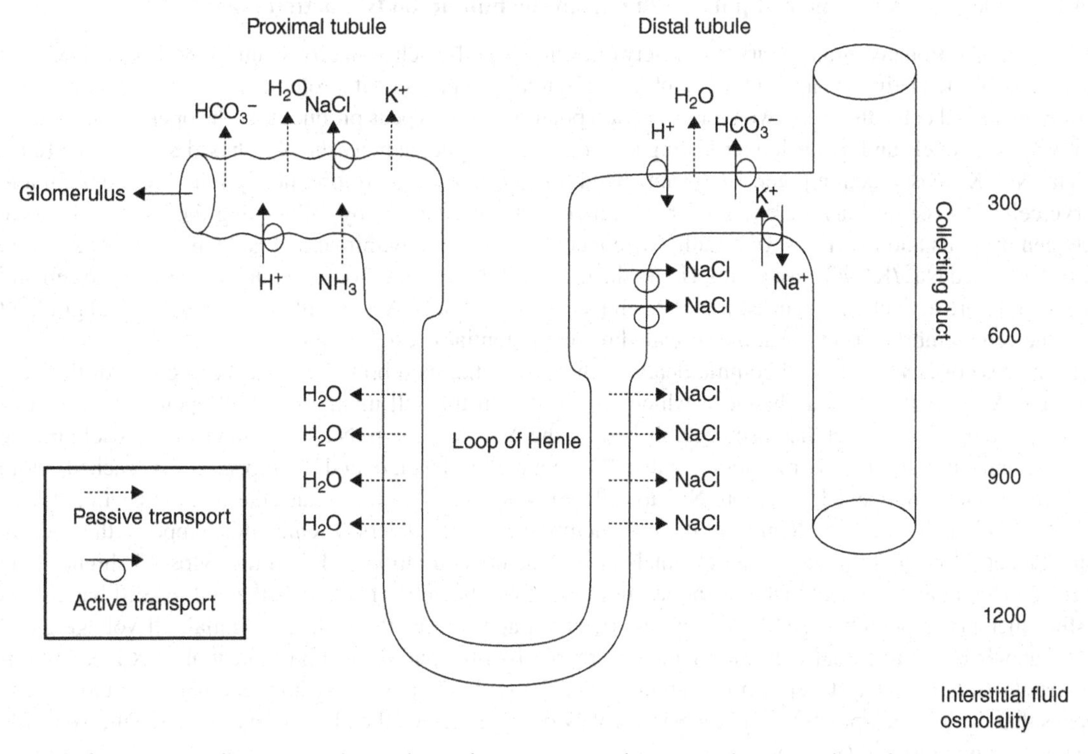

Figure 2.17 Illustration of a nephron showing areas of active and passive electrolyte transport

the maximum salt intake is recommended to be limited to 6 g of NaCl for an adult, whereas intake for children should be significantly lower [12]. Most people exceed this amount on a daily basis, and the high salt plasma levels (hypernatraemia) can result in cardiovascular disorders such as hypertension. Low sodium plasma levels (hyponatraemia), which again can be a result of dysfunction kidneys or sodium loss in the bowels, also cause damage to the human body via osmotic imbalances and if necessary have to be treated. Low blood pressure, dehydration and muscle cramps are signs of a sodium deficiency.

Signs of acute toxicity may be seen after ingestion of 500–1000 mg/kg body weight NaCl. These symptoms can be vomiting, ulceration of the gastrointestinal (GI) tract and renal damage. Also, the increased risk for the formation of kidney stones is believed to be a result of high salt intake [13].

2.4 Potassium and its clinical application

Potassium has atomic number 19 and the chemical symbol K, which is derived from its Latin name 'kalium'. Potassium was first isolated from potash, which is potassium carbonate (K_2CO_3). Potassium occurs in nature only in the form of its ion (K^+) either dissolved in the ocean or coordinated in minerals because elemental potassium reacts violently with water (see Section 2.1.3). Potassium ions are essential for the human body and are also present in plants. The major use of K^+ can be found in fertilisers, which contains a variety of potassium salts such as potassium chloride (KCl), potassium sulfate (K_2SO_4) and potassium nitrate (KNO_3). KCl is also found in table salt, whereas potassium bromate ($KBrO_3$) is an oxidising agent and is used as flour improver. Potassium bisulfite ($KHSO_3$) can be used as a food preservative in wine and beer.

2.4.1 Biological importance of potassium ions in the human body – action potential

The so-called action potential occurs in a variety of excitable cells such as neurons, muscle cells and endocrine cells. It is a short-lasting change of the membrane potential and plays a vital role in the cell-to-cell communication. In animal cells, there are two types of action potential. One type is produced by the opening of calcium ion (Ca^{2+}) channels and this is longer lasting than the second type, namely the Na^+-based action potential.

The Na^+/K^+-based action potential is short-lived (only 1 ms) and therefore mostly found in the brain and nerve cells. Potassium ions are crucial for the functioning of neurons, by influencing the osmotic balance between the cells and the interstitial fluid. The concentration of K^+ within and outside the cells is regulated by the so-called Na^+/K^+-ATPase pump. Under the use of ATP, three Na^+ ions are pumped outside the cell and two K^+ ions are actively transported into the cell (see Section 2.3.2). As a result, an electrochemical gradient over the cell membrane is created; the so-called resting potential is established.

In the case of any cell-to-cell communication, changes of the membrane reach a specific part of the neuron first. As a result, sodium channels, which are located in the cell membrane, will open. As a result of the osmotic gradient, which has been established by the Na^+/K^+-pump, Na^+ ions enter the cytosol and the electrochemical gradient becomes less negative. Once a certain level, called the *threshold*, is reached, more sodium channels are opened, and more Na^+ ions flow inside the cell – this creates the so-called action potential. The electrochemical gradient over the cell membrane is thus reversed. Once this happens, the sodium ion channels close and the potassium channels open. The concentration of K^+ in the cytosol is higher than in the extracellular fluid, and therefore potassium ions leave the cell. This allows the cell to shift back to its resting membrane potential. As the membrane potential approaches the resting potential, all voltage-gated K^+ channels open. In actual fact, the membrane repolarises beyond the resting potential; this is known as *hyperpolarisation*. The last step is now to reintroduce the initial balance of sodium and potassium ions. This means that the Na^+/K^+-pump transports Na^+ actively out of the cytosol and K^+ into it. As a result, the initial steady state is reinstated (Figure 2.18).

2.4.2 Excursus: the Nernst equation

As previously outlined, the electric potential across a cell membrane is created by the difference in ion concentration inside and the outside the cell. The *Nernst equation* is an important equation that allows the calculation of the electric potential for *an individual ion.*

$$\Delta E = -\frac{RT}{zF} \ln \frac{[\text{ion}]_{\text{in}}}{[\text{ion}]_{\text{out}}}$$

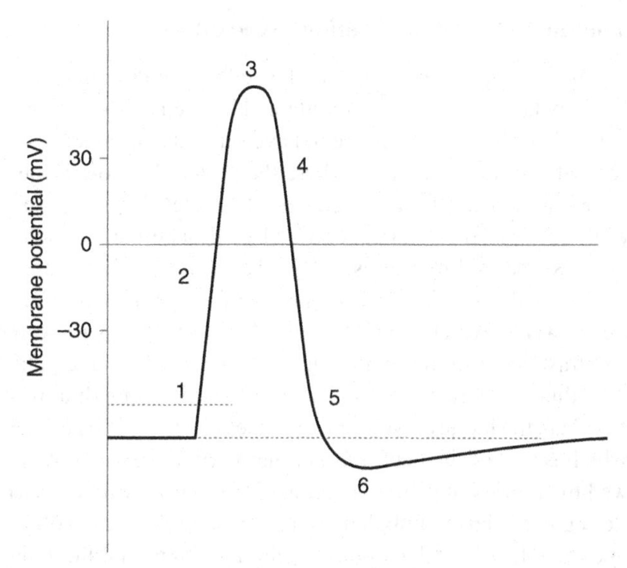

Figure 2.18 Action potential. (1) Threshold of excitation. Na^+ channels open and allow Na^+ to enter the cell. (2) K^+ channels open and K^+ leaves the cell. (3) No more Na^+ enters cell. (4) K^+ continues to leave the cell. This causes the membrane to return to resting potential. (5) K^+ channels close and Na^+ channels resent. (6) Hyperpolarisation

Looking at the Nernst equation, R is the gas constant, T is the temperature in kelvin, F is the Faraday constant and z is the net charge of the ion. $[\text{ion}]_{\text{in}}$ and $[\text{ion}]_{\text{out}}$ are the concentration of the particular ion inside and outside the cell. The equation can also be expressed as $\log_{10}$. Furthermore, at room temperature (25 °C), the term $-RT/F$ can be seen as a constant, and the Nernst equation can be simplified:

$$\Delta E = -\frac{0.059}{zV} \log \frac{[\text{ion}]_{\text{in}}}{[\text{ion}]_{\text{out}}}$$

The Nernst equation can also be used to calculate the overall potential in a redox equation by using the ion concentrations of both half-equations. This also allows the calculation of the reduction potential ΔE of *two different ions* of varying concentration. The Nernst equation can be expanded to the following equation, where [Red] and [Ox] represent the concentrations of the reductant and oxidant ($\text{Ox} + e^- \rightarrow \text{Red}$):

$$\Delta E = E^0 - \frac{0.059}{zV} \log \frac{[\text{Red}]}{[\text{Ox}]}$$

In order to calculate the reduction potential ΔE of *two different ions* of varying concentration, the Nernst equation can be expanded to the following equation:

$$\Delta E = \Delta E^0 - \frac{0.059}{zV} \log \frac{[\text{ion}]_{\text{in}}}{[\text{ion}]_{\text{out}}}$$

It is also possible to calculate the overall electric potential across the cell membrane by taking several different ions into account. The so-called Goldman equation, which will not be further illustrated in this book, can be used to calculate this value.

2.4.3 Potassium salts and their clinical application: hypokalaemia

In the human body, 95% of the K^+ can be found inside the cells, with the remaining 5% mainly circulating in the blood plasma [11]. This balance is carefully maintained by the Na^+/K^+ pump (see Section 2.3.2), and imbalances, such as seen in hypo or hyperkalaemia, can have serious consequences.

Hypokalaemia is a potentially serious condition where the patient has low levels of K^+ in his/her blood plasma. Symptoms can include weakness of the muscles or ECG (electrocardiogram) abnormalities. Mostly, hypokalaemia can be a result of reduced K^+ intake caused by GI disturbance, such as diarrhoea and vomiting, or increased excretion of K^+ caused by diuresis. Hypokalaemia is often found in patients treated with diuretics such as loop diuretics and thiazides. These classes of drugs increase the secretion of Na^+ in the nephrons in order to increase water excretion. Unfortunately, they also increase the excretion of K^+ and lead to hypokalaemia. In contrast, potassium-sparing diuretics actively preserve potassium ions, and patients treated with loop diuretics or thiazides often receive also potassium-sparing diuretics [3a].

Potassium ions are excreted via the kidneys. Within the kidneys, ~150–180 l of plasma is filtered every day through the glomerulus, which is part of the nephron, in order to produce urine. As previously described, the filtration process is followed by a series of processes along the nephron, where a variety of ions are secreted and re-absorbed in order to regulate plasma imbalances and manage the urine volume (Section 2.3.1). K^+ is passively secreted at the proximal tubule and also moves into the interstitial fluid via a counter-flow process to Na^+ mainly at the distal tubule (Figure 2.19).

Oral supplementation in form of potassium salts is especially necessary in patients who take anti-arrhythmic drugs, suffer from renal artery stenosis and/or severe heart failure or show severe K^+ losses due to chronic

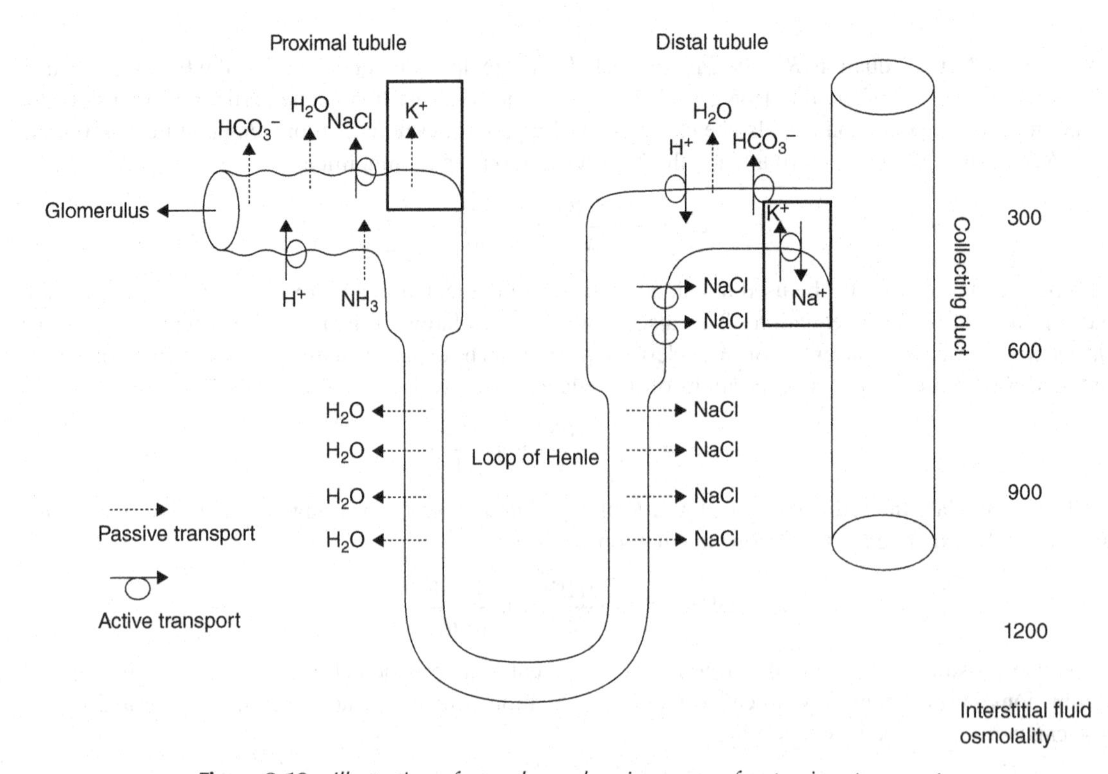

Figure 2.19 *Illustration of a nephron showing areas of potassium transport*

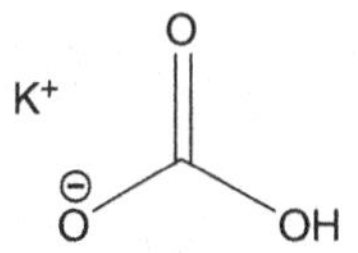

Figure 2.20 *Chemical structure of potassium bicarbonate (KCO_3H)*

diarrhoea or abusive use of laxatives. Regulation of the plasma K^+ level may also be required in the care of elderly patients when the K^+ intake is reduced as a result of changing dietary habits, but special attention has to be given to patients with renal insufficiency because K^+ excretion might be reduced. Potassium salts are preferably given as liquid preparations, and KCl is the preferred salt used. Other potassium-based salts can be used if the patient is at risk of developing hyperchloraemia – increased chloride plasma levels. Typically potassium salts are dissolved in water, but the salty and bitter taste makes them difficult to formulate. Oral bicarbonate solutions such as potassium bicarbonate are typically given orally for chronic acidosis states – low pH of the blood plasma. This can be again due to impaired kidney function. The use of potassium bicarbonate for the treatment of acidosis has to be carefully evaluated, as even small changes of the potassium plasma levels can have severe consequences (Figure 2.20) [3a].

Potassium citrate is used in the United Kingdom as an over-the-counter drug for the relief from discomfort experienced in mild urinary-tract infections by increasing the urinary pH. It should be not given to men if they experience pain in the kidney area (risk of kidney stones) or if blood or pus is present in the urine. Also, patients with raised blood pressure or diabetes should avoid taking potassium citrate without consultation with their general practitioner (GP). Caution is generally advised to patients with renal impairment, cardiac problems and the elderly [3a].

2.4.4 Adverse effects and toxicity: hyperkalaemia

The therapeutic window for K^+ in the blood plasma is very small (3.5–5.0 mmol), and especially hyperkalaemia, an increased level of K^+ in the plasma, can lead to severe health problems [11]. Potassium salts can cause nausea and vomiting and in extreme cases can lead to small bowel ulcerations. Acute severe hyperkalaemia is defined when the plasma potassium concentration exceeds 6.5 mmol/l or if ECG changes are seen. This can lead to cardiac arrest, which needs immediate treatment. Treatment options include the use of calcium gluconate intravenous injections, which minimises the effects of hyperkalaemia on the heart. The intravenous injection of soluble insulin promotes the shift of potassium ions into the cells. Diuretics can also be used to increase the secretion of K^+ in the kidneys, and dialysis can be a good option if urgent treatment is required. Ion-exchange resins, such as polystyrene sulfonate resins, may be used in mild to moderate hyperkalaemia to remove excess potassium if there are no ECG changes present. As previously mentioned, especially in patients suffering from kidney diseases or end-stage renal failure, the potassium levels have to be monitored very carefully and corrected if necessary. Potassium excretion is likely to be disturbed, and a build-up of potassium in the blood plasma may trigger a cardiac arrest [3a].

Potassium salts are also available in the form of tablets or capsules for oral application especially as nonprescription medicine. Usually, their formulation is designed to allow the potassium ions to be slowly secreted, because very high concentrations of K^+ are known to be toxic to tissue cells and can cause injury to the gastric mucosa. Therefore, nonprescription potassium supplement pills are usually restricted to <100 mg of K^+ [6].

2.5 Exercises

2.5.1 Determine the oxidation state for all elements in the following molecules:

(a) H_2O
(b) NaCl
(c) H_2O_2
(d) Fe_2O_3
(e) MnO_4^-
(f) $Cr_2O_7^{2-}$
(g) SiH_4

2.5.2 Complete the reduction and oxidation equation and write the redox equation in the following examples:

(a) $MnO_4^- + Fe^{2+} + ???? \rightarrow Mn^{2+} + Fe^{3+} + ??????$ under acidic conditions

(b) $MnO_4^- + Br^- + OH^- \rightarrow ??? + BrO_3^-$ under basic conditions

(c) $Cr_2O_7^{2-} + Cu^+ \rightarrow Cu^{2+} + ???$ under acidic conditions

(d) $IO_3^- + I^- \rightarrow I_2$ under acidic conditions

2.5.3 Complete the equation and indicate the standard reduction potential E^0_{red} assuming both reactions partners are used in the same concentration.

(a) $Br_2 + 2I^- \rightarrow ????$ $E_0(Br_2/Br^-) = 1.07\,V$
$E_0(I_2/I^-) = 0.54\,V$

(b) $Fe^{2+} + Ce^{4+} \rightarrow ????$ $E_0(Ce^{4+}/Ce^{3+}) = 1.61\,V$
$E_0(Fe^{3+}/Fe^{2+}) = 0.77\,V$

(c) $Br_2 + Fe \rightarrow ????$ $E_0(Br_2/Br^-) = 1.07\,V$
$E_0(Fe^{2+}/Fe) = -0.44\,V$

2.5.4 Calculate E for the following redox pair when $Mn^{3+} = 0.5$ M and $Mn^{2+} = 0.01$ M [$E_0(Mn^{3+}/Mn^{2+}) = 1.51$ V] using the Nernst equation.

2.6 Case studies

2.6.1 Lithium carbonate (Li_2CO_3) tablets

Your pharmaceutical analysis company has been contacted by an important client and asked to analyse a batch of formulated Li_2CO_3 tablets. The description of your brief states that you are supposed to analyse the API in these tablets following standard quality assurance guidelines.

Typical analysis methods used for quality purposes are based on titration reactions. A certain amount of powdered Li_2CO_3 tablets is dissolved in water, and a known amount of HCl is added. The solution is boiled to remove any CO_2. The excess acid is then titrated with NaOH using methyl orange as an indicator [14].

(a) Research the type of titration described. Describe the chemical structure and mode of action of the indicator.
(b) Formulate the relevant chemical equations.
(c) The package states that each tablet contains 250 mg Li_2CO_3. For the experiment, 20 tablets are weighed and powdered (total weight 9.7 g). Powder containing 1 g of Li_2CO_3 is dissolved in 100 ml water, and 50 ml of 1 M HCl is added. After boiling, the solution is titrated against 1 M NaOH using methyl orange as the indicator. For each titration, the following volume of NaOH was used:

35.0 ml	35.5 ml	34.5 ml

Calculate the amount of Li_2CO_3 present in your sample. Express your answer in grams and moles.
(d) Critically discuss your result in context with the stated value for the API.
(e) Research and critically discuss the typically accepted error margins.

2.6.2 Sodium chloride eye drops

Your pharmaceutical analysis company has been contacted by an important client and asked to analyse a batch of eye drops containing a NaCl solution. The description of your brief states that you are supposed to analyse the API in these tablets following standard quality assurance guidelines.

Typical analysis methods used for quality purposes are based on titration reactions. A certain volume of NaCl solution is titrated with silver nitrate ($AgNO_3$). Potassium chromate is used as the appropriate indicator [14].

(a) Research the type of titration described. Describe the chemical structure and mode of action of the indicator.
(b) Formulate the relevant chemical equations.
(c) The package states that the eye drops are a 0.9% w/v aqueous solution of NaCl. For the experiment, a volume containing 0.1 g of NaCl is titrated with a 0.1 M $AgNO_3$ solution. For each titration, the following volume of $AgNO_3$ is used:

16.9 ml	17.0 ml	17.4 ml

Calculate the amount of NaCl present in your sample. Express your answer in grams and moles.
(d) Critically discuss your result in context with the stated value for the API.
(e) Research the typically accepted error margins.

References

1. J. D. Lee, *Concise inorganic chemistry*, 5th ed., Chapman & Hall, London, **1996**.
2. A. D. McNaught, A. Wilkinson, *Compendium of chemical terminology: IUPAC recommendations*, 2nd ed./compiled by Alan D. McNaught, Andrew Wilkinson. ed., Blackwell Science, Oxford, **1997**.
3. (a) G. A. McKay, M. R. Walters, J. L. Reid, *Lecture notes. Clinical pharmacology and therapeutics*, 8th ed., Wiley-Blackwell, Chichester, **2010**; (b) E. R. Tiekink, M. Gielen, *Metallotherapeutic drugs and metal-based diagnostic agents: the use of metals in medicine*, Wiley, Chichester, **2005**.
4. J. F. J. Cade, *Med. J. Aust.* **1975**, *1*, 684–686.
5. J. Levine, K. N. R. Chengappa, J. S. Brar, S. Gershon, E. Yablonsky, D. Stapf, D. J. Kupfer, *Bipolar Disord.* **2000**, *2*, 120–130.
6. *Joint Formulary Committee. British National Formulary*. 60 ed. London: BMJ Group and Pharmaceutical Press; **2010**.
7. N. Farrell, *Uses of inorganic chemistry in medicine*, Royal Society of Chemistry, Cambridge, **1999**.
8. (a) W. C. Drevets, J. L. Price, J. R. Simpson, R. D. Todd, T. Reich, M. Vannier, M. E. Raichle, *Nature* **1997**, *386*, 824–827; (b) D. Ongur, W. C. Drevets, J. L. Price, *Proc. Natl. Acad. Sci. U.S.A.* **1998**, *95*, 13290–13295; (c) G. Rajkowska, J. J. Miguel-Hidalgo, J. R. Wei, G. Dilley, S. D. Pittman, H. Y. Meltzer, J. C. Overholser, B. L. Roth, C. A. Stockmeier, *Biol. Psychiatry* **1999**, *45*, 1085–1098; (d) G. J. Moore, J. M. Bebchuk, K. Hasanat, G. Chen, N. Seraji-Bozorgzad, I. B. Wilds, M. W. Faulk, S. Koch, D. A. Glitz, L. Jolkovsky, H. K. Manji, *Biol. Psychiatry* **2000**, *48*, 1–8.
9. M. J. Berridge, C. P. Downes, M. R. Hanley, *Cell* **1989**, *59*, 411–419.
10. (a) E. S. Emamian, D. Hall, M. J. Birnbaum, M. Karayiorgou, J. A. Gogos, *Nat. Genet.* **2004**, *36*, 131–137; (b) T. Katsu, H. Ujike, T. Nakano, Y. Tanaka, A. Nomura, K. Nakata, M. Takaki, A. Sakai, N. Uchida, T. Imamura, S. Kuroda, *Neurosci. Lett.* **2003**, *353*, 53–56; (c) J. Z. Yang, T. M. Si, Y. S. Ling, Y. Ruan, Y. H. Han, X. L. Wang, H. Y. Zhang, Q. M. Kong, X. N. Li, C. Liu, D. R. Zhang, M. Zhou, Y. Q. Yu, S. Z. Liu, L. Shu, D. L. Ma, J. Wei, D. Zhang, *Biol. Psychiatry* **2003**, *54*, 1298–1301.
11. G. J. Tortora, B. Derrickson, *Principles of anatomy and physiology*, 12th ed., international student/Gerard J. Tortora, Bryan Derrickson. ed., Wiley [Chichester: John Wiley, distributor], Hoboken, N.J., **2009**.
12. NHS, Vol. *2014*, NHS, **2013**.
13. Minerals, E. G. o. V. a., Vol. *2013*, UK Food Standards Agency, **2003**.
14. *British pharmacopoeia*, Published for the General Medical Council by Constable & Co, London.

Further Reading

1. E. Alessio, *Bioinorganic medicinal chemistry*, Wiley-VCH, Weinheim, **2011**.
2. W. Kaim, B. Schwederski, *Bioinorganic chemistry: inorganic elements in the chemistry of life: an introduction and guide*, Wiley, Chichester, **1994**.
3. H.-B. Kraatz, N. Metzler-Nolte, *Concepts and models in bioinorganic chemistry*, Wiley-VCH [Chichester: John Wiley, distributor], Weinheim, **2006**.
4. R. M. Roat-Malone, *Bioinorganic chemistry: a short course*, Wiley, Hoboken, N.J. [Great Britain], **2002**.

3

Alkaline Earth Metals

Members of group 2 of the periodic table (second vertical column) are called *earth alkaline metals*. In this group are included the following elements: beryllium (Be), magnesium (Mg), calcium (Ca), strontium (Sr), barium (Ba) and radium (Ra). Radium is a radioactive element and therefore we will not further discuss it in this chapter (Figure 3.1).

In terms of clinical use, magnesium and calcium are essential ions for the human body and any of their imbalances should be corrected. Strontium is medically used in radiotherapy, and its application is further discussed in Chapter 10. Exposure to excess beryllium can lead to the so-called chronic beryllium disease (CBD), which is discussed later in this chapter. Barium salts are generally highly toxic. Nevertheless, the so-called barium meal is a well-used oral radio-contrast agent.

3.1 Earth alkaline metal ions

Earth alkaline metals together with the alkali metals form the so-called s-block metals. Earth alkaline metals have two electrons in their outer shell which is an s-orbital type. The chemistry of the metals is characterised by the loss of both electrons, which is a result of the relatively low ionisation energy (IE) of both electrons and the subsequent formation of the stable cation M^{2+}, which has a noble gas configuration (Table 3.1).

Group 2 elements are all silvery-white metals with high reactivity, similar to alkali metals, but less soft and not as reactive. Earth alkaline metals can be mostly found in the earth's crust in the form of their cations displayed in minerals and not as the elemental metal, as these are very reactive. For example, beryllium principally occurs as beryl ($Be_3Al_2[Si_6O_{18}]$), which is also known as *aquamarine*.

Magnesium can be found in rock structures such as magnesite ($MgCO_3$) and dolomite ($MgCO_3{\cdot}CaCO_3$), and is the eighth most abundant element in the earth's crust. Calcium is the fifth most abundant element and can be found in minerals such as limestone ($CaCO_3$) and its metamorphs such as chalk and marble.

Earth alkaline metals are harder and have a higher density than sodium and potassium and higher melting points. This is mostly due to the presence of two valence electrons and the resulting stronger metallic bond. Atomic and ionic radii increase within the group, and the ionic radii are significantly smaller than the atomic radii. Again, this is due to the existence of two valence electrons, which are located in the s orbital furthest from the nucleus. The remaining electrons are attracted even closer to the nucleus as a result of the increased

Essentials of Inorganic Chemistry: For Students of Pharmacy, Pharmaceutical Sciences and Medicinal Chemistry,
First Edition. Katja A. Strohfeldt.

Companion website: www.wiley.com/go/strohfeldt/essentials

H																	He
Li	Be											B	C	N	O	F	Ne
Na	Mg											Al	Si	P	S	Cl	Ar
K	Ca	Sc	Ti	V	Cr	Mn	Fe	Co	Ni	Cu	Zn	Ga	Ge	As	Se	Br	Kr
Rb	Sr	Y	Zr	Nb	Mo	Tc	Ru	Rh	Pd	Ag	Cd	In	Sn	Sb	Te	I	Xe
Cs	Ba	La-Lu	Hf	Ta	W	Re	Os	Ir	Pt	Au	Hg	Tl	Pb	Bi	Po	At	Rn
Fr	Ra	Ac-Lr	Rf	Db	Sg	Bh	Hs	Mt	Ds	Rg	Uub						

Figure 3.1 *Periodic table of elements; group 2 elements are highlighted*

Table 3.1 *First, second and third ionisation energies (kJ/mol) of group 2 metals [1]*

	First	Second	Third
Be	900	1 757	14 847
Mg	738	1 450	7 731
Ca	590	1 145	4 910
Sr	550	1 064	4 207
Ba	503	965	3 600

effective nuclear charge. The IEs of the first two valence electrons are similar and relatively low compared to the energy needed to remove the third valence electron, which is part of a fully filled quantum shell. As a result, the dominant oxidation state of earth alkaline metals is +2.

3.1.1 Major uses and extraction

Beryllium is one of the lightest metals and therefore is used in high-speed aircrafts and missiles. Unfortunately, it is highly toxic, and CBD, a scarring of the lung tissue, is often seen in workers from within a beryllium-contaminated work environment.

Calcium is mostly found in limestone and its related forms, such as chalk, and marble and lime (CaO). Ca^{2+} ions are essential for living organisms, as is Mg^{2+}. Magnesium is the only earth alkaline metal that is used on an industrial scale. It is used in ammunition (e.g. tracer bullets and incendiary bombs), as it burns with a very bright white glow. Magnesium alloyed with aluminium results in a low-density and strong material, which is used for lightweight vehicles and aeroplanes.

Magnesium is the only group 2 element that is extracted on a large scale. Its main source is seawater, and the metal is extracted by adding calcium hydroxide. Magnesium hydroxide precipitates, as it is less soluble in water compared to the calcium compound. Magnesium hydroxide is converted into magnesium chloride

Conversion		$2HCl + Mg(OH)_2 \rightarrow MgCl_2 + 2H_2O$
Electrolysis:	at the cathode:	$Mg^{2+}_{(l)} + 2e^- \rightarrow Mg_{(l)}$
	at the anode:	$2Cl^-_{(l)} \rightarrow Cl_{2(g)} + 2e^-$
	Redox:	$2Mg^{2+} + 2Cl^- \rightarrow 2\,Mg_{(l)} + Cl_{2(g)}$

Figure 3.2 *Redox equation for the production of magnesium*

Calcination : Dolomite $[CaMg(CO_3)_2]$ is converted into MgO and CaO

Reduction : $2MgO + 2CaO + FeSi \rightarrow 2Mg + Ca_2SiO_4 + Fe + 1450\,K$

Figure 3.3 *Chemical equation for the production of magnesium*

($MgCl_2$), which can be subsequently electrolysed in a Down's cell (see Section 2.1.1) in order to produce the pure magnesium metal (Figure 3.2).

Alternatively, there is a second method called the *ferrosilicon process* or *pigeon process*. This involves the reduction of magnesium oxide, which is obtained from dolomite, with an iron–silicon alloy. The raw material has to be calcined first, which means the removal of water and carbon dioxide, as these would form gaseous by-products and would reverse the subsequent reduction (Figure 3.3).

3.1.2 Chemical properties

The chemical behaviour of alkaline earth metals is characterised by their strong reducing power, and therefore they very easily form bivalent cations (M^{2+}). The elements within group 2 become increasingly more electropositive on descending within the group.

The **metals** themselves are coloured from grey (Be, Mg) to silver (Ca, Sr, Ba) and are soft. Beryllium and magnesium are passivated and therefore kinetically inert to oxygen or water. The metal barium has to be stored under oil because of its reactivity. Metals such as calcium, strontium and barium react similar to sodium, but are slightly less reactive: All the metals except beryllium form oxides in air at room temperature once the reaction is started. The nitride compound is formed in the presence of nitrogen, and magnesium can burn in carbon dioxide, which means that magnesium fires cannot be extinguished by the use of carbon dioxide fire extinguishers.

$$2M + O_2 \rightarrow 2MO \qquad 3M + N_2 \rightarrow M_3N_2$$

$$2Mg_{(s)} + CO_{2(g)} \rightarrow 2MgO_{(s)} + C_{(s)}$$

The **oxides** of alkaline earth metals have the general formula MO and are generally basic. Beryllium oxide (BeO) is formed by the ignition of beryllium metal in an oxygen atmosphere. The resulting solid is colourless and insoluble in water. Other group 2 oxides (MO) are typically formed by the thermal decomposition of the corresponding metal carbonate or hydroxide.

$$MCO_3 \rightarrow MO + CO_2$$

$$Be(OH)_2 + 2(OH)^- \rightarrow [Be(OH)_4]^{2-}$$
$$Be(OH)_2 + H_2SO_4 \rightarrow BeSO_4 + 2H_2O$$

Figure 3.4 *Chemical equations showing the amphoteric nature of beryllium hydroxide*

Peroxides are known for magnesium, calcium, strontium and barium but not for beryllium. The radius of the beryllium cation (Be^{2+}) is not sufficient to accommodate the peroxide anion.

Beryllium reacts with aqueous alkali (NaOH) and forms beryllium **hydroxide**, which is an amphoteric hydroxide.

The term ***amphoteric*** *describes compounds that can act as an acid and a base.*

Beryllium hydroxide reacts with a base with the formation of the corresponding beryllium salt. Beryllium hydroxide can also be reacted with acids such as sulfuric acid and the corresponding salt, beryllium sulfate, is obtained (Figure 3.4).

Magnesium does not react with aqueous alkali (NaOH). The synthesis of magnesium hydroxide [$Mg(OH)_2$] is based on a metathesis reaction in which magnesium salts are reacted with sodium or potassium hydroxide.

$$Mg^{2+}{}_{(aq)} + 2KOH_{(aq)} \rightarrow Mg(OH)_{2(s)} + 2K^+ \quad (3.1)$$

Calcium, strontium and barium oxides react exothermically with water to form the corresponding hydroxides:

$$CaO_{(s)} + H_2O_{(l)} \rightarrow Ca(OH)_{2(s)}$$
$$SrO_{(s)} + H_2O_{(l)} \rightarrow Sr(OH)_{2(s)}$$
$$BaO_{(s)} + H_2O_{(l)} \rightarrow Ba(OH)_{2(s)} \quad (3.2)$$

Magnesium and calcium hydroxides are sparingly soluble in water and the resulting aqueous solutions are mildly alkaline. In general, group 2 hydroxides, except $Be(OH)_2$, react as bases, and their water solubility and thermal stability increase within the group (Mg→Ba).

Earth alkaline halides (MCl_2) are normally found in their hydrated form. Anhydrous beryllium **halides** are covalent, whereas Mg(II), Ca(II), Sr(II), Ba(II) halides are ionic. As a result of the ionic bond, the later halides have typically high melting points and they are sparingly soluble in water. Additionally, $MgCl_2$, $MgBr_2$, MgI_2 are hygroscopic. Anhydrous calcium chloride also has a strong affinity for water and is typically used as a drying agent.

Hygroscopic compounds *are substances that absorb water from the surrounding air but do not become a liquid.*

3.2 Beryllium and chronic beryllium disease

The element beryllium can be found in the mineral beryl [$Be_3Al_2(SiO_3)_6$] and has minor but important technical applications. Owing to its unique properties, it is used in industrial lightweight systems, for example, turbine rotor blades, automotive parts and electrical contacts. The pure beryllium metal is also used in the nuclear industry.

Beryllium has an exceptionally small atomic radius, and as a result beryllium fluoride, chloride and oxide show evidence of covalent bonds in contrast to the other group 2 oxides or halides. Beryllium halides should be linear if they exhibit the ionic bonding character. This linear form can only be found in the gas phase. In the solid state, the beryllium centre is three or fourfold coordinated, which can be achieved, for example, by polymerisation.

Beryllium and its compounds are extremely poisonous and therefore there is only a very limited potential for their clinical applications. Indeed, even the inhalation of beryllium or its compounds can lead to serious respiratory diseases such as the *chronic beryllium disease*, and soluble beryllium compounds can cause serious skin irritations. Workers within the metal production industry are most likely exposed to beryllium and run the highest risk of developing CBD. But also people working in connected professions such as administrative staff or families are at high risk of beryllium poisoning. Symptoms are not well reported, may occur many years after the exposure and include cough, fatigue and chest pain, whereas nonrespiratory organs can also be affected. However, the introduction of exposure limits and general awareness of the risk have significantly reduced the risk of beryllium exposure and its consequences [2].

3.3 Magnesium: competition to lithium?

The element magnesium (Mg) is a silvery-white and lightweight metal. It is protected by a thin oxide layer, which is very difficult to remove but at the same time removes the need to store it in an oxygen-free environment (see alkali metals). Magnesium reacts with water but much more slowly than its neighbouring earth alkaline metal calcium. Magnesium is a highly flammable metal, and once ignited it burns with a characteristic bright white flame. There are three stable isotopes of magnesium, namely ^{24}Mg (79% occurrence), ^{25}Mg and ^{26}Mg. ^{28}Mg is radioactive with a half-life of 21 h [1].

Most magnesium salts are soluble in water, and given in large amounts they work as a laxative in the human body. Aqueous magnesium ions are sour in taste. Magnesium hydroxide ($MgOH_2$) has only limited solubility in water and the resulting suspension is called *milk of magnesia*, which is commonly used as an antacid and is known to be a mild base. Magnesium is extracted on a large scale using a Down's cell (see Section 2.1.1) or the ferrosilicon process with seawater being the main source (see Section 3.1.1). Mg^{2+} stands in a so-called diagonal relationship to Li^+, which explains why these ions have similar properties and biological activity (see Section 2.2.4).

3.3.1 Biological importance

Mg^{2+} is an essential ion in the human body and is a crucial constituent in numerous enzymatic processes. Indeed, Mg^{2+} is essential to most living cells as a signalling molecule and is involved in nucleic acid biochemistry dealing with the manipulation of ATP (adenosine triphosphate), DNA, RNA and related processes. For example, ATP has to be coordinated to a magnesium ion in order to become biologically active. Mg^{2+} also stabilises DNA and RNA structures, which can be seen in their increased melting points.

Mg^{2+} ions form the redox-active centre in chlorophyll, which facilitates the process of photosynthesis and the connected carbon fixation in green plants. Therefore, green vegetables, as well as milk, whole grain and nuts, are good sources of magnesium. It has to be kept in mind that most magnesium salts are water soluble and therefore processed vegetables, mainly cooked in water, are low in magnesium ion content.

In the human body, Mg^{2+} is the fourth most abundant cation and the second most abundant ion in the interstitial fluid. Mg^{2+} is an essential co-factor dealing with more than 300 cellular enzymatic processes.

On average, the human body contains about 24 g of magnesium ions, with half of it being incorporated into bones and the other half being present in muscles and soft tissue. The majority of Mg^{2+} is absorbed in the ilium and colon, and the kidneys are the major excretory organ. Mg^{2+} is filtered at the glomerulus, and 10–15% is re-absorbed at the proximal tubule, 60–70% at the thick part of the ascending limb of the loop of Henle and 10–15% at the distal tubule [3].

Nevertheless, magnesium salts are generally not well absorbed from the gastrointestinal (GI) tract and therefore are often used as osmotic laxatives. The kidneys regulate the magnesium ion levels in plasma, and as a result high levels of Mg^{2+} are retained when the patient has renal failure. The resulting hypermagnesia can cause muscle weakness and arrhythmia, but it is a rare condition. Hypomagnesia, defined as low magnesium levels in the blood plasma, can be the result of losses in the GI tract, for example, excessive diarrhoea. Magnesium imbalances can also be a result of alcoholism or secondary to treatment with certain drugs. Hypomagnesia is often followed by hypocalcaemia (low calcium ion plasma levels) as well as hypokalaemia and hyponatraemia [3, 4].

3.3.2 Clinical applications and preparations

Magnesium ion imbalances can manifest in a variety of conditions such as hypo- and hypermagnesaemia. Magnesium ion preparations are also used as antacids, mostly in combination with aluminium-based salts (see Section 4.3.5). Additionally, magnesium salts are involved in the treatment of arrhythmia (irregular heart beat) and eclampsia, a life-threatening hypertensive disorder in pregnant women.

Symptomatic hypomagnesaemia is associated with plasma serum Mg^{2+} levels of <0.5–1 mmol/kg for a period of 5 days or more. Mg^{2+} ions are initially given as intravenous (i.v.) or intramuscular injection; the latter is fairly painful and consisting of magnesium sulfate ($MgSO_4$). $MgSO_4$ can also be used as emergency treatment for very serious arrhythmias, a disorder of the heart rate (pulse). In an emergency treatment, it is usually given intravenously as one single dose or with one repeat (Figure 3.5) [4, 5].

Note that the plasma magnesium concentration should be monitored, and the dose has to be reduced in patients with renal impairment as Mg^{2+} is excreted via the kidneys. Magnesium ions can also be given orally to the patient, for example, in the form of magnesium glycerophosphate tablets.

Magnesium hydroxide [$Mg(OH)_2$] is present in antacids because of its laxative properties and is also the main ingredient of the 'milk of magnesia'. The 'milk of magnesia' is a suspension of $Mg(OH)_2$ in water, which has a milk-like appearance because of the low aqueous solubility of $Mg(OH)_2$. It is considered as a strong electrolyte and a weak base and is given to the patient for indigestion and heartburn. The alkaline suspension neutralises any excess stomach acid and therefore works as an antacid. It also stimulates intestinal movement, as the magnesium ions increases the water content in the intestines through its osmotic effect and as a result softens any faeces present.

Magnesium trisilicate ($Mg_2Si_3O_8$) can also be used in antacid preparations especially in the treatment of peptic ulcers. The mode of action includes the increase of the pH of the gastric fluid together with the formation of a colloidal silica precipitate, which forms a protection for the GI mucosa. Most antacids contain a mixture of

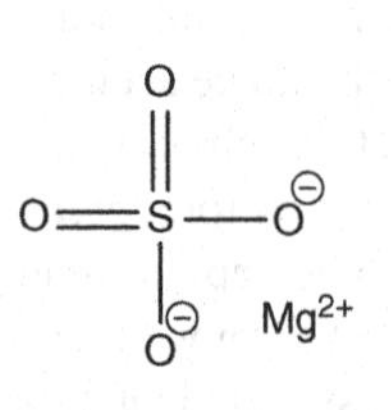

Figure 3.5 *Structure of $MgSO_4$*

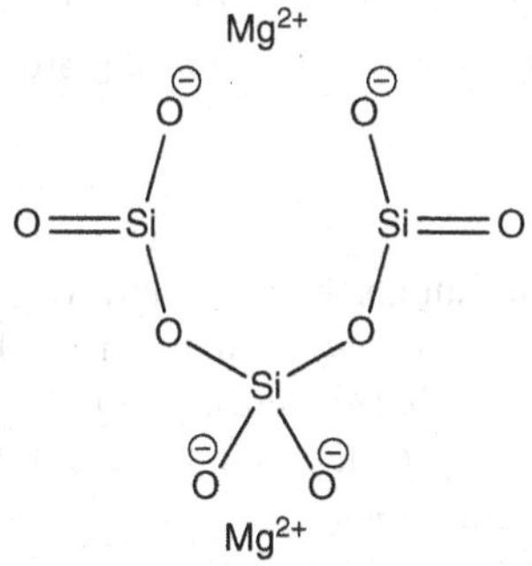

Figure 3.6 *Chemical structure of magnesium trisilicate ($Mg_2Si_3O_8$)*

aluminium hydroxide [$Al(OH)_3$] and magnesium and/or calcium preparations. Therefore, the mode of action will be further discussed in the chapter on aluminium-based drugs (see Section 4.3.5) (Figure 3.6).

Unfortunately, orally taken magnesium salts can show interactions with other drugs taken simultaneously. Magnesium trisilicate reduces the absorption of iron products, certain antibiotics (such as Nitrofurantoin) or antimalarial drugs (such as Proguanil). Magnesium salt preparations, which form part of antacids, are not recommended to be taken at the same time as a variety of drugs such as ACE inhibitors, aspirin and penicillamine. In most cases, antacids reduce the absorption of the simultaneously taken drug. Therefore, before any treatment with antacids, the full medical history of the patient should be taken and possible interactions assessed [4].

3.4 Calcium: the key to many human functions

Calcium is the most abundant inorganic element in the human body and is an essential key for many physiological processes. Ca^{2+} has numerous intra and extracellular physiological roles, for example, a universal role as messenger and mediator for cardiac, skeletal and smooth muscle contractions. Calcium ions are a critical factor in several life-defining biochemical processes as well as in the endocrine, neural and renal aspects of blood pressure homeostasis.

Calcium has the symbol Ca and atomic number 20 and is a soft grey alkaline earth metal. Calcium has four stable isotopes (^{40}Ca and ^{42}Ca–^{44}Ca) and the metal reacts with water with the formation of calcium hydroxide and hydrogen.

$$2Ca + 2H_2O \rightarrow 2CaOH + H_2 \quad (3.3)$$

Calcium salts can be found in everyday life. Limestone, cement, lime scale and fossils are only a few examples where we encounter Ca^{2+}. They also have a wide spectrum of applications spanning from insecticides to clinical applications. Calcium arsenate [$Ca_3(AsO_4)_2$] is extremely poisonous and is used in insecticides. Calcium carbonate ($CaCO_3$) can be found in clinical applications such as antacids, but note that an excessive intake can be hazardous. Calcium chloride ($CaCl_2$) is used in ice removal and dust control on dirt roads, as a conditioner for concrete and as an additive in canned tomatoes. Calcium cyclamate [$Ca(C_6H_{11}NHSO_4)_2$] is used as a sweetening agent, and calcium gluconate [$Ca(C_6H_{11}O_7)_2$] is used as a food additive in vitamin pills. Calcium hypochlorite $Ca(OCl)_2$ can be found in swimming pool disinfectants, in bleaching agents, in deodorants and in fungicides. Calcium permanganate [$Ca(MnO_4)_2$] is used in textile production, as a water-sterilising agent and in dental procedures. Calcium phosphate [$Ca_3(PO_4)_2$] finds applications as a supplement for animal feed, as a fertiliser, in the manufacture of glass and in dental products. Calcium sulfate ($CaSO_4{\cdot}2H_2O$) is the common blackboard chalk.

3.4.1 Biological importance

Calcium ions play important roles in the human body in a variety of neurological and endocrinological processes. Calcium is known as a *cellular messenger* and it has a large intra- versus extracellular gradient (1 : 10 000), which is highly regulated by hormones. This gradient is necessary to maintain the cellular responsiveness to diverse extracellular stimuli. Calcium ions are also involved in the formation of bones and teeth, which act also as a reservoir for calcium ions.

A normal adult body contains ~1000 g of calcium, of which around 99% are extracellular and most of which is stored in bones and teeth. Bones actually serve as a dynamic store for Ca^{2+}. The remaining 1% of Ca^{2+} can be found in the extracellular space, such as plasma, lymph and extracellular water. The intra and extracellular Ca^{2+} concentration is extremely important to many physiological functions and is therefore rigorously controlled (Figure 3.7) [6].

Calcium ions are regulated within the gut, skeleton and kidneys. The Ca^{2+} homeostasis is normally in equilibrium, which means that the amount of Ca^{2+} enters the body is equal to the amount of Ca^{2+} leaving the body. Calcium ion levels are regulated by hormones that are not regulated by the Ca^{2+} level, called *noncalciotropic hormones*, for example, sex hormones and growth factors. In contrast, there are hormones that are directly related to Ca^{2+}, for example, PTH (parathyroid hormone), which are called *calciotropic hormones*. PTH controls the serum plasma level of Ca^{2+} by regulating the re-absorption of Ca^{2+} in the nephron, stimulating the uptake of Ca^{2+} from the gut and releasing Ca^{2+} from the bones which act as a reservoir.

Modified hydroxylapatite, also frequently called *hydroxyapatite* and better known as *bone mineral*, makes up ~50% of our bones. Hydroxylapatite is a natural form of the mineral calcium apatite, whose formula is

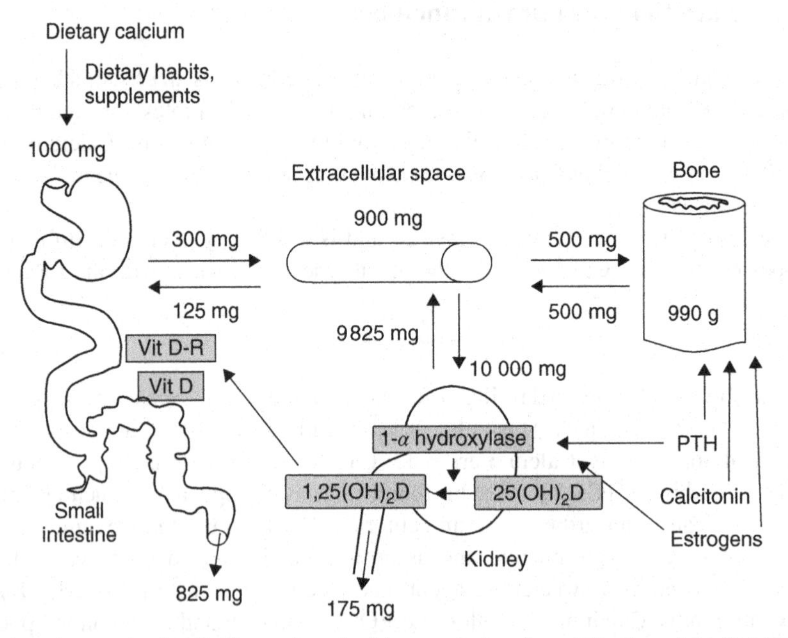

Figure 3.7 Calcium homeostasis [6] (Reproduced with permission from [6]. Copyright © 2005, John Wiley & Sons, Ltd.)

usually denoted as $Ca_{10}(PO_4)_6(OH)_2$. Modifications of hydroxylapatite can also be found in the teeth, and a chemically identical substance is often used as filler for replacement of bones, and so on. Nevertheless, despite similar or identical chemical compositions, the response of the body to these compounds can be quite different.

3.4.2 How does dietary calcium intake influence our lives?

It is believed that an optimal dietary calcium intake can prevent chronic diseases. In the Stone Age, the average calcium intake was 2000–3000 mg Ca^{2+}/day per adult, whereas now-a-days it has decreased to an average of 600 mg/day [7]. This means that we are living in permanent calcium deficiency, and it is believed that there are linkages to various chronic diseases, such as bone fragility, high blood pressure and colon cancer [8].

Ca^{2+} is an essential nutrient, and the required amount varies throughout a person's life time depending on the stage of life. There have been three stages of life identified when the human body needs an increased level of Ca^{2+}. The first one is childhood and adolescence because from birth to the age of ~18 the bones form and grow until they reach their maximum strength. Pregnancy and lactation has also been identified as a time when the human body is in need of an increased level of Ca^{2+}. A full infant accumulates around 30 g of Ca^{2+} during gestation and another 160–300 mg/day during lactation. Ageing has been identified as the third period of life in humans when increased calcium intake is required. This has been associated with several changes to the calcium metabolism in the elderly (Table 3.2).

3.4.3 Calcium deficiency: osteoporosis, hypertension and weight management

Osteoporosis is most commonly associated with calcium deficiency, but an adequate calcium intake should not only be considered as a therapy for bone loss. It should be seen as an essential strategy for the maintenance

Table 3.2 *Optimal daily calcium intake according to NIH Consensus Conference [6]*

Age	mg/d
Neonates	
0–6 mo	400
6–12 mo	600
Children	
1–5 yr	800
6–10 yr	800–1200
Adolescents	
11–24 yr	1200–1500
Male adults	
25–65 yr	1000
Elderly	1500
Female adults	
20–25 yr	1000
Pregnant and nursing	1200–1500
Postmenopausal (>50 yr)	1500
Elderly (>65 yr)	1500

Source: Reproduced with permission from [6].

of health in the ageing human. Ninety-nine percent of Ca^{2+} is found in the bones, as they function as a reservoir. Osteoporosis is known to be the major underlying cause on bone fractures in postmenopausal women. Calcium uptake and plasma concentrations are closely regulated by hormones, as outlined in Section 3.4.1. Nevertheless, there has been no clear and direct relationship between Ca^{2+} intake and bone health established until now. It is believed that a high Ca^{2+} concentration and vitamin D level is essential in the first three decades of life in order to establish an optimum bone density level. These also modify the rate of bone loss, which is associated with ageing.

Studies support the hypothesis that calcium supplementation can reduce blood pressure, being more beneficial to salt-dependent hypertension. The regulation of the cellular calcium metabolism is central to blood pressure homeostasis. It is believed that the higher the level of cytosolic-free calcium ions, the greater the smooth muscle vasoconstrictor tone, which in turn has an effect on the sympathetic nervous system activity and thus on the blood pressure. Nevertheless, studies do not justify the use of calcium supplementation as the sole treatment for patients with mild hypertension.

It has been hypothesised that there exists a link between dietary calcium and weight management in humans. It has been proposed that a low-calorie, high-Ca^{2+} diet helps in supporting the fight against obesity and increase the energy metabolism. The recommended Ca^{2+} intake should be around 1200 mg/day as previously mentioned depending on the age. Available evidence indicates that increasing the calcium intake may substantially reduce the risk of being overweight, although long-term, large-scale prospective clinical trials need to be conducted to confirm or better clarify this association.

3.4.4 Renal osteodystrophy

Renal osteodystrophy, also called *renal bone disease*, is a bone mineralisation deficiency seen in patients with chronic or end-stage renal failure.

Vitamin D is usually activated in the liver to the pro-hormone calcidiol and then in the kidney to calcitriol, which is the active form of vitamin D. Both activation steps are based on a hydroxylation reaction. Pro-vitamin D is hydroxylated in the 25 position in the liver (calcidiol) and then in the kidney at the 1α-position (calcitriol). Calcitriol helps the body to absorb dietary Ca^{2+} (Figure 3.8).

In patients with renal failure, the activation to calcitriol is depressed, which results in a decreased concentration of Ca^{2+} in the blood plasma. Furthermore, the plasma phosphate level increases as a result of the kidney impairment. This, in turn, reduces the amount of free Ca^{2+} in the blood even more, as the phosphate complexes the free Ca^{2+}. The pituitary gland senses the low levels of plasma Ca^{2+} and releases PTH. As previously outlined, PTH increases the re-absorption of Ca^{2+} in the nephron and absorption in the gut, and promotes the release of Ca^{2+} from the bones. In turn, this leads to a weakening of the bone structure.

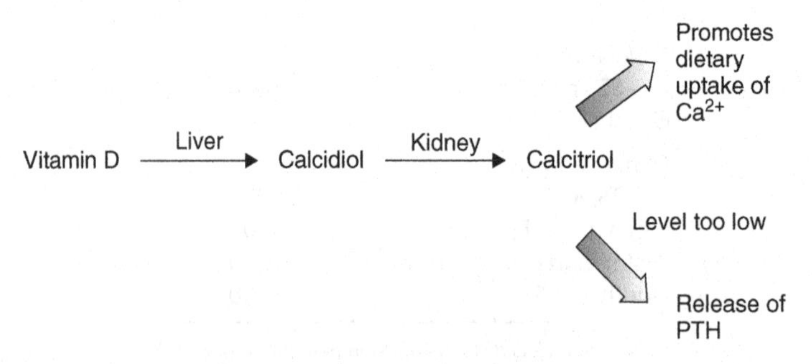

Figure 3.8 *Activation of vitamin D*

Patients can be treated with phosphate binders in order to avoid excess phosphate absorption from the gut. Dialysis will also be helpful in removing excess phosphate from the blood. Furthermore, the patient can be given synthetic calcitriol and potentially calcium supplements.

3.4.5 Kidney stones

Around 20–40% of all kidney stones are associated with elevated Ca^{2+} level in the urine. For a long time, it has been suggested that low dietary calcium intake would be the best method to prevent the recurrence of kidney stones. More recent studies involving patients who suffered from recurring calcium oxalate stones showed that a low calcium diet did not prevent the formation of kidney stones. It was actually found that a higher calcium intake of around 1200 mg/day resulted in a significant reduction of the recurrence of kidney stones by around 50%. It is believed that the restriction of calcium leads to an increase in absorption and excretion of oxalate in the urine and therefore promotes the formation of calcium oxalate stones. Currently, the conclusion is that kidney stone formation in healthy individuals is not associated with calcium supplementation [4].

3.4.6 Clinical application

Calcium supplements are usually required only if the dietary Ca^{2+} intake is insufficient. As previously mentioned, the dietary requirements depend on the age and circumstances; for example, an increased need can be seen in children, in pregnant women and in the elderly where absorption is impaired. In severe acute hypocalcaemia, a slow i.v. injection of a 10% calcium gluconate has been recommended. It has to be kept in mind that the plasma Ca^{2+} level and any changes to the electrocardiogram (ECG) have to be carefully monitored [5].

A variety of calcium salts are used for clinical application, including calcium carbonate, calcium chloride, calcium phosphate, calcium lactate, calcium aspartate and calcium gluconate. Calcium carbonate is the most common and least expensive calcium supplement. It can be difficult to digest and may cause gas in some people because of the reaction of stomach HCl with the carbonate and the subsequent production of CO_2 (Figure 3.9).

Calcium carbonate is recommended to be taken with food, and the absorption rate in the intestine depends on the pH levels. Taking magnesium salts with it can help prevent constipation. Calcium carbonate consists of 40% Ca^{2+}, which means that 1000 mg of the salt contains around 400 mg of Ca^{2+}. Often, labels will only indicate the amount of Ca^{2+} present in each tablet and not the amount of calcium carbonate (Figure 3.10).

$$CaCO_3 + 2HCl \rightarrow CO_2 + CaCl_2 + H_2O$$

Figure 3.9 *Chemical equation showing the synthesis of CO_2 under acidic stomach conditions*

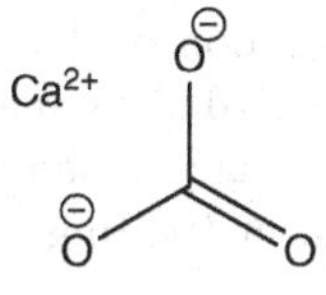

Figure 3.10 *Chemical structure of calcium carbonate*

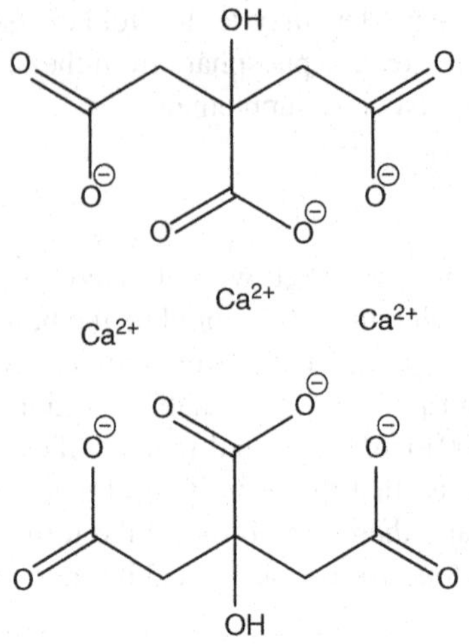

Figure 3.11 Chemical structure of calcium citrate

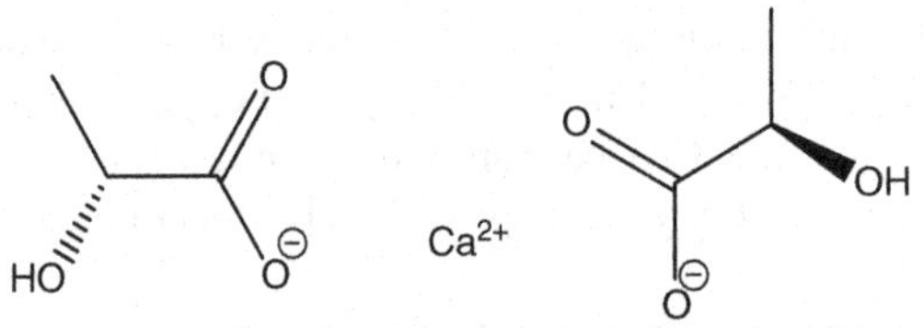

Figure 3.12 Chemical structure of calcium lactate

Calcium citrate is more easily absorbed (bioavailability is 2.5 times higher than calcium carbonate); it is easier to digest and less likely to cause constipation and gas than calcium carbonate. Calcium citrate can be taken without food and is more easily absorbed than calcium carbonate on an empty stomach. It is also believed that it contributes less to the formation of kidney stones. Calcium citrate consists of around 24% Ca^{2+}, which means that 1000 mg calcium citrate contains around 240 mg Ca^{2+}. The lower Ca^{2+} content together with the higher price makes it a more expensive treatment option compared to calcium carbonate, but its slightly different application field can justify this (Figure 3.11).

The properties of calcium lactate are similar to those of calcium carbonate [9], but the former is usually more expensive. Calcium lactate contains effectively less Ca^{2+} per gram salt than, for example, calcium carbonate. Calcium lactate consists of only 18% Ca^{2+}, making it a less 'concentrated' salt (Figure 3.12) [10].

Calcium gluconate is prescribed as a calcium supplement, but it is also used in the urgent treatment of hyperkalaemia (K^+ plasma levels above 6.5 mmol/l). Hyperkaleamia in the presence of ECG changes usually requires immediate treatment, and a 10% calcium gluconate solution intravenously administered is recommended (see Section 2.4.4). Administration of the calcium solution does not lower the plasma K^+ level but protects temporarily against myocardial excitability and therefore temporarily reduces the toxic effects of hyperkalaemia. Calcium gluconate contains effectively the least Ca^{2+} per amount of supplement (only around 9%). That means that in 1000 mg calcium gluconate, only 90 mg is actual Ca^{2+} (Figure 3.13).

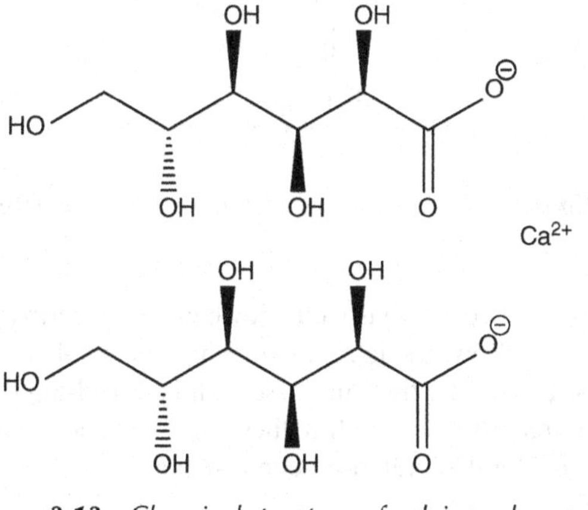

Figure 3.13 Chemical structure of calcium gluconate

3.4.7 Side effects

Several large long-term studies have shown that a daily intake of 1000–2500 mg of calcium salts is safe. Side effects have been observed only at relatively high doses, being manifested in GI disturbances such as constipation and bloating and, in extreme cases, arrhythmia [11]. The GI system normally adjusts after a while, and problems should resolve themselves. Calcium salts are generally better absorbed in an acid environment, so patients with a low production of stomach acid or elderly patients who are on high doses of antiulcer medication might experience problems with absorption. It is then recommended to consume the calcium supplement with a meal [5].

Nevertheless, it is important to note that calcium ions can interfere with the absorption of some drugs, such as antibiotics. For example, tetracycline and quinolone antibiotics can chelate Ca^{2+} ions and form complexes which cannot be absorbed anymore. Therefore, calcium supplements and antibiotics should not be taken together. Patients are typically advised to take antibiotics 1 h before or 2 h after food [5].

3.5 Barium: rat poison or radio-contrast agent?

The element barium (Ba) has the atomic number 56 and is classified as a heavy metal. Barium metal is highly reactive and therefore no elemental barium exists in nature. Natural sources of barium are the water-insoluble minerals barite (barium sulfate) and whiterite (barium carbonate). In order to obtain pure barium compounds, the mineral barite is reacted with carbon, and barium sulfide is formed. Barium sulfide is, in contrast to barium sulfate, water soluble. Subsequently, the pure barium sulfide is treated with sulfuric acid and pure barium sulfate can be obtained.

$$BaSO_4 + 4C \rightarrow BaS + 4CO$$

$$BaS + H_2SO_4 \rightarrow BaSO_4 + H_2S$$

Barium salts can be highly toxic even at low concentrations. Barium carbonate is highly toxic and can be used as rat poison as it readily dissolves in the stomach acid. Barium sulfate is the least toxic barium compound mainly because of its insolubility. Barium sulfate is used in a variety of applications ranging from white paint to X-ray contrast agent.

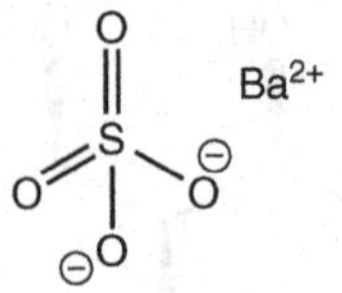

Figure 3.14 *Chemical structure of barium sulfate*

The clinical use of barium sulfate suspension is well known under the term *barium meal*. Patients are given a suspension of barium sulfate to swallow. Using X-ray imaging, the whole oesophagus, the stomach and the intestines can be visualised. Barium sulfate lines the tissue whilst travelling through the digestive tract. The heavy barium ions absorb X-rays readily and therefore these structures become visible in an X-ray screening. Barium sulfate is a well-used and tolerated oral radio-contrast agent. It is also used as radio-contrast agent in enemas (Figure 3.14) [3, 4].

3.6 Exercises

3.6.1 Calcium supplementation

Calcium supplementation is recommended as a dietary supplement especially for menopausal, pregnant or nursing women. There are a variety of calcium salts on the market that can be used for oral administration. For the examples given below, determine the chemical formula, the molecular weight and the Ca^{2+} content (expressed in gram/gram (g/g) and percentage weight/weight (%w/w)).

(a) Calcium carbonate
(b) Calcium lactate
(c) Calcium chloride
(d) Calcium citrate
(e) Calcium gluconate

3.6.2 Complete the redox equation (including the half-equations) and indicate the standard reduction potential assuming that both reaction partners are present in the same concentration.

(a) $Mg + Cl_2 \rightarrow ??$
(b) $Ba + Br_2 \rightarrow ??$

3.6.3 Complete the following redox equation (including the half-equations).

$$Mg + Cl_2 \rightarrow ??$$

Indicate the standard reduction potential assuming a concentration of $[Mg^{2+}] = 0.7$ mol/l and $[Cl^-] = 0.8$ mol/l.

3.6.4 Milk of magnesia

Milk of magnesia is typically an 8.7% w/v aqueous suspension of magnesium hydroxide.

(a) What are the chemical formula and the molecular weight of magnesium hydroxide?
(b) What is the concentration of magnesium hydroxide in gram per litre?
(c) How many moles of such magnesium salts are present in a 100-ml suspension?

3.7 Case studies

3.7.1 Magnesium hydroxide suspension

Magnesium hydroxide mixture is an aqueous oral suspension containing hydrated magnesium oxide. It is indicated for use in constipation in adults and children. Typical analysis methods used for quality purposes are based on titration reactions. A certain volume of the suspension containing hydrated magnesium oxide [$Mg(OH)_2$] is typically reacted with a known amount of sulfuric acid (H_2SO_4). The excess acid is then titrated with sodium hydroxide (NaOH) and methyl orange as an indicator [12].

(a) Research the type of titration described. Describe the chemical structure and mode of action of the indicator.
(b) Formulate the relevant chemical equations.
(c) For the analysis, 10 g of the suspension was reacted with 50 ml of 0.5 M H_2SO_4. The excess H_2SO_4 was titrated with 1 M NaOH using methyl orange as indicator. For each titration, the following volume of NaOH has been used:

11.0 ml	11.2 ml	10.9 ml

Calculate the amount of $Mg(OH)_2$ present in your sample. Express your answer in grams and moles.

3.7.2 Calcium carbonate tablets

Your pharmaceutical analysis company has been contacted by an important client and asked to analyse a batch of injections containing calcium carbonate ($CaCO_3$). The description of your brief states that you are supposed to analyse the active pharmaceutical ingredient (API) in these tablets following standard quality assurance guidelines.

Typical analysis methods used for quality purposes are based on titration reactions. A certain amount of the tablet powder is dissolved in water and hydrochloride acid (HCl). A known amount of disodium edetate is added. After adjustment of the pH, the excess disodium edetate is titrated with zinc chloride (ZnCl) using morbant black II solution as indicator [12].

(a) Research the type of titration described. Describe the chemical structure and mode of action of the indicator. You may want to familiarise yourself with chelation (see Section 11.2).
(b) Formulate the relevant chemical equations.
(c) The package states that each tablet contains 1.5 g of $CaCO_3$. For the experiment, 20 tablets are weighed (total weight 42.6 g) and powdered. An amount of powder containing 50 mg of Ca^{2+} is dissolved in water and HCl and reacted with 50 ml of 0.05 M disodium edetate. After adjusting the pH to 10.9, the excess disodium edetate is titrated with 0.05 M $ZnCl_2$ solution. For each titration, the following volume of $ZnCl_2$ has been used:

25.0 ml	24.8 ml	25.3 ml

Calculate the amount of $CaCO_3$ present in your sample. Express your answer in grams and moles.
(d) Critically discuss your result in context with the stated value for the API.
(e) Research the typically accepted error margins.

References

1. J. D. Lee, *Concise inorganic chemistry*, 5th ed., Chapman & Hall, London, **1996**.
2. B. P. Barna, D. A. Culver, B. Yen-Lieberman, R. A. Dweik, M. J. Thomassen, *Clin. Diagn. Lab. Immunol.* **2003**, *10*, 990–994.
3. G. J. Tortora, B. Derrickson, *Principles of anatomy and physiology*, 12th ed., international student/Gerard J. Tortora, Bryan Derrickson. ed., Wiley [Chichester: John Wiley, distributor], Hoboken, N.J., **2009**.
4. G. A. McKay, M. R. Walters, J. L. Reid, *Lecture notes. Clinical pharmacology and therapeutics*, 8th ed., Wiley-Blackwell, Chichester, **2010**.
5. *British national formulary*, British Medical Association and Pharmaceutical Society of Great Britain, London.
6. E. R. Tiekink, M. Gielen, *Metallotherapeutic drugs and metal-based diagnostic agents: the use of metals in medicine*, Wiley, Chichester, **2005**.
7. S. B. Eaton, D. A. Nelson, *Am. J. Clin. Nutr.* **1991**, *54*, S281–S287.
8. M. J. Bargerlux, R. P. Heaney, *J. Nutr.* **1994**, *124*, S1406–S1411.
9. B. R. Martin, C. M. Weaver, R. P. Heaney, P. T. Packard, D. L. Smith, *J. Agric. Food Chem.* **2002**, *50*, 3874–3876.
10. D. A. Straub, *Nutr. Clin. Pract.* **2007**, *22*, 286–296.
11. (a) M. A. Oconnell, J. S. Lindberg, T. P. Peller, H. M. Cushner, J. B. Copley, *Clin. Pharm.* **1989**, *8*, 425–427; (b) W. F. Caspary, *Eur. J. Gastroen. Hepat.* **1996**, *8*, 545–547.
12. *British pharmacopoeia*, Published for the General Medical Council by Constable & Co, London.

Further Reading

1. E. Alessio, *Bioinorganic medicinal chemistry*, Wiley-VCH, Weinheim, **2011**.
2. W. Kaim, B. Schwederski, *Bioinorganic chemistry: inorganic elements in the chemistry of life: an introduction and guide*, Wiley, Chichester, **1994**.
3. H.-B. Kraatz, N. Metzler-Nolte, *Concepts and models in bioinorganic chemistry*, Wiley-VCH [Chichester: John Wiley, distributor], Weinheim, **2006**.
4. R. M. Roat-Malone, *Bioinorganic chemistry: a short course*, Wiley, Hoboken, N.J. [Great Britain], **2002**.

4

The Boron Group – Group 13

Group 13 (13th vertical column of the periodic table) is called the *boron group* and it consists of boron (B), aluminium (Al), gallium (Ga), indium (In) and thallium (Tl) (Figure 4.1).

All elements within group 13 show a wide variety of properties. It is important to note that boron is a metalloid (semi-metal) whereas aluminium is a metal but shows many chemical similarities to boron. Aluminium, gallium, indium and thallium are considered to be metals of the 'poor metals' group.

> ***Metalloids** are elements that display some properties characteristic for metals and some characteristic for nonmetals.*

In this chapter, the general chemistry of group 13 elements is discussed as well as some clinical applications for boron and aluminium. Further clinical applications for boron as well as applications for thallium can be found in the chapter on radiochemistry (Chapter 10).

4.1 General chemistry of group 13 elements

Group 13 elements are characterised by having three electrons in their valence shell. Therefore, all elements form the stable cation M^{3+}. Most elements (with the exception of B) form additionally the singly positively charged ion M^{+}, which is indeed the more stable oxidation state for Tl.

Boron and aluminium occur only with oxidation number +3 in their compounds, and with a few exceptions their compounds are best described as ionic. The electronic configuration shows three electrons outside a noble gas configuration, two in an s shell and one in a p shell. The outermost p electron is easy to remove as it is furthest from the nucleus and well shielded from the effective nuclear charge. The next two s electrons are also relatively easy to remove. Removal of any further electrons disturbs a filled quantum shell and is therefore difficult. This is reflected in the ionisation energies (Table 4.1).

The main sources of B are the two minerals borax ($Na_2[B_4O_5(OH)_4]\cdot 8H_2O$) and kernite ($Na_2[B_4O_5(OH)_4]$), which are generally used as components in many detergents or cosmetics. Al occurs widely on earth, and it

Essentials of Inorganic Chemistry: For Students of Pharmacy, Pharmaceutical Sciences and Medicinal Chemistry,
First Edition. Katja A. Strohfeldt.

Companion website: www.wiley.com/go/strohfeldt/essentials

H																	He
Li	Be											B	C	N	O	F	Ne
Na	Mg											Al	Si	P	S	Cl	Ar
K	Ca	Sc	Ti	V	Cr	Mn	Fe	Co	Ni	Cu	Zn	Ga	Ge	As	Se	Br	Kr
Rb	Sr	Y	Zr	Nb	Mo	Tc	Ru	Rh	Pd	Ag	Cd	In	Sn	Sb	Te	I	Xe
Cs	Ba	La-Lu	Hf	Ta	W	Re	Os	Ir	Pt	Au	Hg	Tl	Pb	Bi	Po	At	Rn
Fr	Ra	Ac-Lr	Rf	Db	Sg	Bh	Hs	Mt	Ds	Rg	Uub						

Figure 4.1 *The periodic table of elements, group 13 elements are highlighted*

Table 4.1 *Ionisation energy (kJ/mol) for group 13 elements [1]*

	First	Second	Third
B	801	2427	3659
Al	577	1816	2744
Ga	579	1979	2962
In	558	1820	2704
Tl	589	1971	2877

is the most abundant metal and the third most abundant element in the earth's crust. Aluminosilicates, such as clays, micas, feldspar, together with bauxite, are the main sources of Al. Ga, In and Tl occur in traces as their sulfides.

4.1.1 Extraction

Boron (B) can be extracted from borax by converting the latter to boric acid (Equation 4.1) and subsequently to the corresponding oxide (Equation 4.2). Boron of low quality can then be obtained by the reduction of boron oxide with Mg, followed by several steps of washing with bases and acids.

$$Na_2[B_4O_5(OH)_4] \cdot 8H_2O + H_2SO_4 \rightarrow 4B(OH)_3 + Na_2SO_4 + 5H_2O \quad (4.1)$$

$$2B(OH)_3 \rightarrow B_2O_3 + 3H_2O \quad (4.2)$$

Al is extracted from ores such as bauxite or cryolite in the so-called Bayer process. Bauxite contains mainly a mixture of aluminium oxides with Fe_2O_3, SiO_2 and TiO_2 as impurities. In the Bayer process, hot aqueous NaOH is added to the crude ore under pressure and aluminium hydroxide will go into solution. This will result in the separation of Fe_2O_3. The solution is cooled down and seeded with $Al_2O_3{\cdot}3H_2O$ in order to precipitate

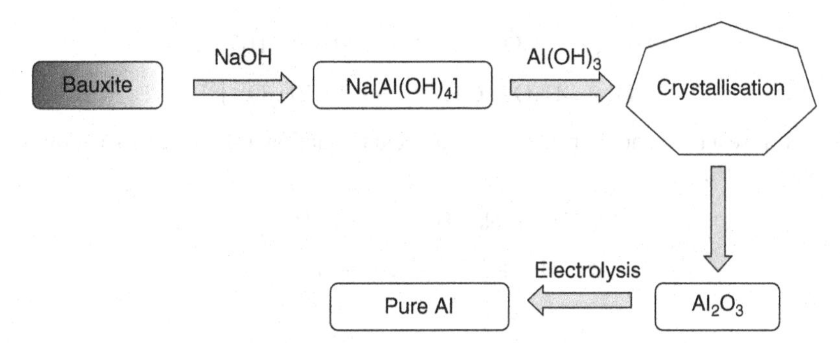

Figure 4.2 *Bayer process*

$Al(OH)_3$. Pure Al can be produced by electrolysis of molten Al_2O_3 (melting point 2345 K), with Al being obtained at the cathode (Figure 4.2).

The main source of Ga is bauxite, but it can also be obtained from the residues from the Zn processing industry. It can be found in the zinc sulfide ore sphalerite. Tl can be obtained as by-product of the processing of Cu and Zn ores. The demand for In and Tl is rather low.

4.1.2 Chemical properties

4.1.2.1 Reactivity

B is chemically unreactive except at high temperatures. Al is a highly reactive metal, which is readily oxidised in air to Al_2O_3. This oxide coating is resistant to acids but is moderately soluble in alkalis. Al itself dissolves in diluted mineral acids (Equation 4.3) and can react with strong alkalis, the product being the tetrahydroxoaluminate ion $[Al(OH)_4{}^-]$ and H_2 (Equation 4.4).

$$Al + 3H_2SO_{4(aq)} \rightarrow Al(SO_4)_3 + 3H_2 \quad (4.3)$$

$$2Al + 2MOH + 6H_2O \rightarrow 2M[Al(OH)_4] + 3H_2 \quad (4.4)$$

Aluminium can be used to reduce metal oxides, the most famous example being the *thermit process.* Al reacts violently with iron(Ill) oxide to produce iron in this highly exothermic process, where Fe is obtained in its liquid form (Equation 4.5).

$$2Al_{(s)} + Fe_2O_{3(s)} \rightarrow 2Fe_{(l)} + Al_2O_{3(s)} \quad (4.5)$$

Ga, In and Tl dissolve in most acids, and as a result the salts of Ga(III), In(III) and Tl(I) are obtained, whereas only Ga reacts with aqueous alkali with the production of H_2.

4.1.2.2 Oxides/hydroxides: amphoteric compounds

Boron oxide (B_2O_3) is an acidic oxide and an insoluble white solid with a very high boiling point (over 2000 K) as a result of its extended covalently bonded network structure. Aluminium oxide (Al_2O_3, Equations 4.6 and 4.7) as well as aluminium hydroxide ($Al(OH)_3$, Equations 4.8 and 4.9) are amphoteric compounds.

***Amphoteric compounds** are substances that can react either as an acid or as a base.*

$$Al_2O_3 + 3H_2O + 2(OH^-) \rightarrow 2[Al(OH)_4]^- \quad (4.6)$$

$$Al_2O_3 + 3H_2O + 6[H_3O]^+ \rightarrow 2[Al(H_2O)_6]^{3+} \quad (4.7)$$

$Al(OH)_3$ can neutralise a base and therefore act as an acid (Equation 4.8); it can also neutralise an acid and act as a base (Equation 4.9).

$$Al(OH)_3 + NaOH \rightarrow NaAl(OH)_4 \quad (4.8)$$

$$Al(OH)_3 + 3HCl \rightarrow AlCl_3 + 3H_2O \quad (4.9)$$

4.1.2.3 Halides

The most important halide of boron is the colourless gas boron trifluoride (BF_3). Aluminium chloride ($AlCl_3$) is a volatile solid which sublimes at 458 K. The vapour formed on sublimation consists of an equilibrium mixture of monomers ($AlCl_3$) and dimers (Al_2Cl_6). It is used to prepare the powerful and versatile reducing agent lithium tetrahydridoaluminate ($LiAlH_4$).

Both boron trichloride (BCl_3) and aluminium trichloride ($AlCl_3$) act as Lewis acids to a wide range of electron-pair donors, and this has led to their widespread use as catalysts. In the important *Friedel–Crafts acylation*, $AlCl_3$ is used as a strong Lewis acid catalyst in order to achieve the acylation of an aromatic ring.

*A **Lewis acid** is defined as a compound that can accept electrons pairs with the formation of a coordinate covalent bond. Any type of electrophile can be a Lewis acid. In contrast, **Brønsted–Lowry acids** are compounds that transfer a hydrogen ion (H^+) and they are the more commonly known type of acids. Analogous definitions apply for a Lewis base (electron donator) and a Brønsted–Lowry base (H^+ acceptor).*

The following sections will describe the clinical application of boron, aluminium and gallium. It is important to note that more information on the clinical use of gallium and thallium can be found in Section 10.4, where radiopharmaceuticals are discussed.

4.2 Boron

4.2.1 Introduction

Boron has the atomic number 5 and the symbol B, and is a so-called metalloid (see Chapter 4). Boron compounds have been known for many centuries and especially used in the production of glass. Boric acid [$B(OH)_3$] is used in the large-scale production of glass. Borosilicate glasses (Pyrex® glass), which are produced by a fusion of B_2O_3 and silicate, are extremely heat resistant and often used in laboratories.

At the beginning of the nineteenth century, it was recognised that boron is an essential micronutrient for plants. A deficiency of boron can lead to deformation in the vegetable growth such as hollow stems and hearts. Furthermore, the plant growth is reduced and fertility can be affected. In general, boron deficiency leads to qualitative and quantitative reduction in the production of the crop. Boron is typically available to plants as boric acid [$B(OH)_3$] or borate $[B(OH)_4]^-$. The exact role of boron in plants is not understood, but there is evidence that it is involved in pectin cross-linking in primary cell walls, which is essential for normal growth and development of higher plants [2].

Borax ($Na_2[B_4O_5(OH)_4]\cdot 8H_2O$) can be applied as a fertiliser and, together with kernite ($Na_2[B_4O_5(OH)_4]\cdot 2H_2O$), forms the two most commercially available borates. Borates find a wide range of practical applications

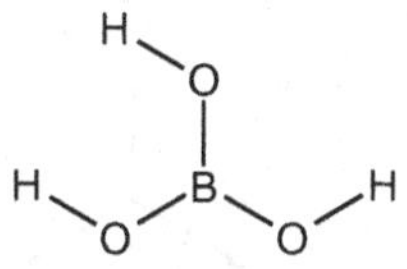

Figure 4.3 Chemical structure of boric acid

such as in detergents, cosmetics, antifungal mixtures as well as components in fibreglass and others. The toxicity of borates in mammals is relatively low, but it exhibits a significantly higher risk to arthropods and can be used as an insecticide.

Boron-based compounds are used in a wide range of clinical applications including their use as antifungal and antimicrobial agent, as proteasome inhibitors and as a noninvasive treatment option for malignant tumours. The latter application will be discussed in the chapter on radiopharmaceuticals (Chapter 10).

4.2.2 Pharmaceutical applications of boric acid

Boric acid is a long-standing traditional remedy with mainly antifungal and antimicrobial effects. For medicinal uses, it has become known as *sal sedativum*, which was discovered by Homberg, the Dutch natural philosopher, in 1702 [3]. Diluted solutions were and sometimes still are used as antiseptics for the treatment of athletes' foot and bacterial thrush, and in much diluted solutions as eyewash (Figure 4.3) [4].

Boric acid can be prepared by reacting borax with a mineral acid:

$$Na_2B_4O_7 \cdot 10H_2O + 2HCl \rightarrow 4B(OH)_3[\text{or } H_3BO_3] + 2NaCl + 5H_2O$$

In general, there are many other health claims around the clinical use of boric acid and boron-containing compounds, but many of those have no supporting clinical evidence.

4.2.3 Bortezomib

Bortezomib belong to the class of drugs called *proteasome inhibitors* and is licensed in the United States and the United Kingdom for the treatment of multiple myeloma. The drug has been licensed for patients in whom the myeloma has progressed despite prior treatment or where a bone marrow transplant is not possible or was not successful. It is marketed under the name Velcade® or Cytomib®. Velcade is administered via injection and is sold as powder for reconstitution (Figure 4.4) [5].

Bortezomib was the first drug approved in the new drug class of proteasome inhibitors and boron seems to be its active element. For the mode of action, it is believed that the boron atom binds with high affinity and specificity to the catalytic site of 26S proteasome and inhibits its action. Therapy with Bortezomib can lead to a variety of adverse reactions, including peripheral neuropathy, myelosuppression, renal impairment and gastrointestinal (GI) disturbances together with changes in taste. Nevertheless, the side effects are in most cases less severe than with alternative treatment options such as bone marrow transplantation [5].

4.3 Aluminium

4.3.1 Introduction

The element aluminium has the atomic number 13 and chemical symbol Al. Aluminium forms a diagonal relationship with beryllium. The name 'aluminium' derives from the salt alum (potassium alum,

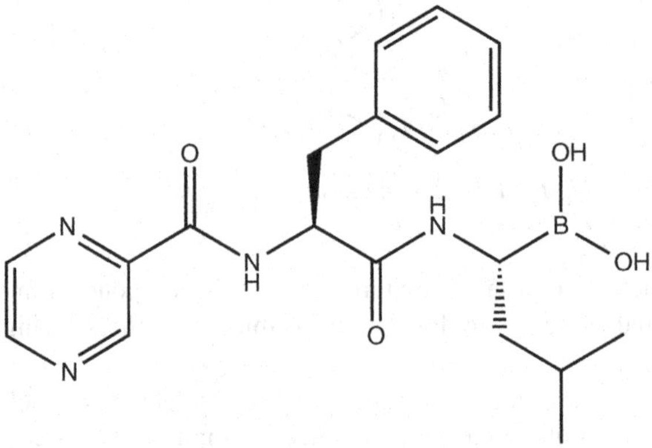

Figure 4.4 Chemical structure of bortzemib

$KAl(SO_4)_2 \cdot 12H_2O$), which was used for medicinal purposes in Roman times. Initially, it was very difficult to prepare pure aluminium and therefore it was regarded as a very precious substance. In the mid-1800s, aluminium cutlery was used for elegant dinners, whereas it is nowadays used as lightweight camping cutlery. In 1886, the manufacture of aluminium by electrolysis of bauxite started, and the price for pure aluminium dropped significantly. Aluminium is a soft, durable and lightweight metal, which makes it attractive to many applications. Nowadays, aluminium is mainly used for the construction of cars and aircrafts and can be found in packaging and construction materials.

4.3.2 Biological importance

The human body contains around 35 mg of Al^{3+}, of which ~50% is found in the lungs and ~50% in the skeleton. There is no known biological role for Al^{3+} and, indeed, the human body has developed very effective barriers to exclude it. Only a minimal fraction of Al^{3+} is taken up from the diet in the gut, and the kidneys fairly quickly excrete most of it. The bones can act as a sink for Al^{3+} if the blood concentration is high and release it slowly over a long period. The brain is vulnerable to Al^{3+} and usually the blood–brain barrier prevents Al^{3+} entering the brain. Al^{3+} can sometimes act as a competitive inhibitor of essential elements such as Mg^{2+}, Ca^{2+} and $Fe^{2+/3+}$ because of their similar ionic radii and charges. It is important to note that at physiological pH, Al^{3+} forms a barely soluble precipitate $Al(OH)_3$, which can be dissolved by changing the pH (see Equations 4.8 and 4.9) [6].

A normal adult diet contains typically between 2.5 mg/day and up to 13 mg/day Al^{3+}, but patients on aluminium-containing medication can be exposed to more than 1000 mg/day. Typically, ~0.001% is absorbed in the digestive tract, but it can be around 0.1–1.0% when it is in the form of aluminium citrate (Figure 4.5) [6b].

Al^{3+} can accumulate in the human body if natural limits are crossed, for example, intravenous administration or patients on dialysis, or when the kidneys are impaired and therefore not able to excrete Al^{3+} sufficiently. Under normal circumstances, Al^{3+} would not accumulate in the human body. Nevertheless, in 1972, Alfrey *et al.* described the new syndrome of progressive dialysis encephalopathy, the so-called dialysis dementia, which was seen in patients being treated with haemodialysis for 15 months or more. The symptoms include speech disorders, problems with the bone mineralisation and general signs of dementia. Investigations showed that brain scans were normal and that there was no connection to the Alzheimer's disease, as neither neurofibrillary tangles nor senile plaques were found. Increased serum and bone concentrations of Al^{3+} were

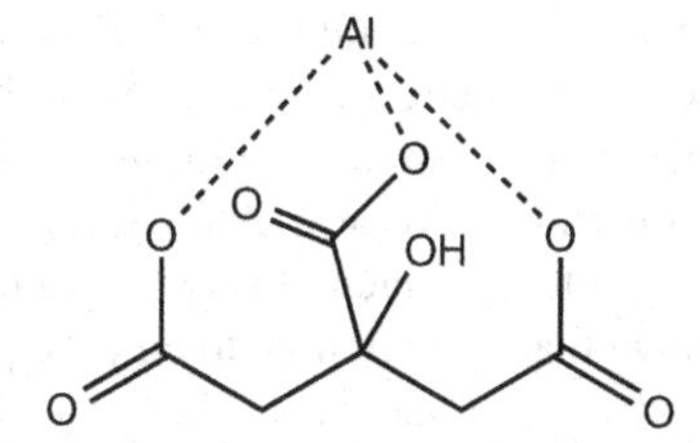

Figure 4.5 Chemical structure of aluminium citrate

found in patients who were on haemodialysis, and the connection was made to the toxicity of the Al^{3+} present in the dialysate solution. Nowadays, the use of modern Al^{3+}-free dialysate solutions or new techniques (e.g. reverse osmosis) prevents 'dialysis dementia' [6a].

4.3.3 Al^{3+} and its use in water purification

Al^{3+} is used in the purification of water. Lime (CaO) and aluminium sulfate $Al_2(SO_4)_3$ are added to waste water in order to accelerate the settling or sedimentation of suspended matter [7]. The addition of lime increases the pH of the water slightly (Equation 4.10). The water becomes more basic, which promotes the precipitation of Al^{3+} as $Al(OH)_3$ (Equation 4.11).

$$CaO_{(s)} + H_2O_{(l)} \rightarrow Ca^{2+}{}_{(aq)} + 2OH^-{}_{(aq)} \tag{4.10}$$

$$Al^{3+}{}_{(aq)} + 3OH^-{}_{(aq)} \rightarrow Al(OH)_{3(s)} \tag{4.11}$$

$Al(OH)_3$ precipitates as a gelatinous precipitate which slowly settles. During this process, it incorporates suspended soil, colloidal material and most bacteria. The water is filtered before leaving the treatment plant in order to remove the flocculate and the vast majority of the Al^{3+}. WHO guidelines allow a maximum concentration for Al^{3+} of 0.2 mg/l [8].

4.3.4 Aluminium-based adjuvants

An adjuvant is an agent or a mixture of agents that possesses the ability to bind to a specific antigen. Adjuvants are added to vaccines in order to increase the antibody responses to the vaccination and/or to stabilise the preparation. Adjuvants can absorb many antigenic molecules over a wide surface area, thus enhancing the interaction of immune cells with the presenting antigens and leading to an increase of the immune response stimulation. Some adjuvants (including aluminium-based ones) can function as a slow-release delivery system. They trap the antigen in a depot created by the adjuvant at the injection site. From there, the antigen is slowly released, which causes a steady stimulation of the immune system.

Aluminium-based adjuvants have a long-standing tradition and have been used for more than 50 years. They are the most widely used adjuvants in human and veterinary vaccines and regarded as safe if applied correctly. Al^{3+} salts are the only kind of adjuvant licensed by the FDA. They are also the only kind of adjuvants used in anthrax vaccines for humans in the United States. Anthrax vaccine contains $Al(OH)_3$, as do the FDA-licensed diphtheria, haemophilus influenzae type B, hepatitis A, hepatitis B, Lyme disease, pertussis and tetanus vaccines. In many countries, vaccines for children contain aluminium-based adjuvants [9].

The adjuvant effect of potassium alum ($KAl(SO_4)_2 \cdot 12H_2O$) was first discovered in 1926. Researchers examined diphtheria toxoids precipitated with alum and were able to show that an injection of this alum precipitate led to a significant increase in immune response. Leading on from this research, alum has found widespread use as an adjuvant. Vaccines containing alum as adjuvant are referred to as *alum-precipitated vaccines*. Unfortunately, it has been shown that alum precipitations can be highly heterogeneous. The homogeneity of the preparation depends on the anions and the conditions present at the point of precipitation [9].

Subsequent research showed that aluminium hydroxide ($Al(OH)_3$) hydrogels can be pre-formed in a standardised manner and be used to absorb protein antigens to form a homologous preparation. Following on from this research, researchers have shown that it is possible to co-precipitate aluminium phosphate ($AlPO_4$) and the diphtheria toxoid in order to form active vaccines. These vaccines are called *aluminium-absorbed vaccines* and, in contrast to alum-precipitated vaccines, the antigens are distributed homogeneously. Nowadays, aluminium-absorbed vaccines have taken over from alum-precipitated ones. Nevertheless, there is a lot of ambiguity found in the literature, where both terms are interchangeably used [9].

In summary, immunisation vaccines containing adjuvants are more effective than those without them. Typical adjuvants are alum [$KAl(SO_4)_2 \cdot 12H_2O$], $Al(OH)_3$, $AlPO_4$, Al_2O_3, but oxides of other metals, such as ZrO_2, SiO_2 and Fe_2O_3, are also under investigation.

The formation of the aluminium hydrogels is generally achieved by reacting Al^{3+} ions (from compound such as $AlCl_3$) under alkaline aqueous conditions. Conditions are strongly regulated, as even smallest changes to parameters such as temperature, concentration and others can influence the quality of the hydrogel. Aluminium phosphate gels are typically produced by reacting Al^{3+} salts in the presence of phosphate ions under alkaline conditions [9].

The mode of action is highly complex and still not fully understood. Initial theories included the physical absorption of the antigen, which is still considered as an important feature, and the gradual release of antigen from the injection side with the adjuvant working as an agglomeration. The latter theory was disproved quickly. Research has shown that antigens need to be adsorbed to the adjuvant before the immunisation reaction. It is believed that the adjuvant will then present the antigen to the immunocomponent of the targeted cell [9].

4.3.5 Antacids

The function of antacids is to neutralise excess stomach acid. They also exhibit cytoprotective effects towards attacks against the gastric mucosa. They are additionally known to heal gastric and duodenal ulcerations; nevertheless, the mechanism is still uncertain.

Antacids have been in use for the past 2000 years, and the initial formulations were based on $CaCO_3$ (coral and limestone). Nowadays, the antacid/anti-gas market is a significant income stream for the pharmaceutical industry and the demand for antacids is expected to grow. The number of people suffering from heartburn increases with an ageing population, more stressful lifestyles and changing eating habits such as eating out more often.

Aluminium hydroxide ($Al(OH)_3$) has several medical applications. It is used as an antacid for treating heartburn as well as acid indigestion (reflux oesophagitis). It is also known to have healing properties of peptic ulcers. In patients suffering from kidney failure, who show elevated serum phosphate levels (hyperphosphataemia), $Al(OH)_3$ is used as a phosphate binder (see Section 4.3.7).

$Al(OH)_3$ is an amphoteric compound (see Section 4.1.2.2), which means it can react as a base or as an acid. In its application as an anti-acid, $Al(OH)_3$ reacts with any excess stomach acid (mainly HCl) with the formation of $AlCl_3$ and water (Equation 4.12).

$$Al(OH)_3 + 3HCl \rightarrow AlCl_3 + 3H_2O \tag{4.12}$$

$Al(OH)_3$ is known to cause constipation, so formulations of anti-acids often include a combination with Mg^{2+} antacids. Usually, oral antifoaming agents, such as simethicone, are added in order to reduce bloating

Table 4.2 *Typical formulation of an antacid/antigas mixture (maximum strength Maalox®, Max®, Norvatis)*

Active ingredient	Quantity (mg)	Purpose
$Al(OH)_3$	400	Antacid
$Mg(OH)_2$	400	Antacid
Simethicone	40	Anti-gas

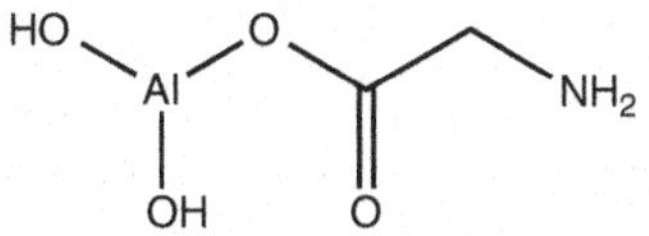

Figure 4.6 *Chemical structure of dihydroxy aluminium glycinate*

and discomfort/pain. Simethicone is a mixture of poly(dimethyl siloxane) and silica gel, which decreases the surface tension of gas bubbles (Table 4.2).

Ancient anti-acid formulations contained sodium bicarbonate (baking soda, $NaHCO_3$), which resulted in a rapid reaction with the gastric acid. The result was an increase in the gastric pH and the production of CO_2 gas as a by-product (Equation 4.13). Large doses of $NaHCO_3$ can cause alkaline urine and this can result in kidney problems. Acid neutralisation using $Al(OH)_3$ does not produce CO_2 and therefore these side effects can be avoided.

$$NaHCO_3 + HCl \rightarrow NaCl + H_2O + CO_2 \tag{4.13}$$

Aluminium glycinate $[Al(NH_2CH_2COO)(OH)_2]$ (Figure 4.6) is also used in anti-acid formulations. For example, Gastralgine® contains, amongst other ingredients, dihydroxy aluminium glycinate $[Al(NH_2CH_2COO)(OH)_2]$, $Al(OH)_3$, magnesium trisilicate and simethicone. It is known to have additionally protective effects from ulcers.

4.3.6 Aluminium-based therapeutics – alginate raft formulations

Heartburn is the primary symptom of the so-called gastro-oesophageal reflux disease (GERD), which is caused by the oesophageal influx of gastric HCl from the stomach. There are also close links to oesophageal cancer, which has a very low survival rate. Relief can be achieved with the use of alginate raft formulations, which typically contain alginic acid, $NaHCO_3$, magnesium trisilicate and $Al(OH)_3$. Alginates are natural polysaccharide polymers which are isolated from brown seaweeds.

In the acidic stomach, alginate salts and alginic acids precipitate to form a low-density viscous gel. When the mixture comes into contact with gastric HCl, the gel matrix formation occurs. HCO_3^-, which is trapped in the gel, leads to the formation of CO_2 gas (Equation 4.13). The gas bubbles trapped in the gel convert it to foam and provide buoyancy, allowing the gel to float on the surface of stomach contents (like a raft on water). $Al(OH)_3$ provides an additional capacity to neutralise any excess stomach acid (Equation 4.12). The raft physically acts as a barrier to gastric reflux and moves into the oesophagus during reflux. It acts as mobile neutralising sealant in the oesophageal space when the gastric pressure is high. Once the pressure reduces, the raft drops back into the stomach and can be digested (Figure 4.7).

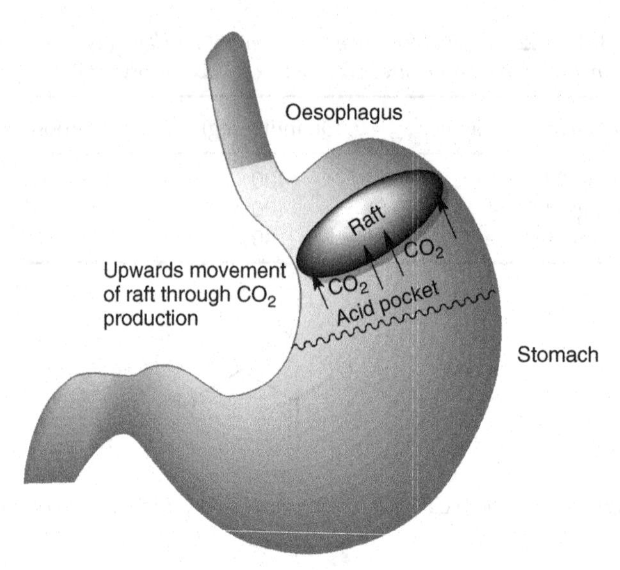

Figure 4.7 Illustration of the stomach, showing the acid pocket and the alginate raft floating on top of it protecting the oesophagus

4.3.7 Phosphate binders

Hyperphosphataemia, that is, increased levels of serum phosphate, is a disorder commonly seen in patients with end-stage renal (kidney) disease where the kidneys are not able to excrete excess phosphate as a result of a low renal clearance rate. This disorder is often seen in patients who are on dialysis treatment. Persistent hyperphosphataemia results in renal osteodystrophy, that is, the weakening of bones due to disturbances in the calcium and phosphate metabolism.

Generally Al^{3+}-containing drugs are given in order to promote the binding of phosphate in the gut. Antacids containing $Al(OH)_3$ can be used as phosphate binders. When $Al(OH)_3$ enters the acidic stomach (pH ~ 1), Al^{3+} ions are formed. Some Al^{3+} ions will be absorbed in the stomach, but the majority is passed to the distal intestines, where the pH is significantly increased (pH 6–8.5). In this high pH range, Al^{3+} freshly precipitates as a colloidal, amorphous $Al(OH)_3$. Its large surface area adsorbs phosphate ions (usually in form of HPO_4^{2-}) and passes them through the remaining intestine without decomposition, as the pH is too high. The Al^{3+}-phosphate complex ($AlPO_4$) is then excreted via the faeces.

Aluminosilicates can also be used as a phosphate binder and is, for example, the active ingredient in Malinal®. In contrast to $Al(OH)_3$, which acts as an efficient PO_4^{3-} binder directly, aluminosilicates need prior exposure to a acid in order to produce free Al^{3+}. Once the free Al^{3+} is formed, it follows the same mode of action.

Initially, aluminium-based phosphate binders were also used in dialysis exchange fluids, especially in patients being treated with haemodialysis. Nevertheless, as a result of the exposure to high concentrations of Al^{3+} salts, relatively high concentrations were found in patients. A significant number of patients developed dementia symptoms after 15 or more months of treatment, which was linked to the high Al^{3+} concentrations in the body including the brain (see Section 4.3.2) [6a, 10].

4.3.8 Antiperspirant

Aluminium trichloride ($AlCl_3$) was the first compound that was used as an antiperspirant. The mechanism of action is still under investigation, but it appears to act by forming a plug of $Al(OH)_3$ within the sweat duct. $AlCl_3$ is a very strong antiperspirant and only advised by doctors if normal antiperspirants do not work. Leading brands of antiperspirants contain usually a ~20% aluminium hexahydrate solution in an alcoholic base. It is thought to work by blocking the openings of the sweat ducts. It tends to work best in the armpits. However, it may also work for sweating of the palms and soles. It can also be applied to the face, taking care to avoid the eyes.

4.3.9 Potential aluminium toxicity

The excessive use of aluminium preparations negatively influences human health. Excessive intake of Al^{3+} has been found to accumulate in sensitive loci and can lead to pathological aberrations and result in dialysis dementia or similar symptoms. It is important to note that Al^{3+} is a major component in over-the-counter drugs such as antacids. Special attention has to be given by the dispensing pharmacist, and the patient has to be made aware of the consequences of overdoses of Al^{3+}-containing products. Al^{3+} is known to have embryonic and foetal toxic effects in humans and animals, causing osteomalacia, which is the softening of the bones due to defective bone mineralisation [5].

Albumin and transferrin bind around 95% of serum aluminium, which is then cleared mainly via the kidneys (a small amount can be found in the faeces). In healthy humans, only 0.3% of orally administered aluminium is absorbed, whereas it has the potential to accumulate when the GI tract is bypassed, for example, in intravenous infusions [10].

4.4 Gallium

4.4.1 Introduction

Gallium has atomic number 31 in the periodic table of elements. It has a silvery-white colour with a melting point of only 29 °C, which means that it melts when held in the hand. It has no known physiological role in the human body, but it can interact with cellular processes and proteins that are normally involved in iron metabolism.

Gallium tartrate has a long research history. Researchers showed in the 1930s that it could be used to treat syphilis in rabbits with no significant toxicity [6a]. In subsequent studies, it has been shown that gallium ions predominantly accumulate in the bone and therefore would be a good candidate for radiotherapy of bone cancer. Unfortunately, the radioactive isotope ^{72}Ga has only a half-life of around 14 h, which is not long enough for effective radiotherapy. Nevertheless, current clinical developments involve the use of radioactive gallium isotopes as tumour imaging reagents (see Chapter 10), gallium nitrate in metabolic bone disease, hypercalcaemia and as anticancer drug, as well as up-to-date research in the area of chemotherapeutic applications.

4.4.2 Chemistry

Gallium exists as the trivalent cation Ga^{3+}, and in aqueous solution it presents as a hydrated complex. Depending on the pH, a variety of hydroxyl species are formed, some of which are insoluble, such as $Ga(OH)_3$. At physiological pH, nearly no free Ga^{3+} is present and the hydroxyl species $Ga(OH_4)^-$ (gallate, the dominant species) and $Ga(OH)_3$ are formed. Gallium hydroxide species are amphoteric, analogous to aluminium hydroxide compounds.

It is important to note that the stability of solutions containing gallium chloride or gallium nitrate for oral administration is affected by the pH. They might not be stable over extended periods and gallium hydroxide precipitates.

4.4.3 Pharmacology of gallium-based drugs

Ga^{3+} has an ionic radius and binding properties similar to those of Fe^{3+} (ferric iron). Unlike Fe^{3+}, it cannot be reduced to its divalent state, which means that it follows a completely different redox chemistry compared to iron. The oxidation and reduction of iron is important in many biological processes, which therefore cannot be mimicked by gallium. One example includes the uptake of Fe^{2+} by the haeme group (see Chapter 8). As Ga^{3+} is not readily reduced to it +II state, it cannot bind to the haeme group.

Transferrin is an important transport protein that controls the level of free Fe^{2+} in the blood plasma. Free iron ions are toxic to most forms of life, and therefore transferrin binds Fe^{2+} and removes it from the blood. There is an excess of transferrin present in the blood, and it has been shown that Ga^{3+} can also bind to this glycoprotein but with a lower affinity than Fe^{3+}. Once the binding capacity of gallium ions to transferrin is exceeded, it is believed to circulate as gallate $[Ga(OH)^{4-}]$ [6a].

The therapeutic action of Ga^{3+} is very much based on the pharmacological activity of Fe^{3+} which it mainly mimics. Ga^{3+} is transported via transferrin to areas of the body that require increased Fe^{3+} levels, including proliferating cancer cells [11]. Ga^{3+} can interrupt the cell cycle and DNA synthesis by competing with iron for the active sites in essential enzymes [12]. Ga^{3+} accumulates in the endosomes mediated by transferrin uptake and transported into the cytosol, where it can bind to the enzyme ribonucleotide reductase. Ribonucleotide reductase has been proposed as the main target for Ga^{3+}. Binding to this enzyme will impair DNA replication and ultimately lead to apoptosis [13]. *In vitro* studies have shown that Ga^{3+} can bind directly to DNA [14].

4.4.4 Gallium nitrate – multivalent use

In clinical trials, gallium nitrate has proved to be highly active as an antitumour agent especially against non-Hodgkin's lymphoma and bladder cancer. The cytotoxic activity of gallium nitrate has been demonstrated as single agent and as part of combination therapy, for example, together with fluorouracil. Gallium nitrate shows a relatively low toxicity and does not produce myelosuppression, which is a significant advantage over other traditional anticancer agents. Furthermore, it does not appear to show any cross-resistance with conventional chemotherapeutic agents (Figure 4.8) [15].

These studies have also shown that gallium nitrate is able to decrease serum calcium levels in patients with tumour-induced hypercalcaemia. Subsequently, several studies have been carried out comparing traditional bisphosphonate drugs with gallium nitrate in their ability to decrease the calcium levels that are elevated as a result of cancer. Based on the clinical efficacy, gallium nitrate injections (Ganite™) was granted approval by the FDA for the treatment of cancer-associated hypercalcaemia. Gallium nitrate is also believed to inhibit the bone turnover and therefore to decrease osteolysis, the active reabsorption of bone material, in patients with bone metastasis secondary to other cancers.

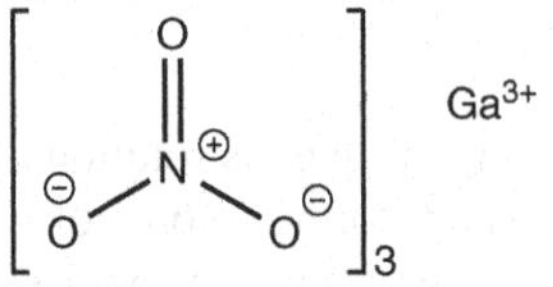

$$\left[O^{-}-N^{+}(=O)-O^{-} \right]_3 \; Ga^{3+}$$

Figure 4.8 Gallium nitrate

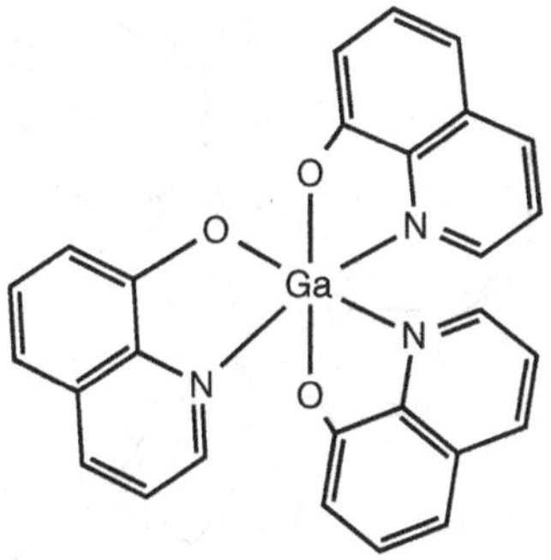

Figure 4.9 Chemical structure of gallium 8-quinolate

4.4.5 Gallium 8-quinolinolate

Gallium 8-quinolinolate is a hexacoordinated Ga^{3+} complex in which the central gallium atom is coordinated by three quinolinolate groups. It was developed as an orally available anticancer agent. It was successfully tested *in vitro* against lung cancer and in transplanted rats against Walker carcinosarcoma [16]. Main side effects were detected in experiments on mice at doses of 125 mg/kg/day. These included leukopaenia and some fatalities. The highest concentrations Ga^{3+} were found in the bone, liver and spleen (Figure 4.9) [6a].

Preclinical studies have established the IC_{50} values for a single-agent activity in the lower micromolar range for a variety of cancer cell lines. These cell lines include human lung adenocarcinoma, where gallium 8-quinolinolate was shown to be 10 times more potent than gallium nitrate. Other cell lines include melanoma and ovarian, colon and breast cancer. The inhibitory effect appears to be dose dependent and not time dependent.

Gallium 8-quinolinolate entered phase I clinical trials under the drug name KP46 in 2004 in order to establish its safety and toxicity profile. KP46 was orally administered as a tablet, containing 10–30%w/w. Dose up to 480 mg/m^2 were given to patients with advanced solid malignant tumours. The drug was well tolerated and preliminary success was seen in patients with renal cell cancer.

4.4.6 Gallium maltolate

Gallium maltolate [tris(3-hydroxy-2-methyl-4*H*-pyran-4-onato)gallium(III)] is a coordination complex containing a central Ga^{3+} ion and three maltolate (deprotonated maltol) groups. Clinical studies have shown that oral administration of gallium maltolate leads to significantly increased bioavailability compared to gallium chloride. The oral bioavailability is estimated to 25–57% in comparison to 2% for gallium chloride (Figure 4.10) [17].

Phase I clinical trials on healthy humans showed that doses were well tolerated up to 500 mg. Furthermore, the results suggested the possibility of a once-per-day treatment option as a result of the half-life of the drug in the blood plasma (17–21 h). Orally administered gallium maltolate is excreted significantly more slowly via the kidneys than gallium nitrate injected intravenously. It has been proposed that rapid intravenous administration leads to the formation of gallate, which is quickly cleared as a small molecule via the kidneys. In contrast, oral administration leads to a slow loading of the blood plasma, and Ga^{3+} is bound to

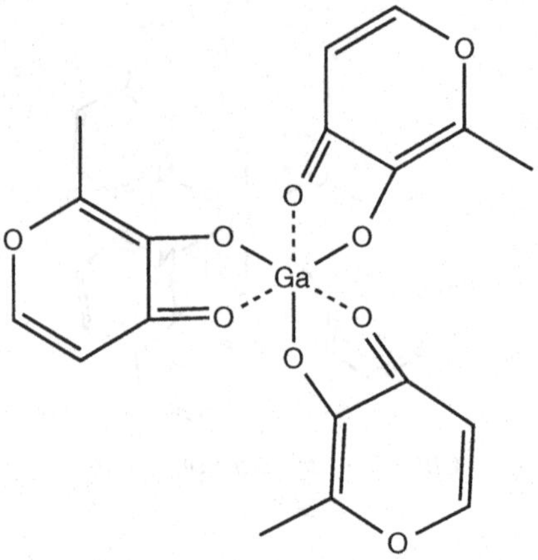

Figure 4.10 Chemical structure of gallium maltolate

transferrin. This may lead to a different mechanism of excretion, leading to a reduction in renal toxicity. Also, the transferrin-bound Ga^{3+} has the potential to be directly transported to the cancer cell without causing significant side effects. Therefore, an oral administration seems to be superior to parenteral administration [17].

4.4.7 Toxicity and administration

Gallium nitrate is usually administered as a continuous intravenous infusion (200 mg/m^2 for 5 days) for the treatment of cancer-induced hypercalcaemia. This dose is well tolerated even by elderly patients. Higher doses are usually used in the treatment of cancer. Renal toxicities being the dose-limiting factor are normally seen when gallium nitrate is administered as a brief intravenous infusion. With the long-term regime as described above, diarrhoea is the most common side effect. Renal toxicity can normally be minimised by adequate hydration of the patient [15].

The advantage of gallium nitrate therapies is that platelet count and white blood cell counts are not suppressed, which means that no myelosuppression takes place, which represents a major advantage over conventional chemotherapeutic agents [15].

4.5 Exercises

4.5.1 Draw the Lewis structure or chemical formula of the following aluminium-based drugs

(a) Aluminium acetate
(b) Aluminium chloride
(c) Aluminium oxide

4.5.2 Research different antacids mixtures, state their content and calculate the weight/volume percentage (%w/v) for each active pharmaceutical ingredient (API).

Drug name	Aluminium hydroxide	%w/v	Magnesium hydroxide	%w/v
Maalox®	220 mg/5 ml	???	195 mg/5 ml	???

4.5.3 A typical antacid capsule contains 475 mg aluminium hydroxide as the active ingredient. How many milligrams of stomach acid (HCl) can be neutralised by one tablet?

4.5.4 An aluminium hydroxide suspension (30 ml) containing 500 mg/5 ml aluminium hydroxide is prescribed to the patient. The prescription states the patient has to take 30 ml four times a day.

(a) What is the chemical formula of aluminium hydroxide?
(b) How many grams of Al^{3+} is given to the patient per single dose?
(c) What is the weight/volume percentage (%w/v) of aluminium hydroxide in the suspension?

4.5.5 Write the chemical equations explaining the amphoteric behaviour of gallium hydroxide.

4.6 Case studies

4.6.1 Boric acid – API analysis

Boric acid is known to have antifungal and antimicrobial properties and therefore has a clinical use. Typical pharmaceutical analysis of boric acid as an API includes its reaction with a known amount of mannitol and subsequent titration of unreacted mannitol.

(a) Research the chemical structure of mannitol including its stereochemistry.
(b) Describe the reaction of boric acid and mannitol, including relevant chemical equations.
(c) For the analysis of boric acid as an API, 1.0 g of the acid is typically dissolved in 100 ml of water, and 15.0 g of mannitol is added. The solution is then titrated with 1 M sodium hydroxide (NaOH) solution using phenolphthalein as indicator.
Calculate the volume of NaOH needed in this titration if the API has a purity of 99%.
(d) Research the typically accepted error margins for the purity of boric acid as an API.

4.6.2 Aluminium hydroxide tablets

Your pharmaceutical analysis company has been contacted by an important client and asked to analyse a batch of aluminium hydroxide tablets. The description of your brief states that you are supposed to analyse the API in these tablets following standard quality assurance guidelines.

Typical analysis methods used for quality purposes are often based on titration reactions, but also a variety of other quantitative analysis methods such as gravimetric analysis can be used. Typically, a certain amount of the tablet powder is dissolved in water, and hydrochloride acid (HCl) is added. An excess of the precipitation reagent is added, and the solution is stirred until precipitation is completed. The precipitate is then filtered, dried to constant weight and weighed.

(a) Draw the chemical formula of aluminium hydroxide.
(b) Research the type of analysis used. Within your studies, you should look at a variety of precipitation reagents and understand how different factors can influence this method.
(c) Formulate the relevant chemical equations.
(d) The package states that each tablet contains 475 mg aluminium hydroxide. For the experiment, 20 tablets are weighed (total weight 12.5 g) and powdered. An amount of powder containing 0.4 g aluminium hydroxide is dissolved in water and HCl, and reacted with excess 8-hydroxy quinolone. After stirring this solution for 2 h near the boiling point, the precipitate is filtered and dried overnight in the oven at 100 °C.

The precipitate weighs 2.0 g.

Calculate the amount of aluminium hydroxide present in your sample. Express your answer in grams and moles.
(e) Critically discuss your result in context with the stated value for the API.
(f) Research the typically accepted error margins.

References

1. J. D. Lee, *Concise inorganic chemistry*, 5th ed., Chapman & Hall, London, **1996**.
2. E.-I. Ochiai, *Bioinorganic chemistry: a survey*, Academic Press, Amsterdam; London, **2008**.
3. H. Kingma, *Can. Med. Assoc. J.* **1958**, *78*, 620–622.
4. R. D. Houlsby, M. Ghajar, G. O. Chavez, *Antimicrob. Agents Chemother.* **1986**, *29*, 803–806.
5. *British national formulary*, British Medical Association and Pharmaceutical Society of Great Britain, London.
6. (a) E. R. Tiekink, M. Gielen, *Metallotherapeutic drugs and metal-based diagnostic agents: the use of metals in medicine*, Wiley, Chichester, **2005**; (b) G. A. McKay, M. R. Walters, J. L Reid, *Lecture notes. Clinical pharmacology and therapeutics*, 8th ed., Wiley-Blackwell, Chichester, **2010**; (c) G. J. Tortora, B. Derrickson, *Principles of anatomy and physiology*, 12th ed., international student/Gerard J. Tortora, Bryan Derrickson. ed., Wiley [Chichester: John Wiley, distributor], Hoboken, N.J., **2009**.
7. N. D. Priest, *J. Environ. Monit.* **2004**, *6*, 375–403.
8. ebrary Inc., 3rd ed., World Health Organization, Geneva, **2004**, p. xix, 515 p.
9. E. B. Lindblad, *Immunol. Cell Biol.* **2004**, *82*, 497–505.
10. C. Ashley, C. Morlidge, *Introduction to renal therapeutics*, Pharmaceutical Press, London, **2008**.
11. L. R. Bernstein, *Pharmacol. Rev.* **1998**, *50*, 665–682.
12. D. W. Hedley, E. H. Tripp, P. Slowiaczek, G. J. Mann, *Cancer Res.* **1988**, *48*, 3014–3018.
13. A. R. Timerbaev, *Metallomics* **2009**, *1*, 193–198.
14. (a) H. A. Tajmirriahi, M. Naoui, R. Ahmad, *Metal. Ions Biol. Med.* **1992**, *2*, 98-101; (b) M. Manfait, P. Collery, *Magnesium-B* **1984**, *6*, 153–155.
15. C. R. Chitambar, *Int. J. Environ. Res. Public Health* **2010**, *7*, 2337–2361.
16. P. Collery, F. Lechenault, A. Cazabat, E. Juvin, L. Khassanova, A. Evangelou, B. Keppler, *Anticancer Res.* **2000**, *20*, 955–958.
17. L. R. Bernstein, T. Tanner, C. Godfrey, B. Noll, *Met.-Based Drugs* **2000**, *7*, 33–47.

Further Reading

E. Alessio, *Bioinorganic medicinal chemistry*, Wiley-VCH, Weinheim, **2011**.

W. Kaim, B. Schwederski, *Bioinorganic chemistry: inorganic elements in the chemistry of life: an introduction and guide*, Wiley, Chichester, **1994**.

H.-B. Kraatz, N. Metzler-Nolte, *Concepts and models in bioinorganic chemistry*, Wiley-VCH [Chichester: John Wiley, distributor], Weinheim, **2006**.

R. M. Roat-Malone, *Bioinorganic chemistry: a short course*, Wiley, Hoboken, N.J. [Great Britain], **2002**.

5

The Carbon Group

Members of group 14 of the periodic table (14th vertical column) are summarised as carbon group consisting of carbon (C), silicon (Si), germanium (Ge), tin (Sn) and lead (Pb) (Figure 5.1).

Group 14 elements have four valence shell electrons and therefore tend to form covalent compounds. Nevertheless, with increasing mass and atomic radius, the elements show increasingly more metallic characteristics and have lower melting and boiling points. Elements within this group show a graduation from nonmetallic elements (C) to elements that are classified as metals (Pb). Silicon is generally seen as nonmetallic, whereas germanium is metallic. Nevertheless, this classification is not definite. Silicon and germanium both form covalent diamond-type structures in the solid state, but their electrical behaviour indicates more a metallic behaviour. Therefore, silicon and germanium are classified as metalloids (see Chapter 4).

Carbon is the essential element to life on earth, and the chemistry related to carbon is classified as organic chemistry and we will therefore not discuss it any further in this book. Organometallic chemistry relates to the interaction of carbon compounds with metals, and the basic concepts will be discussed in Chapter 8.

Tin and lead have been under investigation for use as anticancer and antimicrobial agents, but so far with limited success. This chapter will discuss the pharmaceutical applications of silicon- and germanium-based drugs.

5.1 General chemistry of group 14 elements

5.1.1 Occurrence, extraction and use of group 14 elements

Silicon is, after oxygen, the second most abundant element in the earth's crust. It occurs in a range of minerals and sand (SiO_2, quartz). In contrast, germanium, tin and lead are relatively rare elements, with tin and lead being extracted for thousands of years from their ores. The main source for tin is cassiterite (SnO_2) and for lead galena (PbS). Germanium was first isolated from the mineral argyrodite when it was discovered, but there are no major deposits of this mineral. Germanium is nowadays mainly sourced from zinc and copper ores.

Silicon can be extracted from silicates or sand by reducing SiO_2 with coke at high temperatures at around 3000 °C.

$$SiO_2 + 2C \rightarrow Si + 2CO \tag{5.1}$$

Essentials of Inorganic Chemistry: For Students of Pharmacy, Pharmaceutical Sciences and Medicinal Chemistry,
First Edition. Katja A. Strohfeldt.

Companion website: www.wiley.com/go/strohfeldt/essentials

H																	He
Li	Be											B	C	N	O	F	Ne
Na	Mg											Al	Si	P	S	Cl	Ar
K	Ca	Sc	Ti	V	Cr	Mn	Fe	Co	Ni	Cu	Zn	Ga	Ge	As	Se	Br	Kr
Rb	Sr	Y	Zr	Nb	Mo	Tc	Ru	Rh	Pd	Ag	Cd	In	Sn	Sb	Te	I	Xe
Cs	Ba	La-Lu	Hf	Ta	W	Re	Os	Ir	Pt	Au	Hg	Tl	Pb	Bi	Po	At	Rn
Fr	Ra	Ac-Lr	Rf	Db	Sg	Bh	Hs	Mt	Ds	Rg	Uub						

Figure 5.1 *Periodic table of elements; group 14 elements are highlighted*

Germanium is extracted from zinc ores in a very complicated process as it has aqueous properties similar to those of zinc. Once the germanium/zinc mixture has been sufficiently enriched with germanium, it is heated in HCl with Cl_2 in order to allow the formation of germanium tetrachloride ($GeCl_4$). $GeCl_4$ can be easily separated from $ZnCl_2$ because the former has a significantly lower boiling point, and after hydrolysis germanium dioxide (GeO_2) is obtained. GeO_2 can be reduced to elemental Ge in a stream of hydrogen gas. Elemental Sn is extracted from the ore cassiterite (SnO_2) via reduction with carbon. Lead is obtained from its sulfide (PbS, galena), which is first roasted in the presence of oxygen and then reduced with carbon to give elemental Pb.

$$PbS + 1\tfrac{1}{2}O_2 \rightarrow PbO + SO_2$$

$$PbO + C \rightarrow Pb + CO$$

Silicon is used in a wide variety of applications. In nature, silicon does not exist as the pure metal and most commonly occurs in silica (including sand) and silicates. Silicon dioxide, also known as *silica*, is a hard substance with a high melting temperature and clearly very different from carbon dioxide. Molten silica can be used to make glass, an extremely useful material, which is resistant to attack by most chemicals except fluorine, hydrofluoric acid and strong alkalis. Silicon atoms can also be found in the class of compounds called *silicones*. Silicones are inert synthetic polymers with a wide range of uses including as sealants, cookware, adhesives and medical applications. Silicones contain next to silicon atoms also carbon, hydrogen and oxygen atoms (Figure 5.2).

Pure silicon metal is used in semiconductors, the basis of all electronic devices, and is most well known for its application in solar panels and computer chips. Germanium can be mainly found not only in electrical components and in semiconductors but also in some optical applications and some specialised alloys.

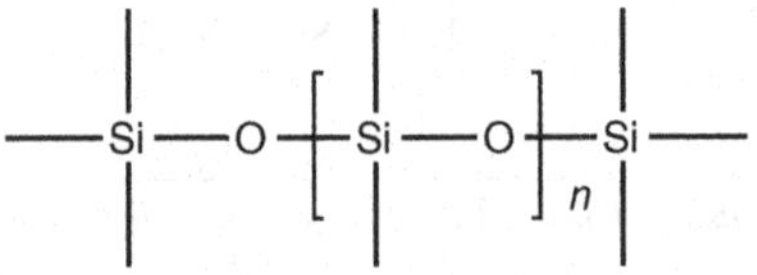

Figure 5.2 *Structure of silicones*

*A **semiconductor** is a crystalline solid that has an electric conductivity between that of a conductor (e.g. copper) and an insulator (e.g. glass). This effect can be a result of impurities or temperature.*

Metallic tin has many applications, but the most well known one is the use as lining for drink and food cans. It is also used in alloys – bronze is an alloy of copper and tin – and organotin compounds find their application in poly(vinyl chloride) (PVC) plastics. Tin is often referred to as a *poor metal* and it has two allotropes: grey tin and white tin.

Lead is a grey metal and most lead is used in batteries. Other major uses, such as in plumbing or as antiknock agent in petrol (tetraethyl lead, $Pb(C_2H_5)_4$), have declined over recent years because of the high toxicity of lead. Pb is a neurotoxin when ingested and many lead compounds are water soluble. Therefore, water lines have been replaced by specialised plastic material, and in most industrialised countries only unleaded petrol is sold.

5.1.2 Oxidation states and ionisation energies

Group 14 elements can have the oxidation state +2 or +4, with the latter one being the dominant one. The stability of the oxidation state +2 increases for elements further down the group, such as Sn and Pb. Group 14 elements have four valence electrons, two in the s orbital and two in the p orbital. Therefore, the first four ionisation energies increase evenly, whereas there is a significant increase to the fifth (Table 5.1).

5.1.3 Typical compounds of group 14 elements

The chemistry of group 14 elements is different from that of its naming element, carbon. The best example is the ability of carbon to form double and triple bonds. None of the remaining group 14 elements shows the same behaviour and there are only very few compounds known that contain silicon–silicon double and triple bonds. Carbon forms double and triple bonds through the overlap of its 2p orbitals. In contrast, the overlap of the 3p orbitals in silicon is weaker, and as a result silicon does not form multiple bonds as readily as carbon.

5.1.3.1 Oxides

Silicon oxides can be very complex structures, which is in contrast to the simple linear molecule carbon dioxide with its C=O bonds. The most basic molecule is SiO_2 (silica), which is an extended array of SiO_4 units where the oxygen atoms bridge the neighbouring silicon atoms. Silicon forms an array of oxides that occur in many minerals. All those are made up of SiO_4 tetrahedra forming rings or chains with an overall

Table 5.1 *Ionisation energy (kJ/mol) for group 14 elements [1]*

	First	Second	Third	Fourth	Fifth
C	1 086	2 352	4 620	6 222	37 827
Si	787	1 577	3 231	4 356	16 091
Ge	762	1 537	3 302	4 410	9 020
Sn	709	1 412	2 943	3 930	6 974
Pb	716	1 450	3 082	4 083	6 640

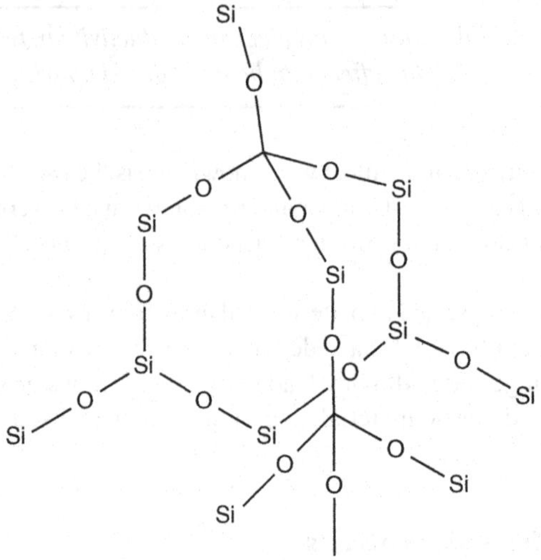

Figure 5.3 Chemical structure of SiO_2

negative charge. The negative charges are balanced by metal ions (e.g. alkali and alkaline earth metals). One of the best known examples is the commercially available synthetic zeolite, which is an aluminosilicate incorporating aluminium ions. Zeolite is used as an absorbent and in the production of laundry products (Figure 5.3).

Germanium dioxide has the chemical formula GeO_2 and forms as a passivation layer on pure germanium when the semi-metal is in contact with atmospheric oxygen. GeO_2 forms structures that are similar to SiO_2 depending on the reaction conditions. Germanium monoxide (GeO) is formed when GeO_2 is heated with powdered germanium at 1000 °C. GeO_2 is used in optical devices such as lenses and optical fibres, amongst other applications.

5.1.3.2 Hydrides

Silicon hydrides are called *silanes* and are structurally similar to their carbon analogues, with the general formula Si_nH_{2n+2}. Silanes are typically colourless and can be gases or volatile oils. Germanium hydrides are accordingly called *germanes* and these are generally less reactive than silanes. A classic example is SiH_4, which ignites spontaneously in air whereas GeH_4 is stable. Tin hydrides (SnH_4), called *stannanes*, are less stable and decompose slowly in air to tin and hydrogen gas. PbH_4 has not been isolated or prepared so far, and only hydrides containing organic substituents have been observed.

5.1.3.3 Halides

Group 14 elements generally form MX_4-type compounds, but germanium, tin and lead also form compounds of the type MX_2. Tetrachlorides of silicon and germanium are important precursors and are used in the synthesis of ultrapure silicon and germanium. These find applications in the electronic industry as materials for semiconductors. Silicon tetrachloride instantly hydrolyses when it comes into contact with water.

5.2 Silicon-based drugs versus carbon-based analogues

Silicon chemistry has been of interest as a source for the design of novel pharmaceutically active compounds. Why is it possible to introduce a silicon group or replace a carbon centre by silicon and what are the resulting changes? Carbon and silicon are both group 14 elements exhibiting similarities and differences:

Valency: Silicon and carbon both possess four valence electrons as they show an analogous electron configuration (C: $[He]2s^2 2p^2$; Si: $[Ne]3s^2 3p^2$).

Coordination number: Unlike that of carbon, the chemistry of silicon is influenced by the availability of its 3d orbitals to be involved in additional bonding interactions. Silicon is therefore capable of increasing its coordination number from 4 to 6 and thus forming isolatable penta- and hexacoordinated silicon-based compounds. Nevertheless, for silicon a coordination number of 4 (sp^3 hybridisation) is favoured especially over coordination numbers 2 (sp hybridisation) and 3 (sp^2 hybridisation). Consequently, the formation of double and triple bonds is disfavoured in contrast to carbon-based reaction centres.

Bond length: Silicon has a larger covalent radius than carbon, resulting in the formation of longer bonds than carbon–carbon bonds (typical C—C bond length = 1.54 Å), silicon–silicon bonds (typical Si—Si bond length = 2.33 Å) and silicon–carbon bonds (typical Si—C bond length = 1.89 Å). As a result, silicon-containing compounds show higher conformational flexibility and therefore steric arrangements different from analogous carbon-based compounds. Differences in interaction with proteins and consequently alterations of pharmacodynamics and pharmacological profiles have been observed.

Electronegativity: Silicon is more positive than the neighbouring carbon (electronegativity according to Pauling: Si = 1.90, C = 2.55), resulting in different bond polarisation of analogous carbon–element and silicon–element bonds. As a result, chemical reactivity and bond strength can differ significantly. This can provide improved or altered potency if carbon moieties are switched to silicon-based ones within pharmacophores, especially if hydrogen bonding is involved in the mode of action (Figure 5.4).

Lipophilicity: In general, silicon-based compounds demonstrate an enhanced lipophilicity in comparison to their carbon-analogous due to their different covalent radii. This provides an interesting opportunity for exploitable pharmacokinetic potential in drug design, for example, for drugs that are prone to hepatic metabolism. Silicon-based compounds involved in hepatic metabolism have been observed to exhibit an increased half-life when compared to their carbon analogues. Increased lipophilicity is also believed to be useful in the design of drugs that are supposed to cross the blood–brain barrier. Therefore silicon analogues with their increased lipophilicity can be very interesting drug candidates.

Currently, there are only a small number of silicon-containing compounds under investigation for pharmaceutical use. Silicones are the only silicon-based compounds widely used in medicine. These oligosiloxanes

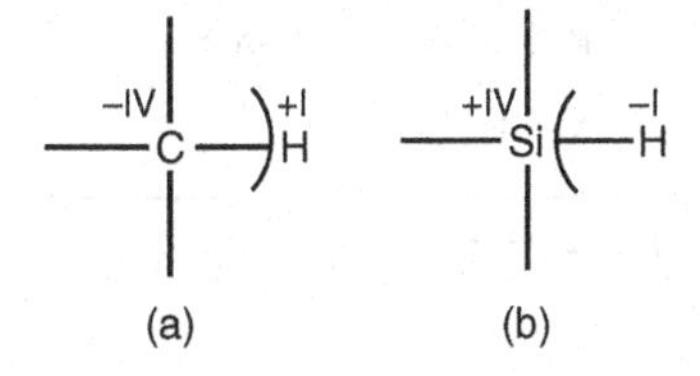

Figure 5.4 *(a) C—H and (b) Si—H bond polarisation*

and polysiloxanes have no carbon analogues and they are widely used for plastics, implants, catheters and many other applications.

The limitations of novel organosilicon compounds are very often attributed to insufficient funding and poor evidence of demonstrated efficacy. There is also an on-going debate about the toxicity of silicon-based therapeutics. So far, no increased systematic toxicity of silicon-containing compounds in comparison to their carbon-analogous has been detected.

Nevertheless, several organosilicon compounds have made it to clinical trials. In the following sections, a couple of interesting examples ranging from steroids being used by bodybuilders to anticancer and antispastic drugs under development are presented.

5.2.1 Introduction of silicon groups

A convenient method to introduce a silicon group is through the so-called silylation. A hydrogen atom that is bonded to a heteroatom (sulfur, nitrogen or oxygen) is exchanged by a silyl group (see Figure 5.5).

Carbon silylation, that is, the introduction of a silicon group next to a carbon centre, is also used for the design of novel drugs. This approach potentially allows changing the properties of the novel drug candidate significantly. It can lead to enhanced blood stability, increased cell penetration and altered pharmacokinetics. Several compounds have entered clinical trials, including the muscle relaxant silperisone (Figure 5.6) [1].

5.2.1.1 Silabolin

The anabolic compound silabolin is an example of a drug in which this silylation approach has been used. Silabolin is an injectable steroid containing a trimethylsilyl group, and was and still is used as an anabolic preparation by bodybuilders. It is known to have a relatively low androgenic activity like the natural anabolic hormone testosterone. Silabolin was officially registered in the (former) USSR as a domestic anabolic drug. It is believed to influence the protein synthesis in humans. Silabolin itself is a white powder, which is sparingly soluble in ethanol but not soluble in water [2]. Its propensity to cause heart and liver defects is under discussion and its effectiveness is being critically discussed amongst bodybuilders (Figure 5.7) [1].

5.2.1.2 Silperisone

Tolperisone is a centrally acting muscle relaxant used, for example, in the treatment of acute muscle spasms in back pain. Previous *in vitro* and *in vivo* studies in mice have demonstrated that silperisone may have the potential to reduce both central nervous system depressing and motor side effects. Phase I clinical trials were

$$R_3C\text{-}X\text{-}H \rightarrow R_3C\text{-}X\text{-}SiR'_3$$
$$X = N, O, S$$

Figure 5.5 *Silylation reaction*

$$R_3C\text{-}H \rightarrow R_3C\text{-}SiR'_3$$

Figure 5.6 *Example for carbon silylation reaction*

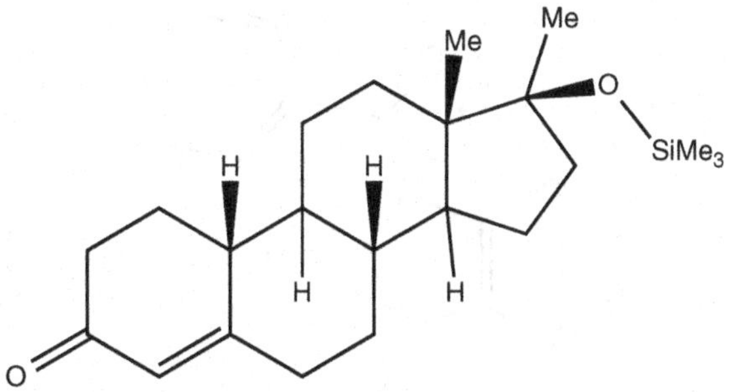

Figure 5.7 Chemical structure of silabolin

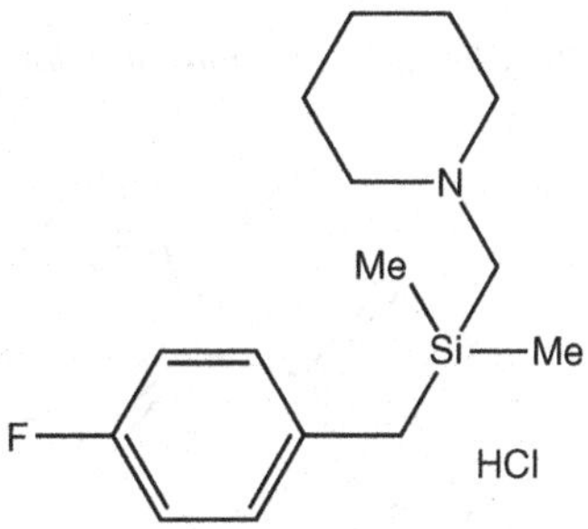

Figure 5.8 Chemical structure of silperisone

conducted with doses up to 150 mg/day. No adverse side effects were detected, and the observed plasma levels were deemed to be effective in preclinical trials. Nevertheless, chronic toxicities were observed in animal studies and the research was discontinued (Figure 5.8) [3].

5.2.1.3 Indomethacin

Indomethacin (see Figure 5.9) is a nonsteroidal anti-inflammatory agent used in pain and moderate to severe inflammation in rheumatic diseases and other musculoskeletal disorders. It is a COX (cyclooxygenase) inhibitor and therefore interrupts the production of prostaglandins [4].

A series of new silicon compounds, based on the structure of indomethacin, have been synthesised and are under investigation as novel anticancer agents. The carboxyl group of indomethacin was reacted with a series of amino-functionalised silanes. The resulting products have been shown to be significantly more lipophilic and more selective to COX-2. Furthermore, *in vitro* testing has shown an increased uptake of the new compounds at the tumour site. The silane-functionalised indomethacin derivatives exhibited a 15-fold increased antiproliferative effect when tested against pancreatic cancer (Figure 5.10) [5].

5.2.2 Silicon isosters

The carbon/silicon switch strategy, meaning the replacement of carbon centres by analogous silicon groups in known biologically active reagents, is currently mainly used for the development of novel silicon-based drug candidates [6].

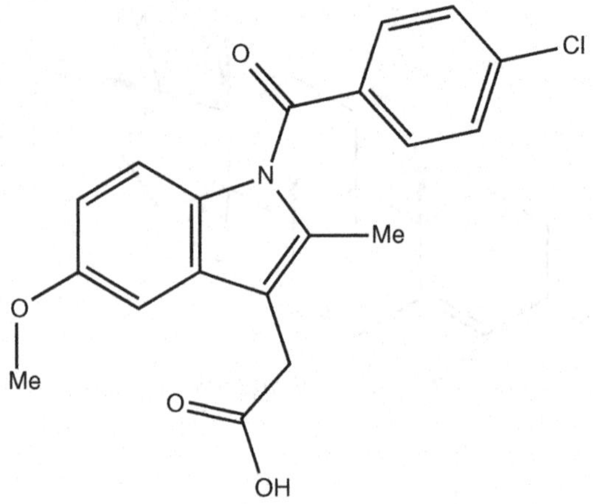

Figure 5.9 Chemical structure of indomethacin

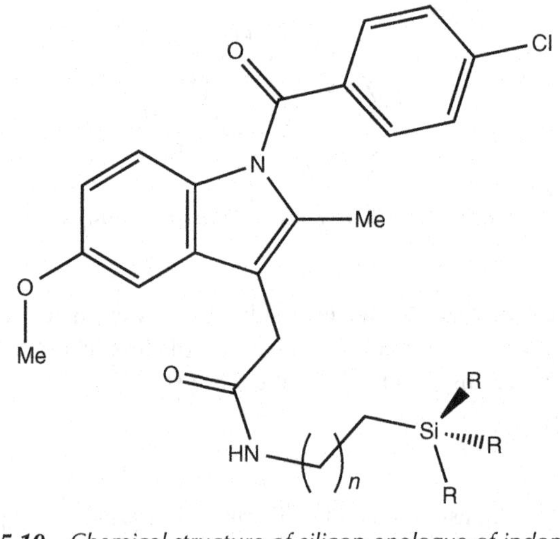

Figure 5.10 Chemical structure of silicon analogue of indomethacin

The idea is that the new silicon-based drug candidates have the same chemical structure, with one carbon atom exchanged by a silicon one. The resulting physiochemical changes include, amongst others, altered bond length and changes in the lipophilicity. These alterations can have a significant effect on the biological activity of these novel silicon-based compounds. A variety of these compounds have been synthesised and tested [1]; two examples are presented in the following:

5.2.2.1 Sila-haloperidol

Haloperidol is an analogue of the dopamine D_2 receptor antagonist and is an older antipsychotic drug. The drug is used in the treatment of schizophrenia, a neuropsychiatric disorder. Schizophrenia is characterised by symptoms such as hallucinations, delusions and disorganised speech. It is believed that schizophrenia is

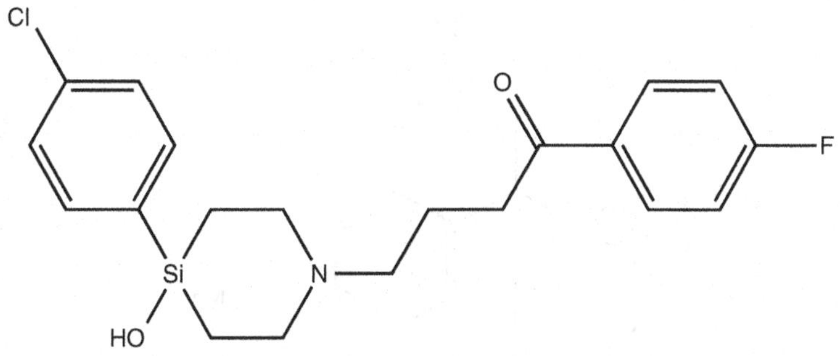

Figure 5.11 Chemical structures of sila-haloperidol

caused by problems involving the dopamine regulation in the brain. In general, antipsychotic drugs work by blocking the dopamine D_2 receptors [7].

Haloperidol is such an antipsychotic drug, which was developed in the 1950s and entered the clinic soon after that. Its use is limited by the high incidence of extrapyramidal symptoms (movement disorders caused by drugs affecting the extrapyramidal system, a neural network which is part of the motor system) [8]. Nevertheless, haloperidol may be used for the rapid control of hyperactive psychotic states and is popular for treating restlessness in the elderly.

The silicon analogue, sila-haloperidol, has been synthesised by a sila-substitution of the quaternary R_3COH carbon atom of the 4-hydroxy-4-(4-chlorophenyl)piperidin-1-yl group of haloperidol (see Figure 5.11). Chemical analyses have shown that haloperidol and sila-haloperidol both exist as two analogous conformers but with a different conformer ratio for the carbon and silicon analogues. Biological studies have also shown large differences between the metabolic pathways of the silicon and carbon analogues. Radiolabelling studies have shown similar potencies of the silicon and the carbon compounds at the human dopamine hD_1, hD_4 and hD_5 receptors. Sila-haloperidol was significantly more potent with the hD_2 receptor, thus giving hope to improved side effects related to the metabolism [9].

5.2.2.2 Sila-venlafaxine

Venlafaxine is a serotonin and noradrenalin reuptake inhibitor (SNRI) and is used as an antidepressant. Compared to tricyclic antidepressants, it lacks the antimuscarinic and sedative side effects. Nevertheless, treatment with venlafaxine can lead to a higher risk of withdrawal symptoms [8].

The silicon analogue, *rac*-1-[2-(dimethylamino)-1-(4-methoxyphenyl)ethyl]-1-silacyclohexan-1-ol, has been synthesised and tested for its biological properties. The hydrochloride salts were examined for their efficacy in reuptake inhibition assays for serotonin, noradrenalin and dopamine. It was concluded that the carbon–silicon switch changed the pharmacological profile significantly in regard to the reuptake inhibition depending on the stereoisomer. (*R*)-Sila-venlafaxine was found to be consistent with selective reuptake inhibition at the noradrenaline inhibitor (Figure 5.12) [10].

5.2.3 Organosilicon drugs

There are several classes of silicon compounds with a clinical use or a proposed biological activity that have no apparent carbon analogues. These compounds use the properties specific to silicon, mainly its ability to form molecules with a penta- and hex-acoordinated silicon centre.

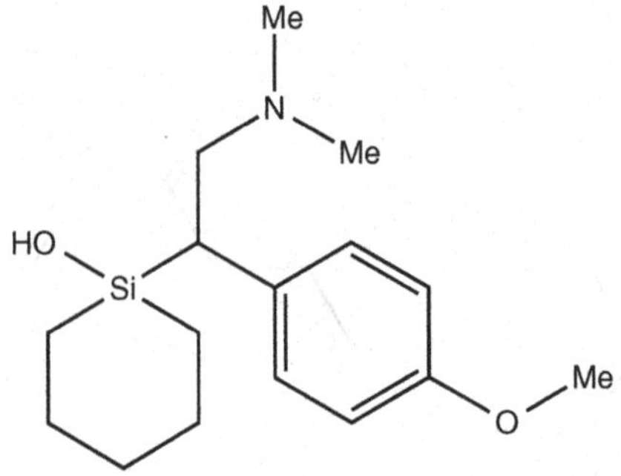

Figure 5.12 Chemical structure of sila-venlafaxine

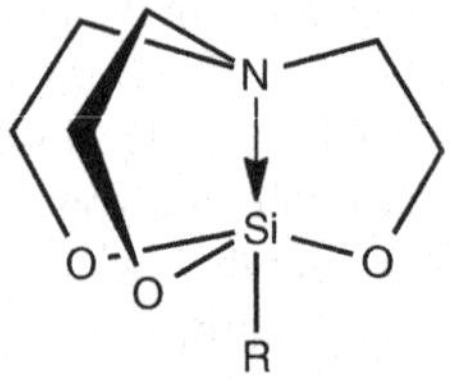

Figure 5.13 Chemical structure of silatrane

Silatranes are silicon compounds in which the central silicon atom is pentacoordinated. Silatranes can be highly toxic depending on the organic rest at the silicon centre. Aryl [11] and 2-thienyl-substituted silatranes [12] have been proposed as rodenticides [1]. These compounds are known for their self-detoxification, resulting in a low hazard for dermal toxicity or long-lasting secondary risk of poisoning (Figure 5.13) [13].

Silatranes substituted with alkyl, alkenyl and other groups are significant less toxic and are under evaluation for a variety of biological or clinical applications ranging from the stimulation of collagen biosynthesis to the proposed use as anticancer agents [1].

Silicones (oligo and polysiloxanes, see Section 5.1.1) are the most widely used class of silicon-based compounds clinically. Silicones can be found in plastics, lubricants, catheters, implants and a variety of other medically used items. Silicone fluids, such as simethicone, are known for their antifoaming properties. Simethicone is an orally administered suspension containing polysiloxanes and silicon dioxide. It is an antifoaming agent and is used to reduce bloating by decreasing the surface tension in bubbles. Excessive formation of gas bubbles in the stomach and intestines can be painful and can also be of hindrance for any ultrasound examination. Simethicone can be found in antacids and in suspensions given to babies against colic.

5.3 Organogermanium compounds: balancing act between an anticancer drug and a herbal supplement

The first organogermanium compound, tetraethylgermane, was synthesised by Winkler *et al.* in 1887, but then it took until the middle of the twentieth century for such compounds to be widely synthesised and examined (Figure 5.14).

The major uses for germanium compounds include their application as optical materials (60%) and semiconductors (10%), as catalysts or in chemotherapy. Some Chinese herbs and vegetables contain a relatively

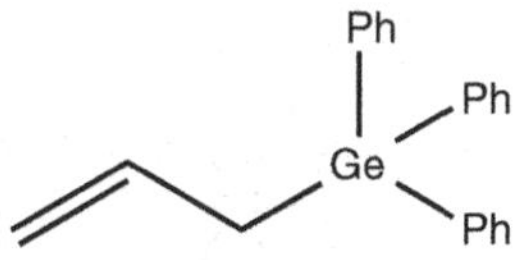

Figure 5.14 *Example for an organogermanium compound*

high amount of germanium, for example, ginseng, oats, soya beans and shiitake mushroom. The germanium is presented in organic form with Ge—O bonds being formed [1].

Germanium dioxide is the oxide of germanium, an inorganic compound, featuring the chemical formula GeO_2. It is formed as a passivation layer on pure germanium after exposure to oxygen. Germanium dioxide generally has a low toxicity, but shows severe nephrotoxicity at higher doses. Germanium dioxide is still offered on the market in some questionable miracle therapies. Exposure to high doses of germanium dioxide can lead to germanium poisoning [1].

In the 1970s, a range of organogermanium compounds were widely marketed as health supplements and became popular because of the therapeutic value of germanium. This encouraged a wide range of research looking into the biological potential of organogermanium compounds. Organogermanium compounds are generally well absorbed after ingestion. Nowadays, mainly compounds with antitumour, immune-stimulating, interferon-reducing and radioprotective properties are being researched. A range of germanium compounds, including germanium sesquioxide, spirogermanium, germatranes, decaphenylgermanocenes, germanium(IV) porphyrins and germyl-substituted heterocycles, have been synthesised and evaluated for their biological activities. Most intensively investigated for a therapeutical application so far have been germanium sesquioxide and spirogermanium (Figure 5.15) [1].

5.3.1 Germanium sesquioxide

2-Carboxyethylgermanium sesquioxide (Ge-132) was investigated in the 1990s to protect the human body from radiation, enrich the oxygen supply, remove heavy metals and scavenge free radicals. Japanese researchers have shown that Ge-132 has a variety of biological activities and could be effective in the treatment of several diseases such as cancer, arthritis and osteoporosis [1].

Ge-132 is a white crystalline powder, which is insoluble in organic solvents and soluble in water when heated. The compound does not melt but decomposes at high temperatures above 320 °C. These properties can be explained by the three-dimensional structure of the Ge-132, which consists of Ge_6O_6 rings. The structure is described as an infinite sheet structure. The carboxylate chains form hydrogen bonds between neighbouring chains and hold these germanium sesquioxide sheets together.

Synthesis starts with the generation of organogermanium trichloride, which can be hydrolysed in several steps to form germanium sesquioxide. Organogermanium trichloride itself can be synthesised by reducing germanium dioxide, a toxic starting material, with sodium hypophosphite. This reaction proceeds via a redox reaction, where sodium hypophosphite is oxidised (oxidation state +1 to +2) whilst GeO_2 is reduced (oxidation state +4 to +2). The resulting trichlorogermane is known to be highly unstable [14] and is therefore reacted *in situ* to the relevant organic germanium trichloride via a so-called hydrogermylation reaction (Figure 5.16) [15].

Germanium sesquioxides are generally not known to be embriotoxic, teratogenic, mutagenic or antigenic. Administration over a short term did not reveal any significant adverse effects. Ge-132 contains relatively stable Ge—C bonds, which prevents its fast hydrolysis to the toxic inorganic compound GeO_2. Ge-132 has good water solubility and is excreted from the human body within 24 h. Side effects are mainly due to impurities of

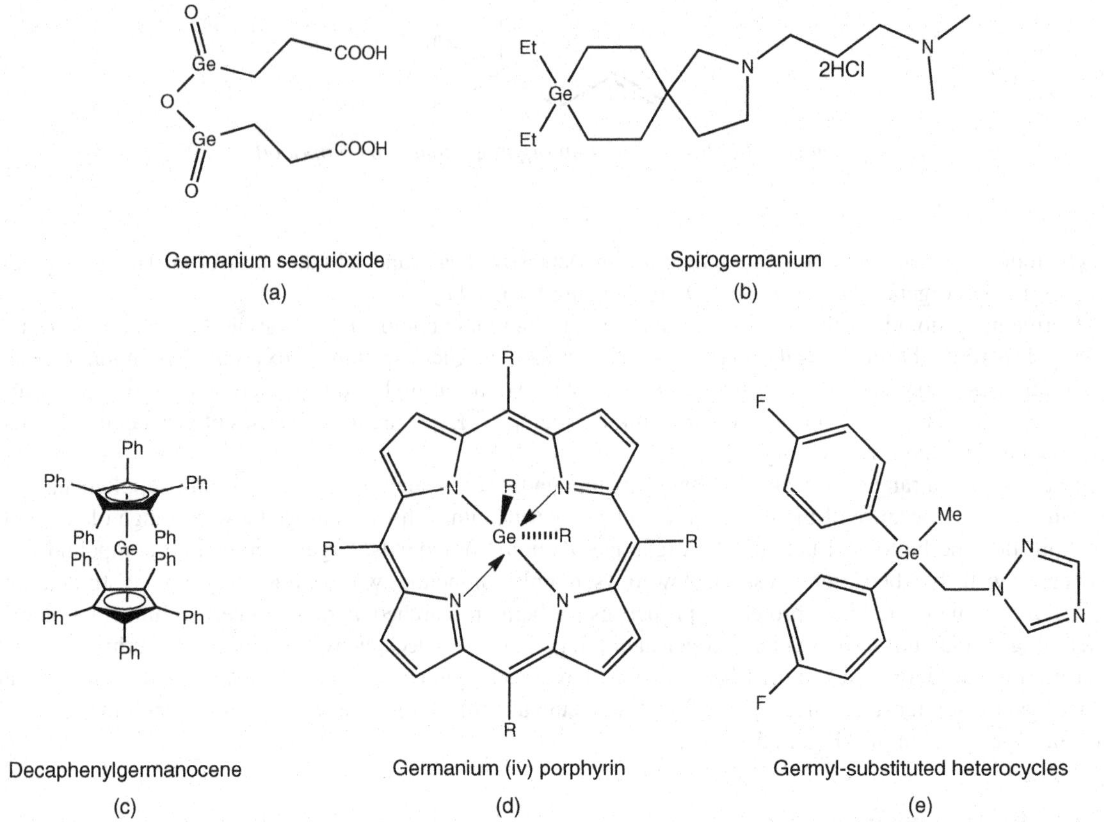

Germanium sesquioxide
(a)

Spirogermanium
(b)

Decaphenylgermanocene
(c)

Germanium (iv) porphyrin
(d)

Germyl-substituted heterocycles
(e)

Figure 5.15 *Chemical structures for germanium compounds investigated for their biological activity. (a) Germanium sesquioxide. (b) Spirogermanium. (c) Decaphenylgermanocene. (d) Germanium(IV) porphyrin. (e) Germyl-substituted heterocycles*

GeO_2 —(i)→ $HGeCl_3$ + acrylic acid —(ii)→ Cl_3Ge-propanoic acid —(iii)→ $(HO)_3Ge$-propanoic acid —(iv)→ germanium sesquioxide

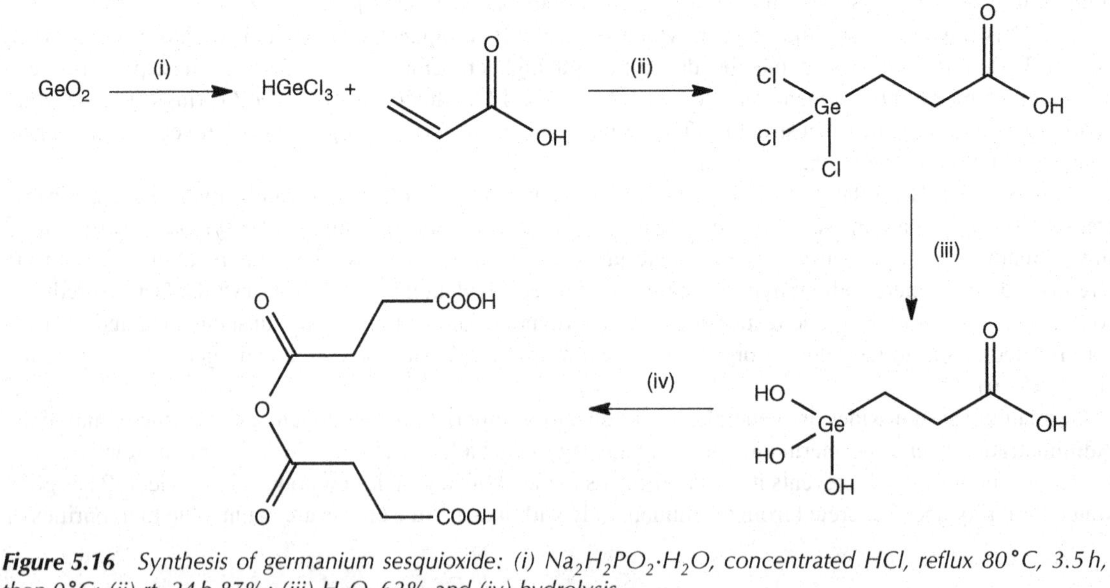

Figure 5.16 *Synthesis of germanium sesquioxide: (i) $Na_2H_2PO_2 \cdot H_2O$, concentrated HCl, reflux 80 °C, 3.5 h, then 0 °C; (ii) rt, 24 h 87%; (iii) H_2O, 62% and (iv) hydrolysis*

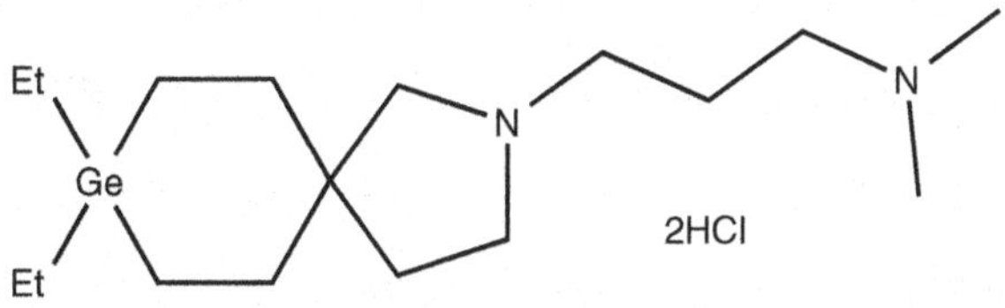

Figure 5.17 *Chemical structure of spirogermanium*

the pharmaceutical product with GeO_2, which can induce renal damage and accumulate in the kidneys, liver and spleen [16].

Lately, the antitumour activity of Ge-132 has been studied. It has been revealed that it possesses antitumour and immune-modulating activity. The first anticancer activity was reported when tested on Ehrlich Ascites tumour. Furthermore, studies were carried out on Lewis lung carcinoma and other cancer types. Oral treatment of pulmonary spindle cell carcinoma with Ge-132 showed complete remission of the cancer [1].

Interestingly, no cytotoxicity was proven when the studies were carried out *in vitro*, and it was concluded that the mechanism works via a stimulation of the host-mediated immunopotentiating mechanism. Nevertheless, the precise mechanism of the anticancer activity of Ge-132 is still not fully understood.

Ge-132 has been scrutinised for a range of biological activities, and studies suggest that the germanium compound may also exhibit antiviral, cardiovascular, antiosteoprotic and antioxidant activities [1, 15b, 17]. For example, studies have shown that Ge-132 is able to avert the decrease in bone strength and loss of bone mineral resulting from osteoporosis [17].

5.3.2 Spirogermanium

2-(3-Dimethylaminopropyl)-8,8-diethyl-2-aza-8-germaspiro[4,5]decane (spirogermanium) was the first organogermanium compound tested as an anticancer agent on a wide variety of human cancer cell lines, such as ovarian, cervix, breast, renal cell cancers and others. Preclinical toxicological evaluation in white mice confirmed the lack of bone marrow toxicity [18].

Spirogermanium entered clinical trials and showed good drug tolerance in phase I clinical trials. Phase II clinical trials revealed consistent neurotoxicity as well as pulmonary toxicities and only moderate activity against ovarian cancer. The mode of action involved is believed to be based on the inhibition of protein synthesis and a secondary suppression of RNA and DNA synthesis (Figure 5.17) [19].

5.4 Exercises

5.4.1 Draw the Lewis structures of the following silicon compounds and compare them with their carbon analogues.

(a) Silicon dioxide
(b) Silicon tetrachloride
(c) Silane
(d) Trichlorosilane

5.4.2 Calculate the ΔEN for all bonds in the following silicon compounds and compare these values with the corresponding carbon bonds.

(a) Silicon tetrachloride
(b) Silane
(c) Trichlorosilane

5.4.3 Research and calculate the bond length in the following compounds and compare them with their carbon analogues.

(a) Silane
(b) Germane
(c) Germanium tetrachloride
(d) Silicon tetrachloride
(e) Silicon dioxide

5.5 Cases studies

5.5.1 Simethicone

Simethicone is a silicon-based antifoaming agent that can be found in a variety of formulations.

(a) Describe the chemical structure of simethicone.
(b) Research its mode of action and route through the human body.
(c) Identify clinical applications for simethicone.
(d) Typical oral suspension used for infants contain simethicone 40 mg/ml. Calculate the weight/volume percentage (%w/v).

5.5.2 Germanium supplements

Organic germanium is sold as a dietary supplement and contains mainly bis-carboxyethyl germanium sesquioxide. Despite many positive health claims, several severe side effects including kidney failure have been reported. These side effects are often explained by toxic impurities, and therefore information on the testing procedures can often be found on the packaging. Typical information would read as follows: 'Analysis performed includes identification by infrared spectroscopy and solubility testing. The purity has been identified by acid group titration and the absence of inorganic germanium has been confirmed by a colour limit test'.

(a) Draw the chemical structure of bis-carboxyethyl germanium sesquioxide.
(b) What is the toxic impurity typically found in these preparations?
(c) What is the solubility of bis-carboxyethyl germanium sesquioxide?
(d) Describe the results you would expect from the infrared analysis.
(e) Research a method for the acid base titration mentioned.
(f) Research the colour limit test for inorganic germanium.

References

1. E. R. Tiekink, M. Gielen, *Metallotherapeutic drugs and metal-based diagnostic agents: the use of metals in medicine*, Wiley, Chichester, **2005**.
2. A. A. Shishkina, T. I. Ivanenko, N. A. Zarubina, O. N. Volzhina, V. G. Angarskaya, K. K. Pivnitsky, *Khim. Farm. Zh.* **1986**, *20*, 232–237.
3. S. Farkas, *CNS Drug Rev.* **2006**, *12*, 218–235.
4. G. T. Bikzhanova, I. Toulokhonova, G. Stephen, W. Robert, *Silicon chemistry, Vol. 3*, Springer, **2007**.
5. S. Gately, R. West, *Drug Dev. Res.* **2007**, *68*, 156–163.
6. (a) G. A. Showell, J. S. Mills, *Drug Discovery Today* **2003**, *8*, 551–556; (b) G. Showell, *Chem. Ind.-London* **2003**, 17.
7. G. A. McKay, M. R. Walters, J. L. Reid, *Lecture notes: Clinical pharmacology and therapeutics*, 8th ed., Wiley-Blackwell, Chichester, **2010**.
8. *British national formulary* British Medical Association and Pharmaceutical Society of Great Britain, London.
9. R. Tacke, F. Popp, B. Muller, B. Theis, C. Burschka, A. Hamacher, M. U. Kassack, D. Schepmann, B. Wunsch, U. Jurva, E. Wellner, *ChemMedChem* **2008**, *3*, 152–164.
10. (a) J. O. Daiss, C. Burschka, J. S. Mills, J. G. Montana, G. A. Showell, J. B. H. Warneck, R. Tacke, *J. Organomet. Chem.* **2006**, *691*, 3589–3595; (b) G. A. Showell, M. J. Barnes, J. O. Daiss, J. S. Mills, J. G. Montana, R. Tacke, J. B. H. Warneck, *Bioorg. Med. Chem. Lett.* **2006**, *16*, 2555–2558; (c) J. O. Daiss, C. Burschka, J. S. Mills, J. G. Montana, G. A. Showell, J. B. H. Warneck, R. Tacke, *Organometallics* **2006**, *25*, 1188–1198.
11. M. G. Voronkov, *Vestn. Akad. Nauk. SSSR* **1978**, 97–99.
12. E. Lukevits, S. Germane, O. A. Pudova, N. P. Erchak, *Khim. Farm. Zh.* **1979**, *13*, 52–57.
13. C. B. Beiter, M. Schwarcz, G. Crabtree, *Soap Chem. Special* **1970**, *46*, 38–46.
14. F. Glocking, *The chemistry of germanium*, Academic Press, London, **1969**.
15. (a) L. T. Chang, H. L. Su, *Vol. U.S. Patent* 4420430, **1982**; (b) Z. Rappoport, *The chemistry of organo-germanium, tin and lead compounds, Vol. 2*, Wiley, Chichester, **2002**; (c) M. Lesbre, *The organic compounds of germanium*, Interscience, London, **1971**.
16. S. H. Tao, P. M. Bolger, *Regul. Toxicol. Pharmacol.* **1997**, *25*, 211–219.
17. Y. Wakabayashi, *Biosci. Biotechnol. Biochem.* **2001**, *65*, 1893–1896.
18. (a) M. Slavik, L. Elias, J. Mrema, J. H. Saiers, *Drug Exp. Clin. Res.* **1982**, *8*, 379–385; (b) J. E. K. Mrema, M. Slavik, J. Davis, *Int. J. Clin. Pharmacol. Ther.* **1983**, *21*, 167–171.
19. (a) T. E. Lad, R. R. Blough, M. Evrard, D. P. Shevrin, M. A. Cobleigh, C. M. Johnson, P. Hange, *Invest. New Drugs* **1989**, *7*, 223–224; (b) C. Trope, W. Mattsson, I. Gynning, J. E. Johnsson, K. Sigurdsson, B. Orbert, *Cancer Treat. Rep.* **1981**, *65*, 119–120.

Further Reading

E. Alessio, *Bioinorganic medicinal chemistry*, Wiley-VCH, Weinheim, **2011**.

C. Ashley, C. Morlidge, *Introduction to renal therapeutics*, Pharmaceutical Press, London, **2008**.

W. Kaim, B. Schwederski, *Bioinorganic chemistry: inorganic elements in the chemistry of life: an introduction and guide*, Wiley, Chichester, **1994**.

H.-B. Kraatz, N. Metzler-Nolte, *Concepts and models in bioinorganic chemistry*, Wiley-VCH [Chichester: John Wiley, distributor], Weinheim, **2006**.

J. D. Lee, *Concise inorganic chemistry*, 5th ed., Chapman & Hall, London, **1996**.

R. M. Roat-Malone, *Bioinorganic chemistry: a short course*, Wiley, Hoboken, N.J. [Great Britain], **2002**.

G. J. Tortora, B. Derrickson, *Principles of anatomy and physiology*, 12th ed., international student/Gerard J. Tortora, Bryan Derrickson. ed., Wiley, [Chichester: John Wiley, distributor], Hoboken, N.J., **2009**.

6
Group 15 Elements

Members of group 15 of the periodic table (15th vertical column) are summarised as group 15 elements (or the nitrogen group) consisting of nitrogen (N), phosphorus (P), arsenic (As), antimony (Sb) and bismuth (Bi) (Figure 6.1).

The appearance of group 15 elements varies widely, reflecting the changing nature of the elements when descending within the group from nonmetal to metal. This trend can be seen both in their structures as well as in their chemical and physical properties. Nitrogen is a colourless and odourless gas. Phosphorous exists as white, red and black solids, whereas arsenic is found as yellow and grey solids. Antimony presents itself in a metallic grey form, and bismuth is a white crystalline metal.

Nitrogen atoms are included in a variety of organic drugs, and their application will not further be discussed in this book. Phosphorus is also an essential element for human life, and some of its biochemical uses as well as clinical applications will be discussed in Section 6.2. The clinical use of arsenic is known as the *start of chemotherapy*. Arsenic, despite its known toxicity, is still clinically used to combat a variety of diseases including cancer (see Section 6.3).

6.1 Chemistry of group 15 elements

6.1.1 Occurrence and extraction

Nitrogen makes up 78% (by volume) of air, whereas phosphorus can be found in several minerals and ores. Phosphorus is an essential constituent of plants and animals, being present in deoxyribonucleic acid (DNA), bones, teeth and other components of high biological importance. Phosphorus does not occur in its elemental state in nature, as it readily oxidises and therefore is deposited as phosphate rock. The remaining elements of group 15 are mostly obtained from minerals, but can also be found in their elemental form in the earth's crust. Arsenic is mostly presented in nature as mispickel (FeAsS), realgar (As_4S_4) and orpiment (As_2S_3). Bismuth occurs as bismuthinite (Bi_2S_3) as well as in its elemental form.

Dinitrogen (N_2) is extracted by fractional distillation of liquid air. By-products such as dioxygen (O_2) are removed by addition of H_2 and the use of a Pt catalyst. Elemental phosphorus is extracted from phosphate

Essentials of Inorganic Chemistry: For Students of Pharmacy, Pharmaceutical Sciences and Medicinal Chemistry,
First Edition. Katja A. Strohfeldt.

Companion website: www.wiley.com/go/strohfeldt/essentials

H																	He
Li	Be											B	C	N	O	F	Ne
Na	Mg											Al	Si	P	S	Cl	Ar
K	Ca	Sc	Ti	V	Cr	Mn	Fe	Co	Ni	Cu	Zn	Ga	Ge	As	Se	Br	Kr
Rb	Sr	Y	Zr	Nb	Mo	Tc	Ru	Rh	Pd	Ag	Cd	In	Sn	Sb	Te	I	Xe
Cs	Ba	La-Lu	Hf	Ta	W	Re	Os	Ir	Pt	Au	Hg	Tl	Pb	Bi	Po	At	Rn
Fr	Ra	Ac-Lr	Rf	Db	Sg	Bh	Hs	Mt	Ds	Rg	Uub						

Figure 6.1 *Periodic table of elements. Group 15 elements are highlighted*

rock by reacting with sand and coke in an electrically heated industrial oven. Phosphorus vapour is isolated and condensed under water – white phosphorus is extracted.

$$2Ca_3(PO_4)_2 + 6SiO_2 + 10C \rightarrow P_4 + 6CaSiO_3 + 10CO$$

Elemental arsenic is extracted mainly from FeAsS by heating and subsequent condensation of arsenic in the absence of air.

$$FeAsS \rightarrow FeS + As \quad \text{(in absence of air)}$$

Antimony is obtained from stibnite (Sb_2S_3) after reduction with iron. Bismuth is extracted from its sulfide or oxide ores via a reduction with carbon.

$$Sb_2S_3 + 3Fe \rightarrow 2Sb + 3FeS$$

6.1.2 Physical properties

The physical properties of group 15 elements vary widely, from nitrogen being a gas to the remaining elements being solids with increasing metallic character. Nitrogen exists as a diatomic molecule N_2 and is a colourless and odourless gas (condensation at 77 K). Nitrogen forms relatively strong and short bonds, resulting in the formation of a triple bond in the N_2 molecule. Furthermore, nitrogen has an anomalously small covalent radius and therefore can form multiple bonds with N, C and O atoms. Group 15 elements follow the general trend showing an increasing covalent radius when descending within the group.

Phosphorus has several allotropes, with white, red and black phosphorus being the main ones.

> ***Allotropes*** *are defined as the two or more physical forms of one element. Allotropes of carbon are graphite, carbon and diamond. These allotropes are all based on carbon atoms but exhibit different physical properties, especially with regard to hardness.*

White phosphorus is a solid consisting of tetrahedral P_4 molecules with single bonds. White phosphorus is the standard state of the element, but it is metastable, potentially due to the strained 60° bond angles (Figure 6.2).

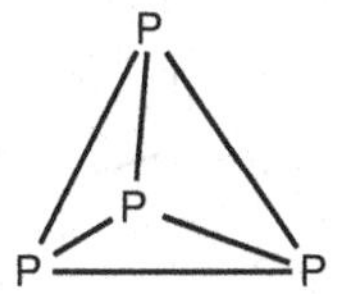

Figure 6.2 *Structure of white phosphorus*

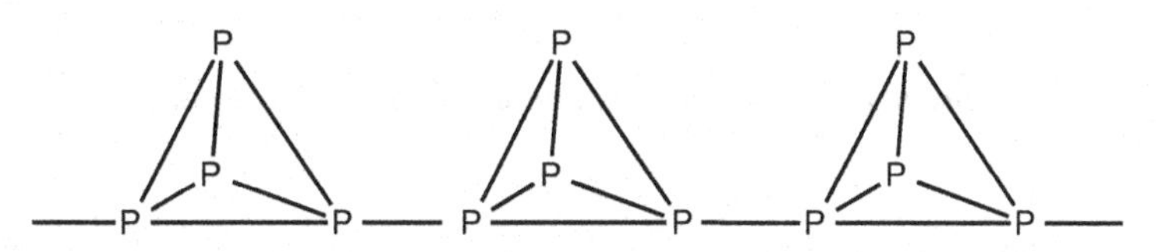

Figure 6.3 *Structure of red phosphorus*

Heating of white phosphorus in an inert gas atmosphere results in red phosphorus, which is an amorphous solid (several crystalline forms are known) with an extended covalent structure (Figure 6.3).

Black phosphorus is the most stable allotrope of phosphorus and can be obtained by heating white phosphorus under high pressure. In contrast to white phosphorus, black phosphorus does not ignite spontaneously in air. The reactivity of red phosphorus lies between those of the white and black allotropes. White phosphorus is insoluble in water and is therefore stored under water to prevent oxidation.

Arsenic and antimony vapour consists of As_4 or Sb_4 molecules, respectively. In the solid state, arsenic, antimony and bismuth are grey solids with a lattice structure similar to that of black phosphorus.

6.1.3 Oxidation states and ionisation energy

The general electron configuration for group 15 elements is ns^2np^3 and all elements form the oxidation states of +3 and +5. Nitrogen is more versatile and shows a range of oxidation states ranging from −3 to +5.

The ionisation energies relating to the removal of the first five electrons (two s and three p electrons) are relatively low. There is a significant increase in the ionisation energy necessary for the removal of a sixth electron, as this will be removed from an inner complete quantum shell (Table 6.1).

Table 6.1 *Ionisation energies (kJ/mol) for group 14 elements [1]*

	First	Second	Third
N	1 403	2857	4578
P	1 012	1897	2910
As	947	1950	2732
Sb	834	1596	2440
Bi	703	1610	2467

Source: Reproduced with permission from [1]. Copyright © 1996, John Wiley & Sons, Ltd.

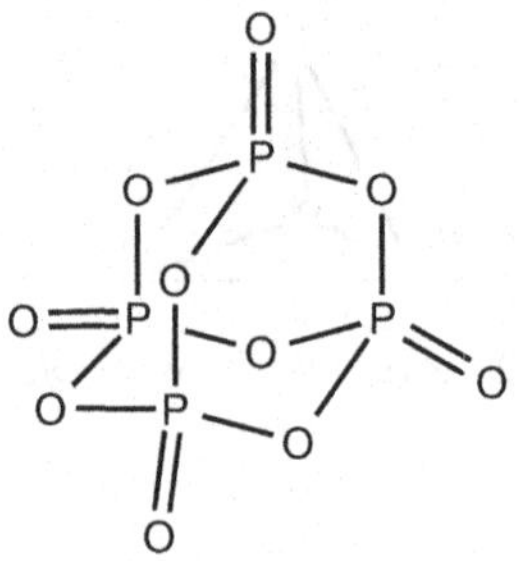

Figure 6.4 Chemical structure of phosphorus(V) oxide

6.1.4 Chemical properties

Nitrogen is relatively unreactive because the bond enthalpy of the nitrogen–nitrogen triple bond is very high (944 kJ/mol). N_2(g) is usually used as an inert atmosphere for reactions that cannot be carried out in oxygen. Only lithium reacts directly with nitrogen with the formation of Li_3N. *Nitrogen fixation* is an important mechanism developed by some microorganism in order to directly incorporate nitrogen gas into proteins. This process is an important step in the early food chain.

There are five *oxides* of nitrogen known – N_2O, NO, N_2O_3, NO_2 and N_2O_5 (oxidation numbers ranging from +1 to +5, respectively). Nitric(III) acid (nitrous acid) HONO and nitric(V) acid (nitric acid) HNO_3 are the most important oxoacids of nitrogen. HNO_3 is a highly reactive oxidising and nitrating agent.

In general, phosphorus is more reactive than nitrogen. White phosphorus ignites spontaneously in air and forms phosphorus(V) oxide. Phosphoric acid (H_3PO_4) is the most important oxoacid of phosphorus and its main use is in the manufacture of fertilisers (Figure 6.4).

Hydrogen-containing compounds of nitrogen and phosphorus, namely NH_3 and PH_3, both act as a Lewis base because of their lone pair. Phosphine (PH_3) is less water soluble than NH_3 as it does not form hydrogen bonds. NH_3 (ammonia) is produced in the so-called Haber Bosch process. This industrial process uses finely divided iron as catalyst and a reaction temperature of around 450 °C at a pressure of 50 atm. Ammonia is used to produce fertilisers, nitric acid, nylon and many more products important to our modern life style.

$$N_2(g) + 3H_2(g) \rightarrow 2NH_3(g)$$

For clinical applications, nitrogen and phosphorus compounds are mostly used as heteroatoms in organic compounds or counter-ions in inorganic salts with no specific therapeutic effect. Arsenic differs because it exhibits its own typical therapeutic and toxic properties, which has resulted in its long-standing use in clinical applications and in the invention of chemotherapy. Therefore, the following clinical discussion will concentrate on arsenic-based drugs.

6.2 Phosphorus

Phosphorus (P) is a nonmetal of the nitrogen group. As previously outlined, +3 and +5 are the preferred oxidation states, forming a variety of allotropes, with black phosphorus being the most stable one. Phosphorus is one of the most abundant elements in the human body and is often found in conjunction with calcium, because together they are the building materials for bones and teeth. Phosphorus is also involved in the building of our genetic material as well as in the energy supply of cells and many biochemical processes.

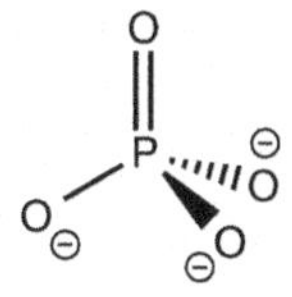

Figure 6.5 *Chemical structure of phosphate*

Within the human body, phosphate is the main phosphorus-containing compound. Phosphate is an inorganic compound and is the salt of phosphoric acid. It can form organic esters with a variety of compounds and these are important in many biochemical processes. Phosphate has the empirical formula PO_4^{3-}. It is a tetrahedral molecule, where the central phosphorus atom is surrounded by four oxygen atoms (Figure 6.5).

The phosphate ion PO_4^{3-} is the conjugated base of the hydrogen phosphate ion (HPO_4^{2-}). HPO_4^{2-} is the conjugated base of the dihydrogen phosphate ion ($H_2PO_4^-$). The latter is the conjugated base of phosphoric acid (H_3PO_4).

A ***conjugated base*** *is formed from an acid by the removal of a proton. This means that the conjugate base of an acid is this acid without a proton. An analogous definition applies to a conjugate acid.*

$$\text{A conjugate base (of the acid)} + H^+ \rightarrow \text{Acid}$$

$$\text{A conjugate acid (of the base)} \rightarrow \text{Base} + H^+$$

In biological systems, phosphate is often found either as the free ion (inorganic phosphate) or as an ester after reaction with organic compounds (often referred to as *organic phosphates*). Inorganic phosphate (mostly denoted as P_i) is a mixture of HPO_4^{2-} and $H_2PO_4^-$ at physiological pH.

6.2.1 Adenosine phosphates: ATP, ADP and AMP

Adenosine phosphates are organic-phosphate-containing compounds that are responsible for the energy flow in many biochemical processes in living cells. Adenosine phosphates consist of three parts: a sugar molecule (ribose) as the backbone, to which a nucleobase adenine and a varying number of phosphate groups are connected. Adenine is bonded to C-1 of the sugar, whilst the phosphate groups are connected to each other and then are attached to the C-5 atom of the ribose backbone. There is a series of adenosine phosphates depending on the number of phosphate groups present. Adenosine triphosphate (ATP) contains three phosphate groups, whilst adenosine diphosphate (ADP) contains two and adenosine monophosphate (AMP) contains one phosphate group (Figure 6.6) [2].

Within living cells, energy is transferred by dephosphorylation of ATP, which results in a transfer of energy to biochemical processes and the production of ADP. The enzyme ATPase is used to cleave off the phosphate group.

6.2.2 Phosphate in DNA

DNA is a major macromolecule in all living organism responsible for the encoding of genetic material. Mostly, DNA consists of a double-stranded helix. Each strand has a backbone of alternating sugar moieties (deoxyribose) and phosphate groups. The phosphate group is attached to 5′ position of the deoxyribose (Figure 6.7).

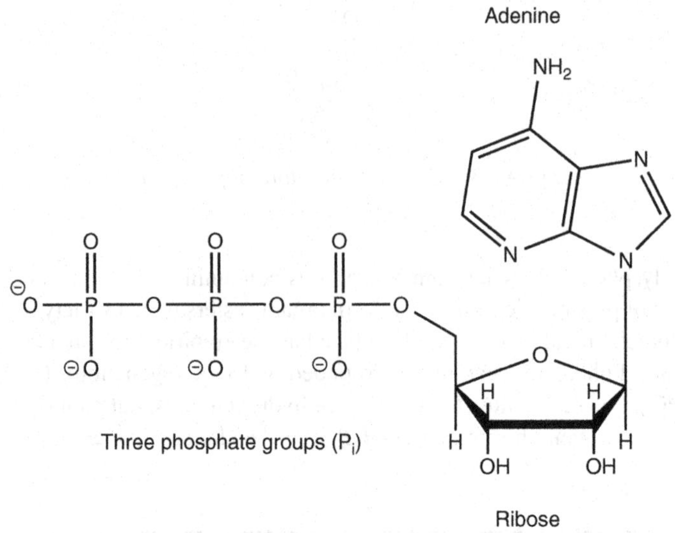

Adenine

NH2

Three phosphate groups (Pi)

Ribose

Figure 6.6 Chemical structures of ATP

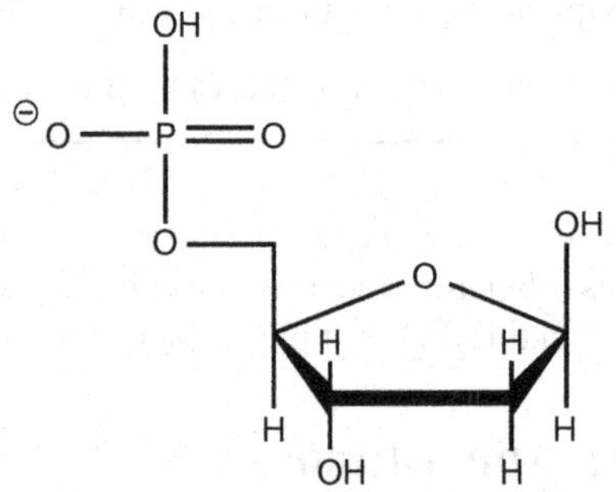

Figure 6.7 DNA backbone

There is also an organic base attached to the sugar moiety in the 1′ position. In DNA, these organic bases are thymine (T), cytosine (C), adenine (A) or guanine (G) (Figure 6.8). This unit, the deoxyribose with the phosphate group and a base, is called a *nucleotide* (Figure 6.9).

These nucleotides are then joined together by a condensation reaction of the phosphate and the hydroxyl group in 3′ position (Figure 6.10).

As previously mentioned, DNA mostly exists as a double-stranded helix. The double strand is formed by the interaction between the base pairs. Adenine and thymine form two hydrogen bonds, whereas guanine and cytosine are held together by three hydrogen interactions (Figure 6.11).

It is important to note that the two strands run in opposite directions. Whilst one strand runs in the 5′–3′ direction, the other one runs in the 3′–5′ direction.

6.2.3 Clinical use of phosphate

Phosphorus-containing compounds, mainly phosphates, are usually present in abundance in the human diet. They are mostly found in milk, meat (protein-rich food), grains, dried fruits and carbonated soft drinks.

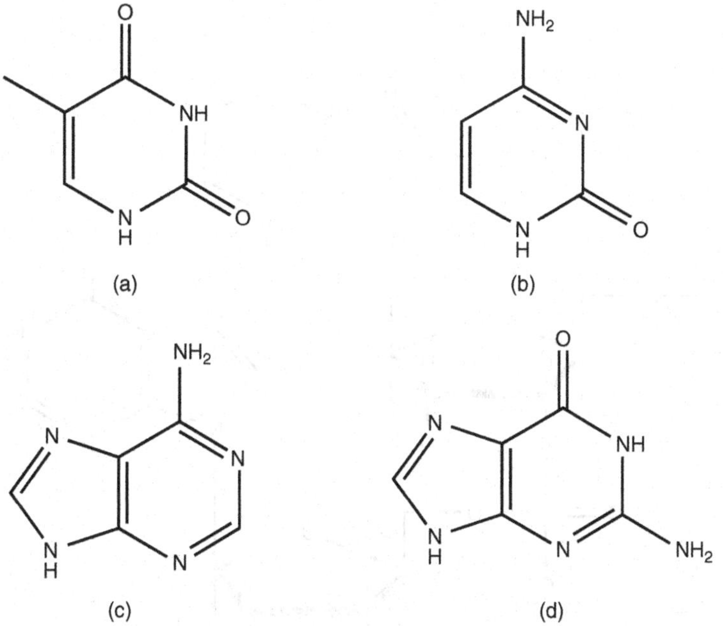

Figure 6.8 Chemical structures of bases: (a) thymine, (b) cytosine, (c) adenine and (d) guanine

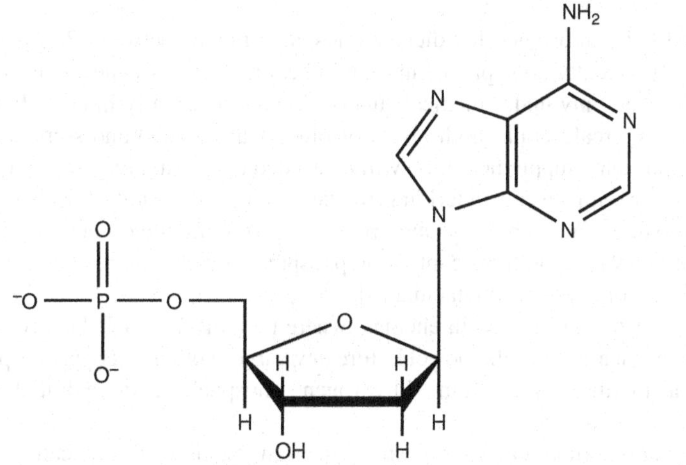

Figure 6.9 Chemical structure of the nucleotide adenine

Hypophosphataemia (low levels of phosphate in the serum) is rare and is often caused by some underlying illness, extreme lifestyle situations such as starvation or alcoholism or drug interactions; for example, some diuretics may cause low phosphate levels. Hyperphosphataemia, in contrast, is more common and often caused by kidney problems (reduced clearance) or dietary behaviour (increased intake). Phosphate and calcium ions work closely together and therefore an imbalance of either ion can have serious consequences for bone health or can even lead to cardiovascular problems due to hardening of the soft tissue [2, 3].

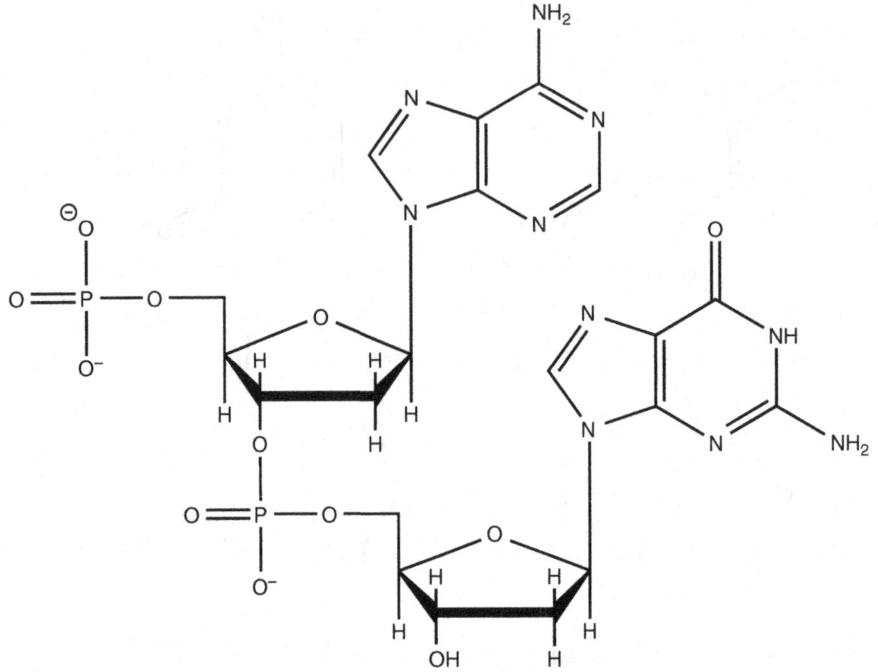

Figure 6.10 Result of the condensation reaction of the phosphate and the hydroxyl group in 3′ position

The recommended daily allowance for dietary phosphate ranges between 700 and 1250 mg depending the circumstances, and typically no supplementation in healthy humans is necessary [4]. Phosphate supplementation should be taken only under medical supervision and is usually indicated in the following cases: Hypophosphataemia, hypercalcaemia (high levels of blood Ca^{2+} levels) and sometimes for calcium-based kidney stones. Oral phosphate supplementation will be needed only in a minority of patients, often in patients dependent on alcohol or with severe underlying medical conditions. Oral phosphate tablets and solutions typically contain a mixture of monobasic sodium phosphate, sodium dihydrogen phosphate and/or disodium phosphate. Intravenous (IV) preparations containing phosphate together with potassium and sodium ions can be used in extreme cases of hypophosphataemia [5].

Phosphate solutions can also be used in enemas, where they display their laxative properties. Phosphate enemas are used for the clearance of the bowel before any surgery or endoscopy. A typical phosphate enema solution contains a mixture of sodium dihydrogen phosphate and hydrated disodium phosphate (Figure 6.12) [5].

Hyperphosphataemia is a more common problem seen in patients either because of excessive intake of phosphate or owing to reduced renal clearance such as in patients with renal failure. In this case, phosphate binding agents are used to treat the patients to manage high blood phosphate levels. Mostly, calcium preparations (e.g. calcium citrate tablets or capsules) are used as phosphate binding agents. The aim is to bind any excess phosphate in the gut before absorption. In patients on dialysis, Sevelamer or lanthanum salts (see Chapter 11) can be used to maintain normal phosphate blood levels. Sevelamer is a polymer containing amine groups. These protonated amine groups can react with the negatively charged phosphate groups via ionic binding and prevent absorption of phosphate from the gut. In the past, aluminium preparations were used, but they are not recommended anymore because of potential aluminium accumulation (see Chapter 4) [5].

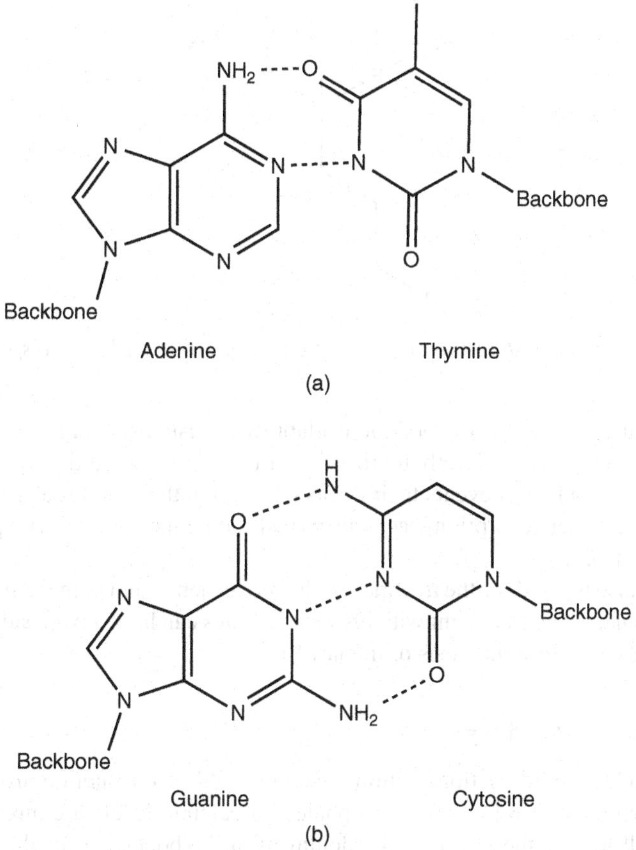

Figure 6.11 *Illustration of base pairing. (a) A–T base pair and (b) G–C base pair*

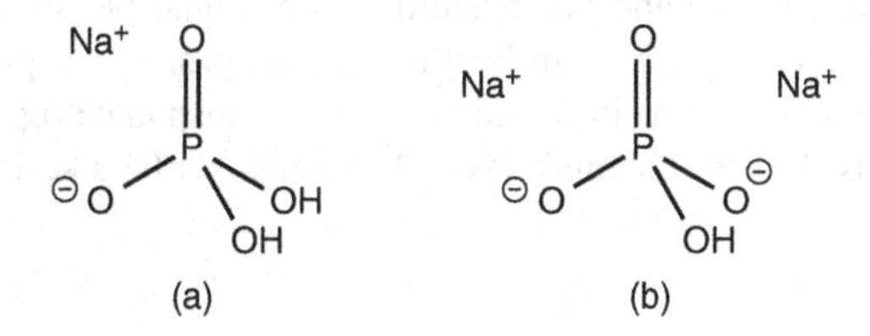

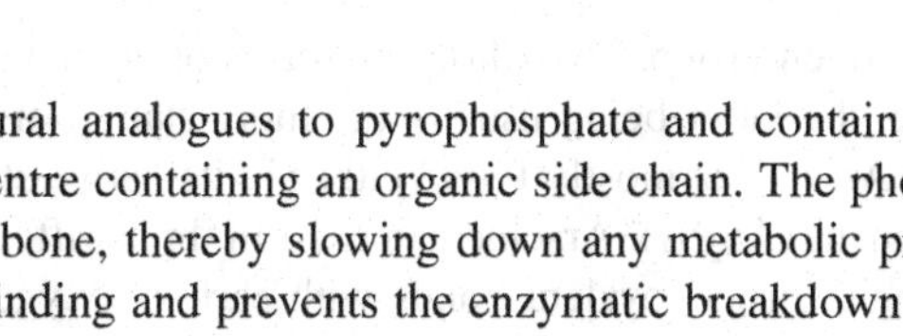

Figure 6.12 *Chemical structures of sodium dihydrogen phosphate (a) and disodium phosphate (b)*

Bisphosphonates are structural analogues to pyrophosphate and contain two phosphate groups linked together by a central carbon centre containing an organic side chain. The phosphonate is absorbed onto the hydroxyapatite crystals in the bone, thereby slowing down any metabolic processes in the bone. The side chain influences the skeletal binding and prevents the enzymatic breakdown in the gastrointestinal tract by phosphatases.

Bisphosphonates are mainly licensed for the treatment of osteoporosis, and alendronic acid is mostly the drug of choice. Administration typically is by mouth as tablets once a day or once a week for post-menopausal

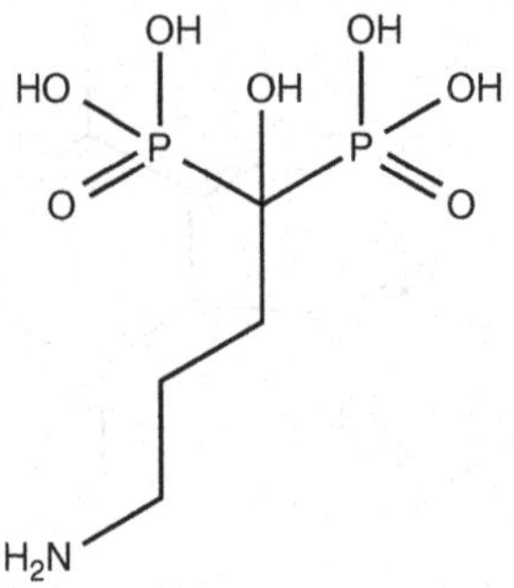

Figure 6.13 Chemical structure of alendronic acid as an example of a bisphosphonate

women [5]. It is interesting to note that the oral bioavailability of bisphosphonates, which are typically large hydrophilic molecules, is very low with only up to 1.5% of the administered dose [6]. Bisphosphonates are large hydrophilic molecules, which prevents their diffusion through the transcellular route [7]. Additionally, the negative charges hinder other absorption mechanisms and encourage reaction with positively charged ions and molecules (Figure 6.13).

Bisphosphonates can also be used for the treatment of hypercalcaemia and pain from metastatic bone cancer in patients with breast cancer. The treatment with bisphosphonates can have severe side effects, ranging from gastrointestinal disturbances to osteonecrosis of the jaw [5].

6.2.4 Drug interactions and toxicity

Excess of phosphate can lead to interactions within the human body with calcium, iron and magnesium, and can lead to diarrhoea and may even be toxic. Phosphate and calcium levels are directly connected, and an excess of phosphate will lead to the removal of calcium from the bones and teeth. This will cause osteoporosis and problems with the health of teeth and gums. Athletes often use phosphate supplementation, but a healthcare specialist should monitor this application.

Interactions for phosphate preparations with several over-the-counter and prescription drugs are known. Antacids containing aluminium, calcium and magnesium ions can bind phosphate in the digestive tract and prevent phosphate from being absorbed. This can lead in extreme cases to hypophosphataemia. Potassium-sparing diuretics and potassium supplements in combination with phosphate preparations may lead to elevated levels of blood potassium levels (hyperkalaemia). Hyperkalaemia can be a serious life-threatening problem (see Chapter 2).

6.3 Arsenic

Arsenic is a metalloid of the nitrogen group. Two allotrope forms of elemental arsenic have been reported: yellow arsenic and grey arsenic, the latter being usually the more stable form. Arsenic readily oxidises in air to arsenic trioxide (As_2O_3). Arsenic is mostly found either in its native state or as arsenic sulfide in the form of realgar (As_4S_4) or orpiment (As_2S_3). Arsenic can exist in three different valence states (zerovalent, trivalent and pentavalent). Arsenic forms covalent bonds with carbon, oxygen and hydrogen. The toxicity varies widely and depends on the physical state of the compound and its absorption/elimination rate.

Trivalent arsenics (As(III)) are derivatives of the arsenous acid (H_2AsO_3 – arsenite) and arsenic trioxide (AsO_3). Examples of pentavalent arsenic (As(V)) include derivatives of the arsenic acid (H_3AsO_4 – arsenate).

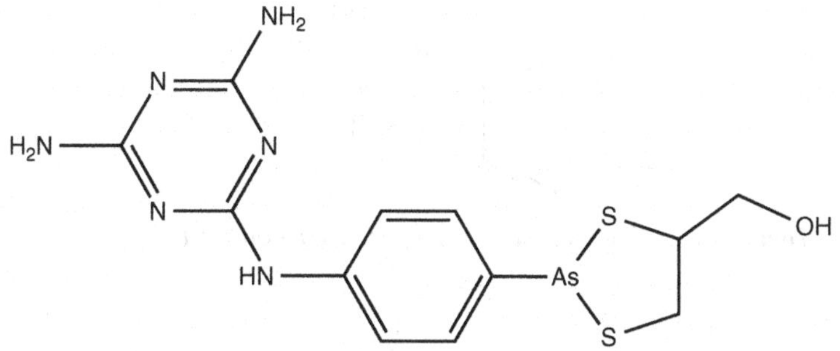

Figure 6.14 Chemical structure of Melarsoprol

Organic arsenic-based compounds, that is, compounds containing arsenic–carbon bonds, are usually less toxic than their inorganic counterparts. This is mainly due to their quicker excretion from the human body.

Arsenic is known to be one of the most toxic heavy metals. Compounds containing arsenic have a long history of use as poisons, but they also have a long historical medicinal use. As_2O_3, As_2S_3 (orpiment) and As_2S_2 (realgar) have been used as early as 2000 BC as drugs, for example, to cure cancerous tumours, ulcers and other diseases of the time. Nevertheless, the therapeutic use of arsenic-based compounds continued; for example, Galen (130–200 AD) recommended the application of a paste of arsenic sulfide against ulcers. Paracelsus 'ignored' any kind of formulation and recommended the clinical use of elemental arsenic. Fowler's solution (1% potassium arsenite) was applied in a variety of clinical applications. Interestingly, it was the main treatment option for chronic myelogenous leukaemia (CML) until it was replaced by radiation and chemotherapy in the twentieth century. Again, until the twentieth century, arsenic-based drugs were, for example, mostly used to combat trypanosomal infections. Indeed, Melarsoprol is currently approved for the treatment of late-stage African trypanosomiasis (Figure 6.14).

6.3.1 Salvarsan: the magic bullet – the start of chemotherapy

Salvarsan was a synthetic arsenic-based drug discovered in 1909 by Ehrlich and his team. In 1910, Ehrlich introduced Salvarsan (3-amino-4-hydroxyphenylarsenic), also known as *arsphenamine* or *compound 606*, to the market as a cure for syphilis caused by the bacterium *Treponema pallidum*.

6.3.1.1 Historical developments

Early in his studies, Ehrlich believed in the search for the 'magic bullet' – a treatment that would result in 'the use of drugs to injure an invading organism without injury to the host' [8]. This can be regarded as the start of chemotherapy. Early in his research career, Ehrlich became interested in bacteriology and the use of aniline and other dyes to selectively stain bacteria. In 1904, Ehrlich used trypan red to selectively stain trypanosomas (protozoa responsible for the African sleeping sickness). He discovered that no other cells took up the dye, and he got the idea of selectively targeting single cells from this experiment – the early start to chemotherapy. Ehrlich and his team managed to show that mice infected with trypanosomas could be cured with trypan red, but human experiments failed. Even today, trypan blue as staining agent is used to distinguish between living and dead cells because living cells do not take up the dye [8].

In 1905, the bacterium *T. pallidum* was identified by Schaudinn and Hoffmann as the cause syphilis. This discovery inspired Ehrlich to search for a cure using his targeted approach. 'We must search for magic

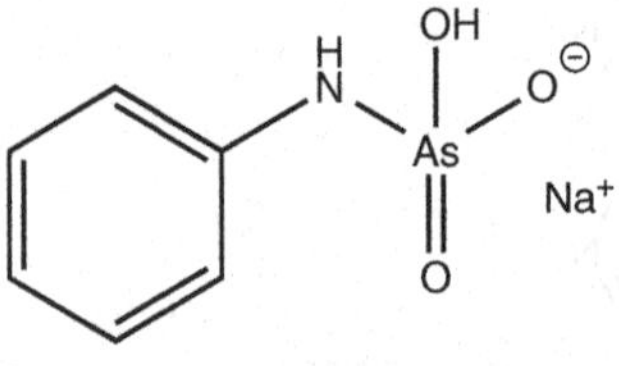

Figure 6.15 Chemical structure of Atoxyl according to Béchamp

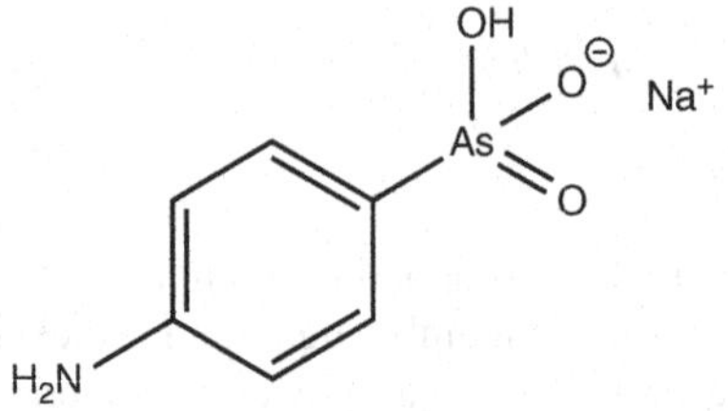

Figure 6.16 Chemical structure of Atoxyl according to Ehrlich and Bertheim

bullets' [9], Ehrlich commented during his research. 'We must strike the parasites and the parasites only, if possible, and to do this, we must learn to aim with chemical substances' [9].

Béchamp, teaching medicinal chemistry at the University of Montpellier, synthesised in 1863 a compound from aniline and arsenic acid, which became known later as *Atoxyl*. The name Atoxyl stems from its decreased toxicity. Béchamp characterised his compound as an anilide, and its structure is shown in Figure 6.15 [10].

In 1905, Thomas and Breinl showed that Atoxyl was effective in the treatment of trypanosomas, mainly *Trypanosoma brucei* gambiense – the cause of the African sleeping sickness, which was the main health problem around that time in Africa. Nevertheless, very high doses were required to show any pronounced effect, and as a result severe side effects such as blindness and damage to the optical nerve were common issues [10].

Inspired by this research, Ehrlich hired the chemist Bertheim in 1905. Bertheim revised the structure of Atoxyl and the correct chemical formula was assigned. Atoxyl was identified as an *p*-anilinyl arsenic acid derivative on the basis of its properties to reduce Tollen's reagent $[Ag(NH_3)_2]^+$ to metallic silver and its potential use to synthesise the corresponding diazo dye. Diazotisation is possible only for primary aromatic amines and therefore it could be concluded that Atoxyl had to be an arsenic acid rather than an anilide derivative; the correct structure according to Ehrlich and Bertheim is shown in Figure 6.16 [10].

Ehrlich's coworker, Hata, discovered a way to infect rabbits with *T. pallidum*. No one before had been able to produce syphilis in an animal, and in 1909 the first successful *in vivo* experiments in rabbits were performed. Having identified the correct structure of Atoxyl, Ehrlich and his team were inspired to search for a huge number of derivatives. Eventually, compound 606 was synthesised and introduced as an agent against syphilis. The compound was later marketed as Salvarsan, receiving its name from the Latin word 'salvare', which means to preserve, to heal. In 1909 and 1910, the first human tests on patients with syphilis and relapsing fever were extremely successful, and Salvarsan was marketed from 1910. For the first time, an infectious and fatal disease in humans could be treated with a man-made molecule, and Salvarsan brought Ehrlich world-wide fame (Figure 6.17) [10].

However, Ehrlich and his team did not stop with the discovery of Salvarsan. In particular, formulation issues encouraged them to search for a derivative which was easier to administer in order to make an injectable

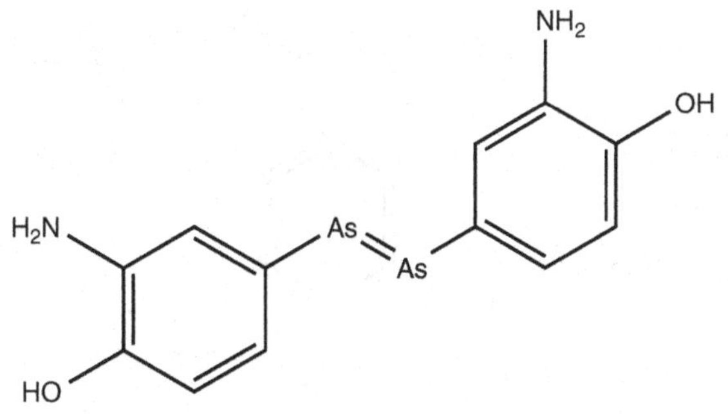

Figure 6.17 Chemical structure of Salvarsan

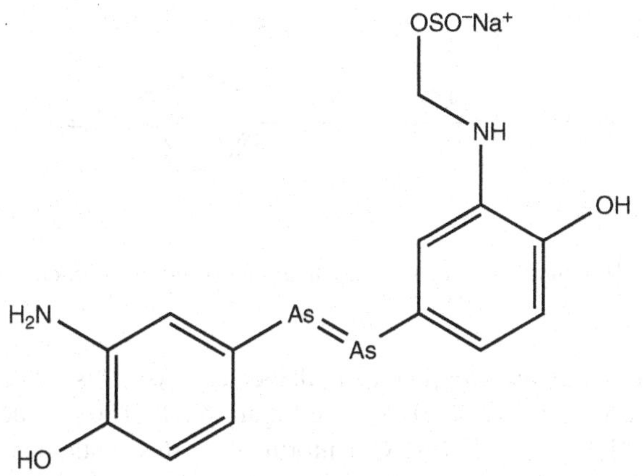

Figure 6.18 Chemical structure of Neosalvarsan

solution. Neosalvarsan (compound 914) is a salt derivative of Salvarsan and is water soluble, which showed reduced side effects (Figure 6.18) [10].

Around a decade later, doubts arose about the stability of an As=As double bond, as analysis of the arsenic content of the samples never conformed to the structure stated. Later work showed that neither Salvarsan nor Neosalvarsan was the active pharmaceutical ingredient (API). In 1930, the oxidised compound Oxophenarsine, containing an As=O unit, was identified as the active ingredient and was later marketed under the trade name Mapharsen. Mapharsen was used until the 1940s when it was replaced by Penicillin. Mapharsen was actually synthesised in Ehrlich's laboratory as compound number 5, but it was believed to be too toxic for any clinical application (Figure 6.19) [10].

Generally, the use of arsenic-based drugs has ceased, especially as a result of the development of Penicillin. Nevertheless, Melarsoprol and an arsenic-based drug closely related to Atoxyl are licensed to treat sleeping sickness.

6.3.1.2 Synthesis and structural analysis of Salvarsan

In Ehrlich's time, it was very reasonable to formulate the structure of Salvarsan as he did. But the As=As double bonds are not stable under the reaction conditions chosen by Bertheim and Ehrlich. Their proposed synthetic route was based on the reaction of 3-nitro-4-hydroxyphenyl-arsonic acid with dithionite. As a result,

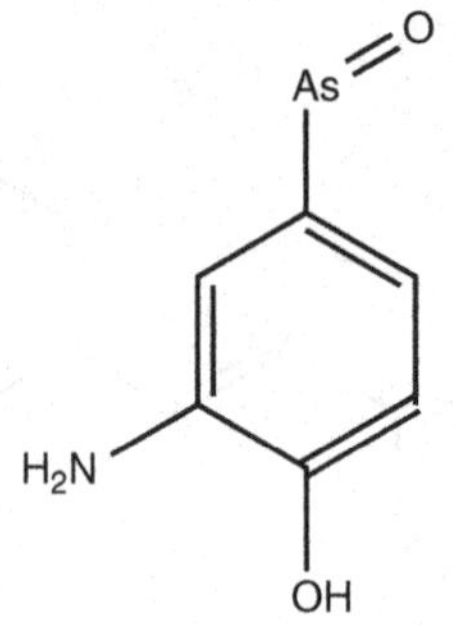

Figure 6.19 Chemical structure of Mapharsen

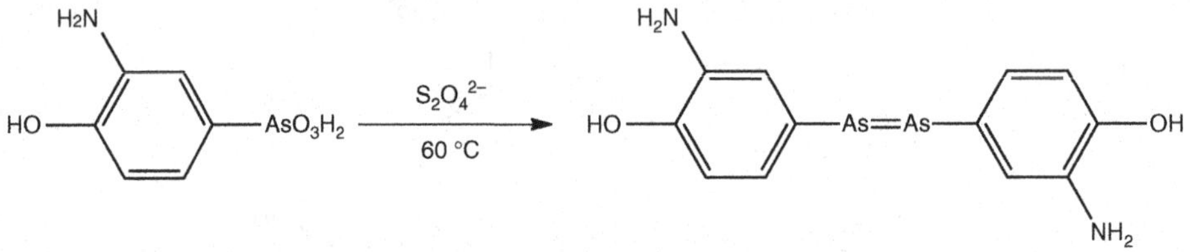

Figure 6.20 The initial synthesis of Salvarsan according to Ehrlich and Bertheim

the nitro group is reduced to an amine group and simultaneously As(V) is reduced to As(I), resulting in a compound with the formula 3-H_2N-4-HOC_6H_3As (see Figure 6.20). The product was then isolated as the hydrochloride salt 3-H_2N-4-$HOC_6H_3As{\cdot}HCl{\cdot}H_2O$. Unfortunately, this synthetic route was not always reproducible [11].

Christiansen *et al.* published in 1920 a two-step process leading to the sulfur-free product. The reaction involved the initial reduction of the nitro group with sodium dithionite and the subsequent reduction of the As(V) with hypophosphorous acid (Figure 6.21) [11].

Nevertheless, subsequent research has shown that dimeric arsenic-based structures exist only in sterically crowded molecules. The real structure of Salvarsan is not dimeric. Research published in 2005 by Lloyd *et al.* using different mass spectroscopic techniques showed that Salvarsan in solution consisted of small cyclic species with ring sizes of three and five arsenic atoms (see Figure 6.22). Nevertheless, this final structure of Salvarsan has still not been entirely identified [11].

6.3.2 Arsenic trioxide: a modern anticancer drug?

Arsenic trioxide, often denoted as As_2O_3 but more correctly stated as As_4O_6, is an inorganic compound mainly used as the precursor for organoarsenic compounds. It can be obtained by the oxidation of arsenic-containing minerals in the air, such as roasting of orpiment.

$$2As_2S_3 + 9O_2 \rightarrow As_4O_6 + 6SO_2$$

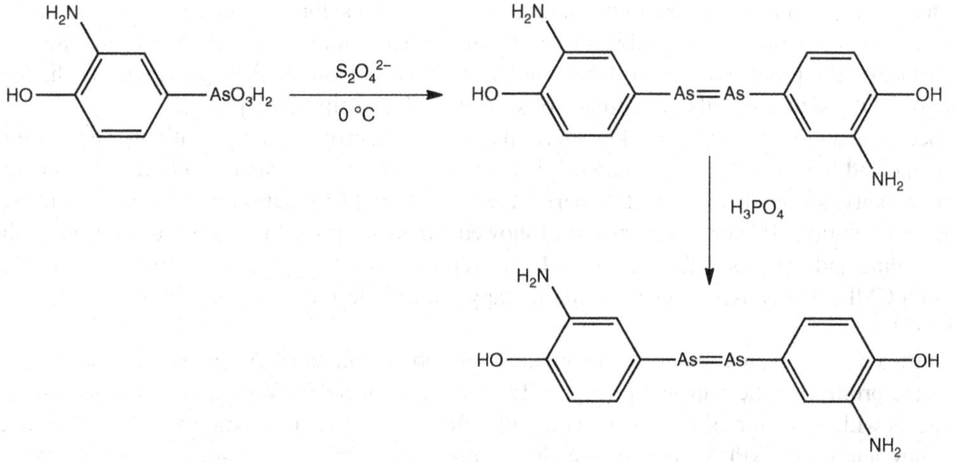

Figure 6.21 Synthesis of Salvarsan according to Christiansen

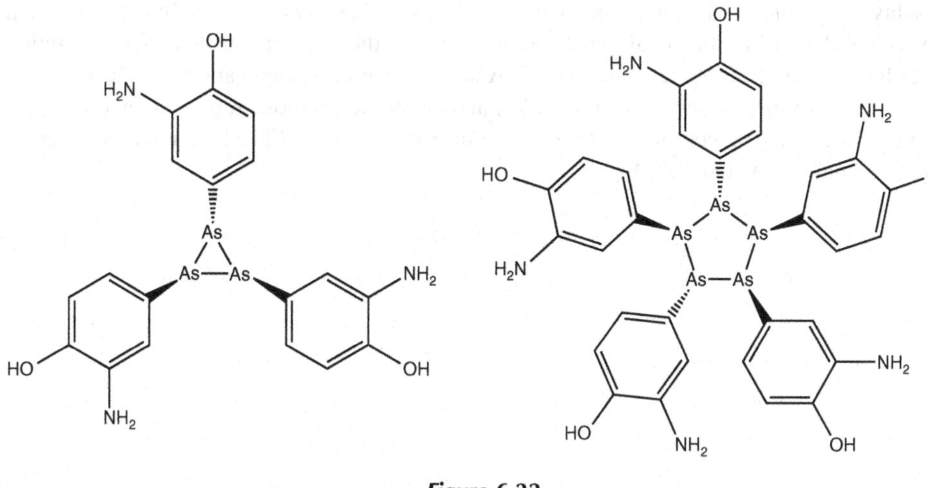

Figure 6.22

As_4O_6 is sparingly soluble in water and is an amphoteric compound. It reacts with alkali with the formation of arsenates, and arsenite trichlorides are synthesised in the presence of an acid.

$$As_4O_6 + 12NaO \rightarrow 4Na_3AsO_3 + 6H_2O$$

$$As_4O_6 + 12HCl \rightarrow 4AsCl_3 + 6H_2O$$

Arsenic trioxide is highly toxic. It is readily absorbed in the digestive system, through inhalation and skin contact. With a half-life of 1–2 days, elimination occurs rapidly at first via a methylation reaction and excretion in the urine. Around 30–40% of arsenic trioxide is incorporated into bones, muscles, hair and nails.

This means that elimination can take months, and any arsenic poisoning is detectable for the same period. Arsenic poisoning is characterised by digestive problems such as vomiting, diarrhoea and abdominal pain, as well as cardiovascular problems. Lower doses can lead to liver and kidney damage as well as changes in the pigmentation of the skin and nails (occurrence of so-called Mees stripes) [12].

Nevertheless, arsenic trioxide is long known for its therapeutic properties especially in the traditional Chinese medicine and homeopathy. In the latter, it is known as *arsenicum album* (dilution of arsenic trioxide). Despite its toxicity, arsenic trioxide and its derivatives have found application in the treatment of cancer. In 1878, Fowler's solution (1% potassium arsenite) showed a reduction of white blood cells when administered to healthy people and patients with leukaemia. It was reported in 1930 that arsenic trioxide was effective in patients with CML. It was used after radiation therapy until modern chemotherapy replaced this treatment approach [12].

Arsenic trioxide – marketed under the trade name Trisenox – gained FDA approval in 2000 for the treatment of acute promyelocytic leukaemia (APL). Trisenox, an injectable formulation, has been licensed for use in patients with induction of remission with APL after all-*trans* retinoic acid (ATRA) and anthracycline chemotherapy, and where APL is characterised by the presence of the t(15;17) translocation or PML/RAR-α (promyelocytic leukemia/retinoic acid receptor-alpha) gene expression. Trisenox was also approved in 2002 by the European Agency for the Evaluation of Medical Products for the treatment of adults with relapsed APL [13].

APL is a subtype of acute myelogenous leukaemia (AML), which is a cancer of the blood and bone marrow. The disease is caused by a chromosomal translocation involving the RAR-α gene and therefore unique compared to other forms of AML in its response to ATRA therapy. Unfortunately, about 20–30% of patients do not achieve remission from the combination of ATRA and cytotoxic chemotherapy, or they relapse. Trisenox was reported to achieve a 70% complete response rate in patients with APL which relapsed after treatment with cytotoxic chemotherapy and ATRA [13].

6.4 Exercises

6.4.1 Oral phosphate preparations contain typically monobasic sodium phosphate, and/or disodium phosphate. Determine the chemical formulae of these three compounds.

6.4.2 Draw the Lewis structure of the following molecules and determine the oxidation state of phosphorus:

(a) H_3PO_4
(b) PCl_3
(c) Na_2HPO_4

6.4.3 Determine the oxidation state of arsenic in the following molecules:

(a) As_2O_3
(b) H_3AsO_4
(c) AsF_5

6.4.4 A typical IV solution for the treatment of hypophosphataemia contains 100 mmol/l phosphate, 19 mmol/l potassium ions and 162 mmol/l sodium ions. Calculate the amount of phosphate, potassium and sodium ions present in 500 ml. Express your answer in grams.

6.5 Case studies

6.5.1 Phosphate solution for rectal use

Your pharmaceutical analysis company has been contacted by an important client and asked to analyse a batch of phosphate solutions for use in enemas. The description of your brief states that you are supposed to analyse the APIs in this solution following standard quality assurance guidelines.

Typical analysis methods used for quality purposes are based on titration reactions. A certain amount of solution is diluted with water. A known amount of this solution is then titrated with sodium hydroxide or hydrochloric acid depending on the API. Phenolphthalein and methyl red solutions are typically used as indicators [14].

(a) Research the APIs typically present in a solution used for enemas and describe their chemical structure. Describe the type of titration suggested. Describe the chemical structure and mode of action of the indicators.
(b) Formulate the relevant chemical equations.
(c) Phosphate enemas typically contain 12.8 g sodium dihydrogen phosphate dihydrate and 10.24 g disodium phosphate dodecahydrate and water made up to a 128 ml solution.

For the determination of the sodium dihydrogen phosphate dehydrate content, 20 ml of this solution is diluted with 80 ml of water and titrated with 0.5 M sodium hydroxide solution using phenolphthalein as indicator. The following amounts of sodium hydroxide are used:

25.0 ml	25.6 ml	25.4 ml

For the determination of the disodium hydrogen phosphate dodecahydrate content, 50 ml of this solution is titrated with 0.5 M hydrochloric acid using methyl red as indicator. The following amounts of hydrochloric acid are used:

22.0 ml	22.3 ml	22.4 ml

Calculate the amount of sodium dihydrogen phosphate and disodium hydrogen phosphate present in your sample. Express your answer in grams and moles.
(d) Critically discuss your result in context with the stated value for the API.
(e) Research the typically accepted error margins.

6.5.2 Forensic test for arsenic

Arsenic compounds are known for their toxicity and are often used in poisons. In order to obtain forensic evidence, specific tests were developed for the qualitative analysis of arsenic-based compounds. The so-called Marsh test was published in 1836 and is based on the reduction of As^{3+} in the presence of Zn.

(a) Research the conditions needed for the Marsh test.
(b) Formulate the relevant reduction and oxidation equations.
(c) Is it possible to use the Marsh test as a quantitative test?

References

1. J. D. Lee, *Concise inorganic chemistry*, 5th ed., Chapman & Hall, London, **1996**.
2. G. J. Tortora, B. Derrickson, *Principles of anatomy and physiology*, 12th ed., international student/Gerard J. Tortora, Bryan Derrickson. ed., Wiley [Chichester: John Wiley, distributor], Hoboken, N.J., **2009**.
3. G. A. McKay, M. R. Walters, Reid J. L. *Lecture notes. Clinical pharmacology and therapeutics*, 8th ed., Wiley-Blackwell, Chichester, **2010**.
4. U.S. Department of Health and Human Services and U.S. Department of Agriculture (ed. U.S. Department of Health and Human Services and U.S. Department of Agriculture), Rockville, M.D., **2005**.
5. *British national formulary*, British Medical Association and Pharmaceutical Society of Great Britain, London.
6. (a) S. Khosla, J. P. Bilezikian, D. W. Dempster, E. M. Lewiecki, P. D. Miller, R. M. Neer, R. R. Recker, E. Shane, D. Shoback, J. T. Potts, *J. Clin Endocrinol. Metab.* **2012**, *97*, 2272–2282; (b) S. Cremers, S. Papapoulos, *Bone* **2011**, *49*, 42–49.
7. A. G. Porras, S. D. Holland, B. J. Gertz, *Clin. Pharmacokinet.* **1999**, *36*, 315–328.
8. S. Riethmiller, *Abstr. Pap. Am. Chem. S* **1998**, *216*, U896–U896.
9. S. Lehrer, *Explorers of the body*, 1st ed., Doubleday, Garden City, N.Y., **1979**.
10. F. Winau, O. Westphal, R. Winau, *Microbes Infect.* **2004**, *6*, 786–789.
11. N. C. Lloyd, H. W. Morgan, B. K. Nicholson, R. S. Ronimus, *Angew. Chem. Int. Ed.* **2005**, *44*, 941–944.
12. K. H. Antman, *Oncologist* **2001**, *6*, 1–2.
13. M. H. Cohen, S. Hirschfeld, S. F. Honig, A. Ibrahim, J. R. Johnson, J. J. O'Leary, R. M. White, G. A. Williams, R. Pazdur, *Oncologist* **2001**, *6*, 4–11.
14. *British pharmacopoeia*, Published for the General Medical Council by Constable & Co., London.

Further Reading

E. Alessio, *Bioinorganic medicinal chemistry*, Wiley-VCH, Weinheim, **2011**.

C. Ashley, C. Morlidge, *Introduction to renal therapeutics*, Pharmaceutical Press, London, **2008**.

W. Kaim, B. Schwederski, *Bioinorganic chemistry: inorganic elements in the chemistry of life: an introduction and guide*, Wiley, Chichester, **1994**.

H.-B. Kraatz, N. Metzler-Nolte, *Concepts and models in bioinorganic chemistry*, Wiley-VCH [Chichester: John Wiley, distributor], Weinheim, **2006**.

J. D. Lee, *Concise inorganic chemistry*, 5th ed., Chapman & Hall, London, **1996**.

E. R. Tiekink, M. Gielen, *Metallotherapeutic drugs and metal-based diagnostic agents: the use of metals in medicine*, Wiley, Chichester, **2005**.

R. M. Roat-Malone, *Bioinorganic chemistry: a short course*, Wiley, Hoboken, N.J. [Great Britain], **2002**.

7

Transition Metals and d-Block Metal Chemistry

7.1 What are d-block metals?

The elements in groups 3–12 as shown in the schematic periodic table below are defined as the so-called d-block metals. The term *transition metal* is also often used to describe this group of elements. However, the IUPAC (International Union of Pure and Applied Chemistry) defines transition metals as elements with an incomplete d subshell or elements that can form a cation with an incomplete d subshell. Therefore, the group 12 metals zinc (Zn), cadmium (Cd) and mercury (Hg) are not typically classified as transition metals (Figure 7.1) [1].

Elements in the f-block, the so-called lanthanides and actinides, have been in the past called *inner transition metals*. Nowadays, they are more often referred to as *f-block elements* and selected examples will be discussed in Chapter 11.

Each group of d-block metals is formed by three members and is therefore called a *triad*. Sometimes, elements are grouped according to their chemical behaviour. One example is the group of platinum group metals, which encompasses ruthenium (Ru), osmium (Os), palladium (Pd) and platinum (Pt). Sometimes, you can find the term *heavier d-block metals*, which refers to d-block metals of the second and third row.

7.1.1 Electronic configurations

Generally, the ground-state electronic configurations of the first, second and third-row d-block metals follows the progressive filling of the 3d, 4d and 5d atomic orbitals, respectively. Nevertheless, there are exceptions, such as the ground state of chromium, which is $[Ar]4s^1 3d^5$ rather than $[Ar]4s^2 3d^4$. The reasons are fairly complicated and will not be further discussed in this book (Table 7.1).

d-Block metals can show several oxidation states as their valence electrons can be present in more than one atomic orbital. M^{2+} and M^{3+} ions of the first-row d-block metals follow the general formula $[Ar]3d^n$. The electronic configurations for second- and third-row d-block metals are again more complicated and will not be further discussed in this book.

Essentials of Inorganic Chemistry: For Students of Pharmacy, Pharmaceutical Sciences and Medicinal Chemistry,
First Edition. Katja A. Strohfeldt.

Companion website: www.wiley.com/go/strohfeldt/essentials

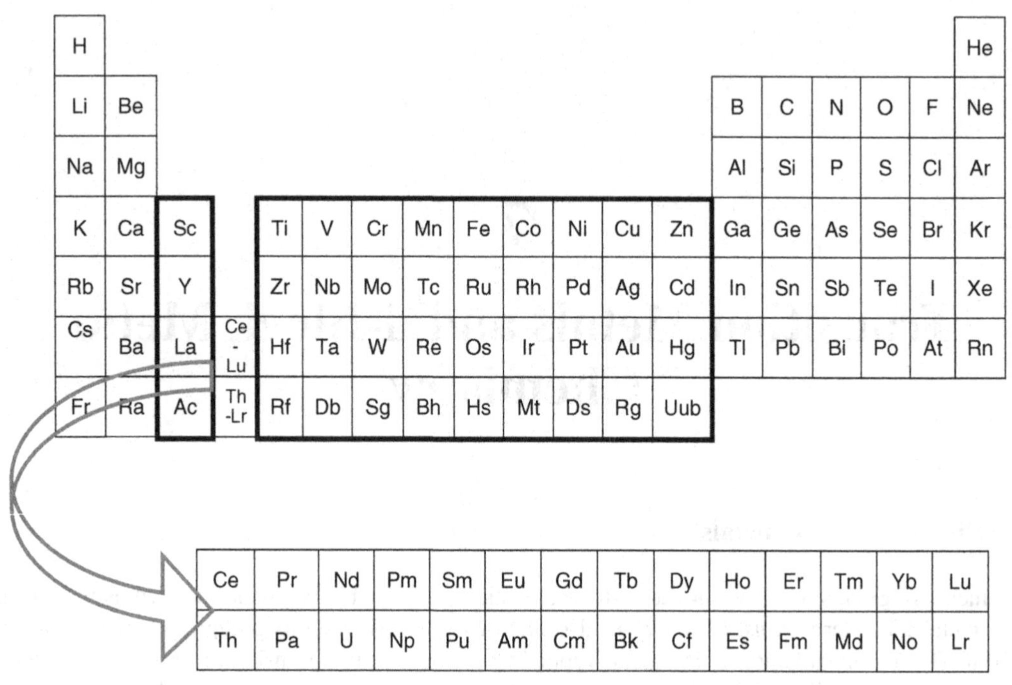

Figure 7.1 *Periodic table of elements; d-block elements are highlighted*

Table 7.1 *Examples of ground-state electronic configurations*

Sc	Ti	V	Cr	Mn	Fe	Co	Ni	Cu	Zn
d^1s^2	d^2s^2	d^3s^2	d^5s^1	d^5s^2	d^6s^2	d^7s^2	d^8s^2	$d^{10}s^1$	$d^{10}s^2$

7.1.2 Characteristic properties

Nearly all d-block metals are hard, malleable and ductile, and conduct electricity and heat. Most of them will form one of the typical metal structures. Exceptions are manganese (Mn), Zn, Cd and Hg. Only three d-block elements, namely iron (Fe), cobalt (Co) and nickel (Ni), are known to produce a magnetic field.

d-Block metals tend to readily form complexes with a characteristic colour if their ground-state electronic configuration is different from d^0 or d^{10}. Complex formation is often characterised by a colour change. For example, $[CoF_6]^{3-}$ is green, whereas $[Co(NH_3)H_2O]^{3+}$ is red and $[Co(H_2O)_6]^{3+}$ is blue.

> ***Coordination complexes*** *are defined as chemical structures that consist of a central atom or metal ion, and the surrounding molecules or anions called the ligands.*

Paramagnetism is a phenomenon that is often observed for d-block metal compounds. This is a result of the presence of unpaired electrons and can be investigated using electron paramagnetic resonance (EPR) spectroscopy. As a result, abnormalities in the NMR spectra can be observed such as broadening of the signals or unusual chemical shifts.

***Paramagnetism** is defined as the phenomenon whereby some materials show magnetic properties only once they are exposed to a magnetic field. Outside this magnetic field, no magnetic properties are seen. This is in contrast to ferromagnets, which show magnetic properties independent of the environment.*

7.1.3 Coordination numbers and geometries

d-Block metal compounds readily form complexes displaying different coordination numbers and geometries. In this section, we will restrict our discussion to complexes with only one metal centre (mononuclear complex). Also, it is important to note that the discussed geometries are regular geometries, which in practice can often be distorted as a result of steric hindrances. Additionally, in reality, fluxional behaviour in solution can be observed if the energy difference between the different structures is small enough, but we will restrict the following discussion to the solid state of complexes.

In general, steric and electronic factors dictate the coordination number. Sterically demanding ligands are more likely to form complexes with a low coordination number. In contrast, complexes containing small ligands and a large metal centre favour high coordination numbers (Figure 7.2).

The *Kepert model* is typically used to describe the shape of d-block metal complexes. The metal is defined to be the centre of the complex, and the ligands are arranged freely on a sphere around the centre. Only ligands are taken into consideration when determining the geometry of the complex. This is in contrast to the valence shell electron pair repulsion (VSEPR) model, which is used to determine the structure of p-block element compounds, where also nonbonding electrons are considered.

***The η-nomenclature for ligands** is used in organometallic chemistry and describes the number of atoms in a ligand that directly interact with the metal. The prefix η (eta) is accompanied by a number, which equals the number of atoms coordinating to the metal. This is called **hapticity** of a ligand.*

7.1.3.1 Coordination number 2: linear

A coordination number of 2 can be typically found for the metals Cu(I),[1] Ag(I), Au(I) and Hg(II). The metal forms the centre of the complex, and the two ligands are arranged at 180° to each other (Figure 7.3).

7.1.3.2 Coordination number 3: trigonal planar or trigonal pyramidal (less common arrangement)

In general, trifold coordination of a metal centre is not very common. There are examples of metals with a full d-orbital (d^{10} metals), which form trigonal planar structures. This means three ligands are arranged around the metal centre in one plane with 120° angle to each other. Examples include complexes of Cu(I), for example, in $[Cu(CN)_3]^{2-}$ and Ag(I), Au(I), Hg(II) and Pt(0) in $[Pt(PPh_3)_3]$ (Figure 7.4).

7.1.3.3 Coordination number 4: tetrahedral or square planar

The coordination number 4 is extremely common for d-block metal complexes. Most frequently, a tetrahedral arrangement can be observed. Examples include $[MnO_4]^{2-}$, $[FeCl_4]^-$ and $[CrO_4]^{2-}$ (Figure 7.5).

The square planar arrangement in which all four ligands are arranged around the metal centre in one plane is less commonly observed and often connected to d^8 metals. Nevertheless, some of these complexes are very important as a result of their medical application. For example, the square planar complex $[PtCl_4]^{2-}$

[1] Oxidation states for d-block elements are often denoted in Roman numbers.

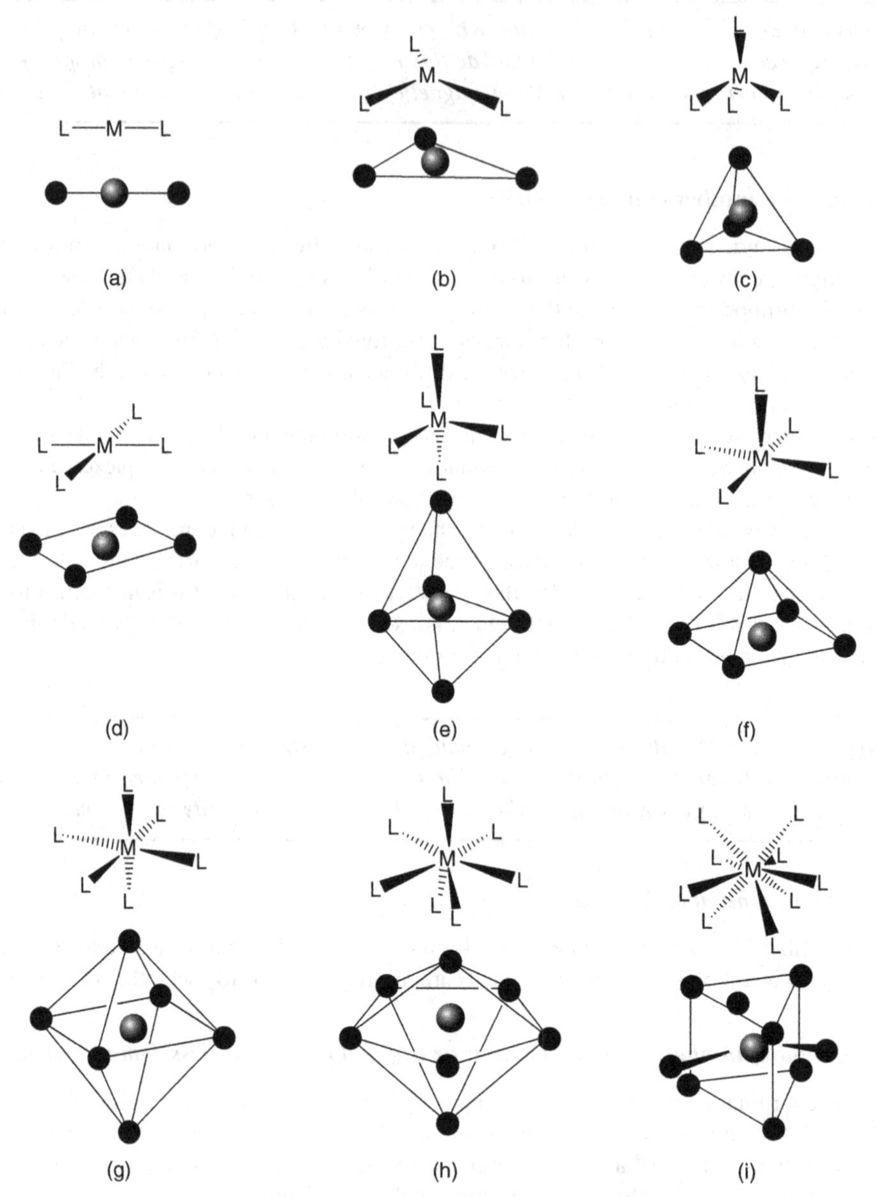

Figure 7.2 *Common geometries of metal complexes [2]. (a) Linear, (b) trigonal, (c) tetrahedran, (d) square planar, (e) trigonal bipyramid, (f) square pyramid, (g) octahedron, (h) pentagonal bipyramid and (i) tricapped trigonal prism (Reproduced with permission from [2]. Copyright © 2009, John Wiley & Sons, Ltd.)*

F—Be—F

Figure 7.3 *Example of a linear complex (BeF_2 in the gas phase)*

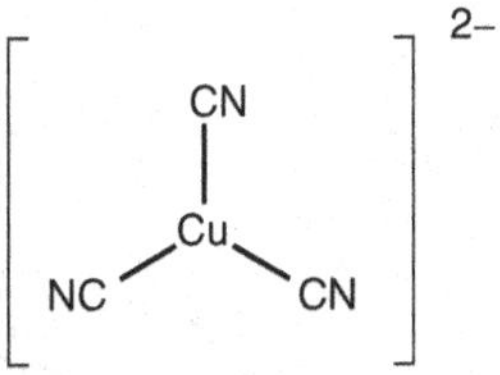

Figure 7.4 Example of a trigonal planar complex

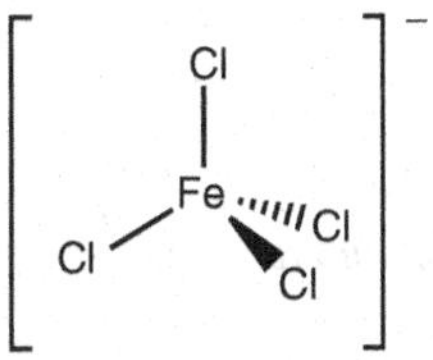

Figure 7.5 Example of a tetrahedral complex

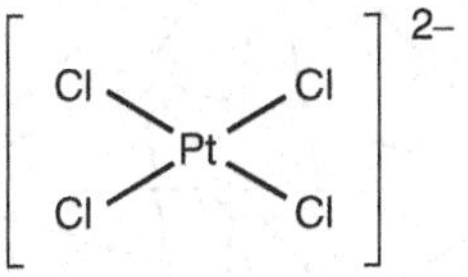

Figure 7.6 Example of a square planar complex

is commonly used as a precursor to the known chemotherapeutic agent cisplatin, whereas square planar complexes $[PdCl_4]^{2-}$, $[AuCl_4]^-$ and $[RhCl(PPh_3)_3]$ are all under investigation for their use in medicine (Figure 7.6).

7.1.3.4 Coordination number 5: trigonal bipyramidal or square-based pyramidal

The energy difference between the trigonal bipyramidal structure and the square-based pyramidal structure is usually fairly small and therefore many structures lie in reality between those two. Examples for simple trigonal bipyramidal structures include $[CdCl_5]^{3-}$ and $[CuCl_5]^{3-}$, whereas $[WCl(O)]^-$ and $[TcCl_4(N)]^-$ form square-based pyramidal structures typical for a series of oxo and nitrido complexes (Figure 7.7).

7.1.3.5 Coordination number 6: octahedral or trigonal prismatic (less common geometry)

Octahedral geometry is most commonly observed for the coordination number 6. Metals of all kinds of electronic configurations form octahedral complexes, for example, $[Mn(OH_2)_6]^{3+}$, $[V(OH_2)_6]^{3+}$, $[Fe(CN)_6]^{3-}$ and $[Fe(OH_2)_6]^{2+}$ (Figure 7.8).

It was long believed the octahedral geometry is the only geometry for coordination number 6 in existence, but eventually examples of trigonal prismatic coordination have been confirmed by X-ray analysis. Examples are $[ZrMe_6]^{2-}$ and $[ReMe_6]$ (Figure 7.9).

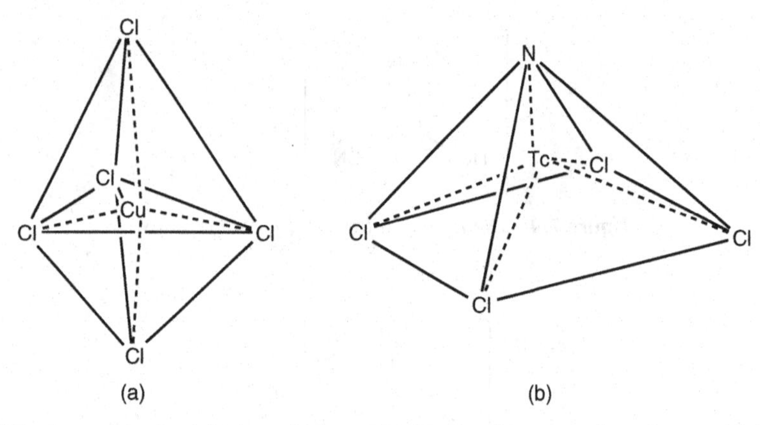

Figure 7.7 *Examples of a (a) trigonal bipyramidal and a (b) square-based pyramidal complex*

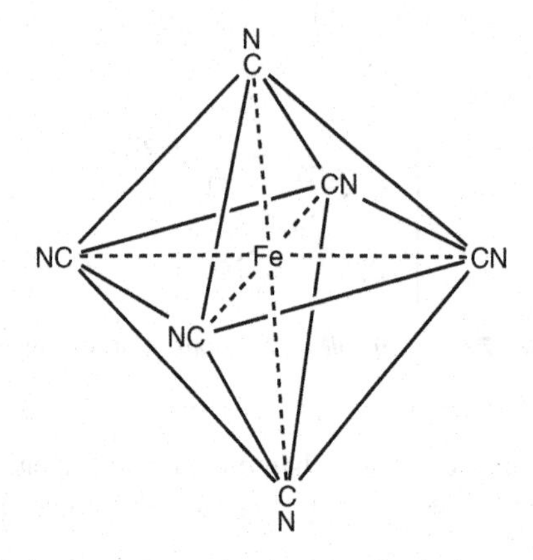

Figure 7.8 *Example of an octahedral complex*

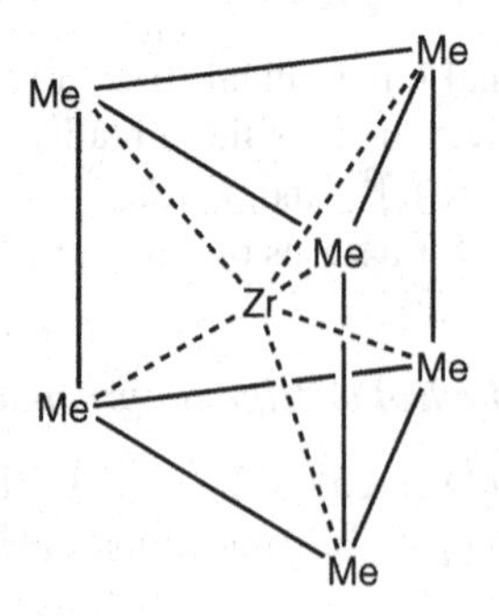

Figure 7.9 *Example of a trigonal prismatic complex*

7.1.3.6 Coordination number 7 and higher

Coordination numbers of 7 and higher are most commonly observed for d-block metals of the second and third row. Geometries can become fairly complicated and are not further discussed here.

7.1.4 Crystal field theory

Many transition-metal complexes are coloured. This and other spectroscopic properties, such as magnetism and hydration enthalpies, can be explained with the so-called crystal field theory (CFT).

> *The CFT describes the degeneration of the d- and f-orbitals in transition-metal complexes. It does not attempt to describe any type of chemical bonds.*

CFT is based on the interaction of a positively charged cation and the nonbinding (negatively charged) electrons of the ligand. The general principle is that the five d orbitals are degenerated, meaning that they do not occupy the same energy level anymore. Once the ligands approach the central positively charged cation, the electrons of the ligands will become closer to some of the d orbitals of the metal. This results in the degeneration of the d orbitals. Electrons in d orbitals that are closer to the ligands will occupy a higher energy level as the negatively charged electrons will repel each other (Figure 7.10).

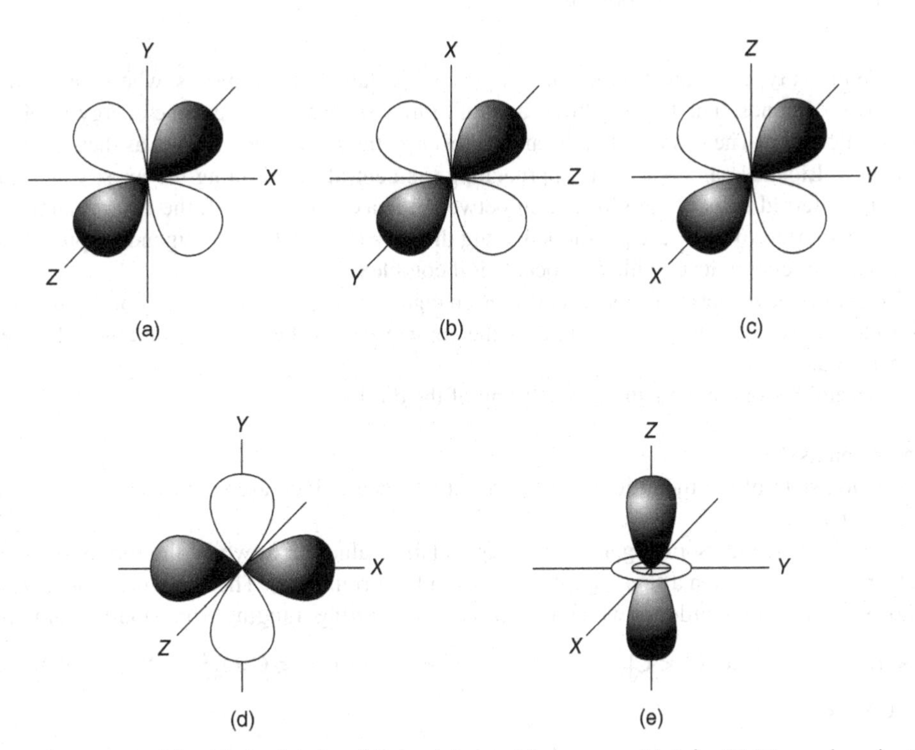

Figure 7.10 Geometry of d orbitals: (a) d_{xy}, (b) d_{xz}, (c) d_{yz}, (d) $d_{x^2-y^2}$ and (e) d_{z^2} [2] (Reproduced with permission from [2]. Copyright © 2009, John Wiley & Sons, Ltd.)

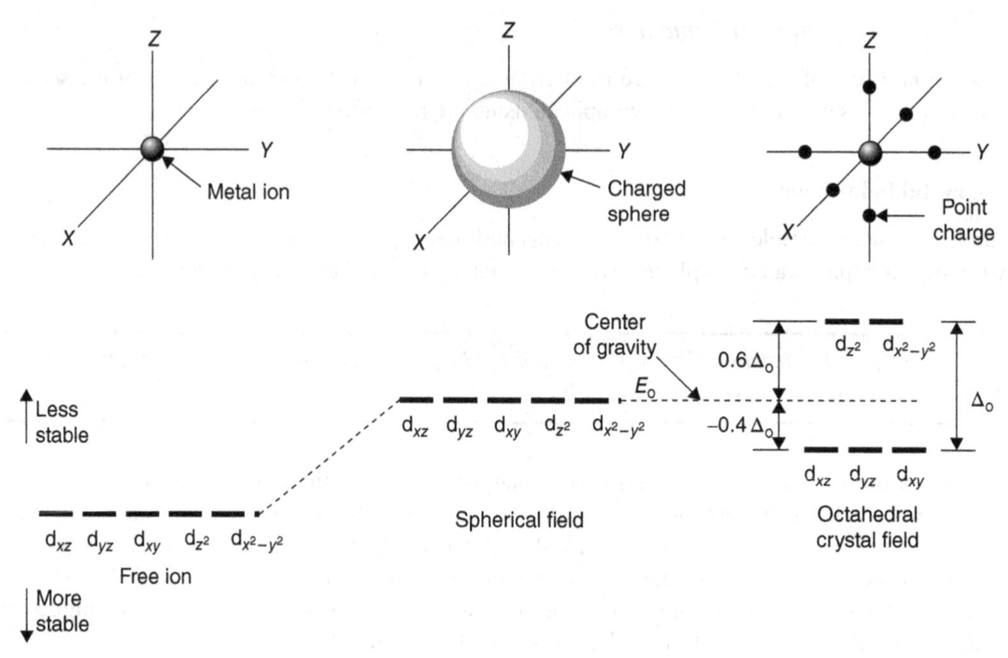

Figure 7.11 Generation of the octahedral crystal field from the free ion [2] (Reproduced with permission from [2]. Copyright © 2009, John Wiley & Sons, Ltd.)

The most common type of transition-metal complexes is octahedral complexes, where the central metal is coordinated by six ligands. The five d orbitals are split into two sets with the same energy level, the energy difference is called Δ_{oct}. The orbitals d_{xy}, d_{xz} and d_{yz} occupy the lower energy level as they are further away from the ligands. Ligands can be seen as approaching the central metal along the axes, where the orbitals d_{xy}, d_{xz} and d_{yz}, referred to as t_{2g}, are located in between the axes. In contrast, the two remaining d orbitals $d_{x^2-y^2}$ and d_{z^2}, referred to as e_g, are positioned along the axes (from where the ligands approach) and therefore occupy a higher energy level within an octahedral complex.

Figure 7.11 shows how energy levels of d orbitals change, ranging from the energy of d orbitals of the free metal ion, followed by the energy of d orbitals of the complex in a spherical field and the split energy levels observed in an octahedral field.

There are several factors influencing this splitting of the d orbitals:

- The metal ion itself.
- The oxidation state of the metal ion: The energy difference Δ increases with increasing oxidation state for any given metal.
- The nature of the ligand: Some ligands encourage a large value for Δ, whilst the formation of complexes with other ligands results in a small splitting of t_{2g} and e_g orbital sets. The *spectrochemical series* is a list that shows ligands in the order in which they produce a splitting, ranging from small to large Δ:

$$I^- < Br^- < S^{2-} < SCN^- < Cl^- < NO_3^- < N_3^- < F^- < OH^- < C_2O_4^{2-} < NH_3 < NO_2^- < PPh_3 < CN^- < CO$$

- The geometry of the transition-metal complex: The arrangement of energy level changes and is dependent on the geometry of the complex.

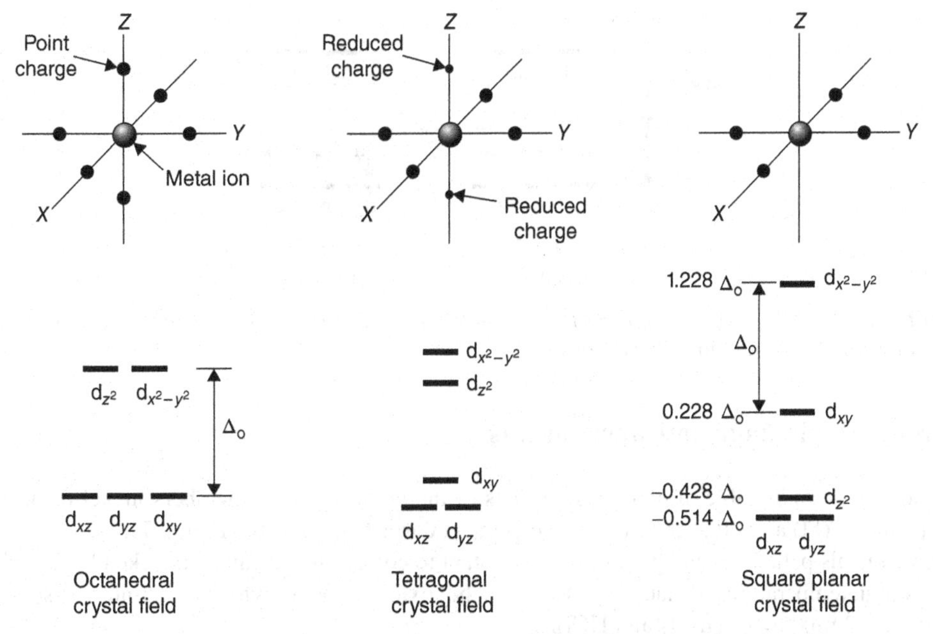

Figure 7.12 *Octahedral, tetragonal and square planar crystal fields [2] (Reproduced with permission from [2]. Copyright © 2009, John Wiley & Sons, Ltd.)*

Figure 7.11 shows the energy diagram for octahedral complexes. In Figure 7.12, the CFT splitting diagrams for other geometries, including tetrahedral, square planar and trigonal bipyramidal complexes are shown.

It is interesting to observe what happens when these orbitals are filled with electrons. The lowest energy levels will be filled first, and Hund's rule states that orbitals are filled with one electron first, before the electrons are paired. The *Pauli Exclusion Principle* states that two electrons in the same orbital are not allowed to have the same spin; nevertheless, it consumes energy to change the spin of an electron. Looking at octahedral complexes and the energy scheme shown in Figure 7.11, filling the orbitals with one, two or three electrons is straightforward. Each of the t_{2g} orbitals will be filled with one electron. The addition of the next electron leads now to two possibilities. Electron number 4 can either be added to a vacant e_g orbital, which means the energy Δ_{Oct} is needed to promote it to this level. These types of complexes are called *high-spin complexes*. Alternatively, the fourth electron can be added to one of the already occupied orbitals of t_{2g} once the electron spin has changed. These complexes are called as *low-spin complexes*. Depending on the size of Δ_{Oct}, different types of complex are formed. If Δ_{Oct} is very large, it consumes less energy to fill a t_{2g} orbital with a second electron; vice versa, if Δ_{Oct} is relatively small, it is energetically favourable to promote the fourth electron to the e_g level.

Examples of a low- and high-spin complex with four electrons in the d orbitals are shown in Figure 7.13. Both complexes have the same metal ion in the same oxidation state as the centre, and Δ_{Oct} depends only on the ligands. The same discussion follows for the addition of a fifth electron.

As mentioned at the beginning of this chapter, CFT can be used to explain the colour of these often brightly coloured transition metals. When a molecule absorbs a photon, one or more electrons are temporarily promoted within the set of split d orbitals from the lower to the higher energy level. This leads to a complex in an excited state, and the energy difference to the ground state equals the energy of the absorbed photon. The latter energy is inversely related to the wavelength of the light absorbed. Therefore the transition-metal complex can be seen in the complementary colour.

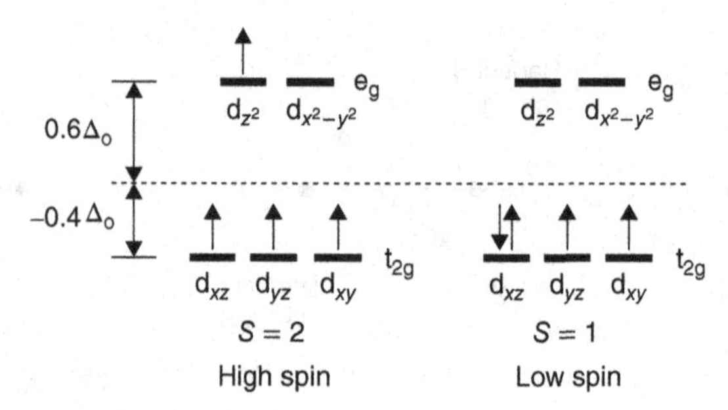

Figure 7.13 High- and low-spin possibilities for d^4 in an octahedral crystal field [2] (Reproduced with permission from [2]. Copyright © 2009, John Wiley & Sons, Ltd.)

7.2 Group 10: platinum anticancer agents

Group 10 of the periodic table of elements consists of the nonradioactive members nickel (Ni), palladium (Pd) and platinum (Pt) as well as the radioactive element darmstadtium (Ds) (Figure 7.14).

The noble metals palladium and platinum are resistant to corrosion and can be attacked by O_2, F_2 and Cl_2 only at very high temperatures. Palladium dissolves in hot oxidising acids, whereas platinum dissolves in only 'aqua regia' (1 : 3 mixture of HNO_3 and HCl).

Palladium is used as a hydrogenation catalyst, for $H_2/D_2/T_2$ separation and purification as well as a catalyst in the Wacker process. The Wacker process, which facilitates the oxidation of ethylene to acetaldehyde, was the first organopalladium reaction that was applied on an industrial scale. Platinum is also intensively used as a catalyst, for example, in HNO_3 production, as oxidation catalysts, in petroleum reforming, in hydrogenations and in many more chemical processes, as well as for jewellery.

All three nonradioactive elements of group 10 show a high diversity in their electronic configuration (Figure 7.15).

Most stable oxidation states are +II for all three nonradioactive elements, whereas Pt(II) and Pt(IV) are not only stable but also kinetically inert. The bromide and iodide salts of Pt(II) and Pd(II) are insoluble. Pt(II) has

H																		He
Li	Be												B	C	N	O	F	Ne
Na	Mg												Al	Si	P	S	Cl	Ar
K	Ca	Sc		Ti	V	Cr	Mn	Fe	Co	Ni	Cu	Zn	Ga	Ge	As	Se	Br	Kr
Rb	Sr	Y		Zr	Nb	Mo	Tc	Ru	Rh	Pd	Ag	Cd	In	Sn	Sb	Te	I	Xe
Cs	Ba	La	Ce - Lu	Hf	Ta	W	Re	Os	Ir	Pt	Au	Hg	Tl	Pb	Bi	Po	At	Rn
Fr	Ra	Ac	Th -Lr	Rf	Db	Sg	Bh	Hs	Mt	Ds	Rg	Uub						

Figure 7.14 Periodic table of elements; group 10 elements are highlighted

Ni : $[Ar]4s^2 3d^8$
Pd : $[Kr]4d^{10}$
Pt : $[Xe]4f^{14}5d^9 6s^1$

Figure 7.15 *Electronic configuration of group 10 metals*

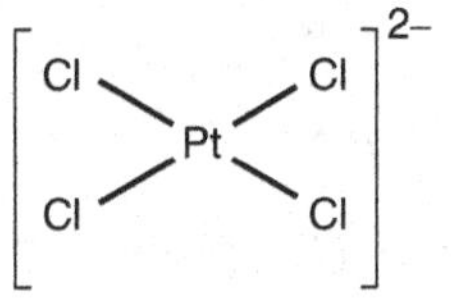

Figure 7.16 *Structure of $[PtCl_4]^{2-}$*

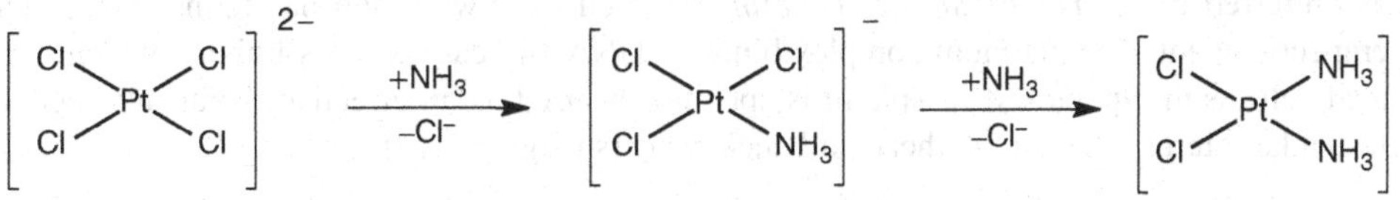

Figure 7.17 *Synthesis of cisplatin explaining the translabellising effect*

an electron configuration of d^8, and the square planar geometry is the dominant structure. The $[PtCl_4]^{2-}$ anion is an example where this square planar geometry is adapted. It is a stable anion, and indeed most platinum(II) chemistry starts with $K_2[PtCl_4]$.

$[PtCl_4]^{2-}$ is also the starting material for the synthesis of *cis*-diamminedichloroplatinum(II) (cisplatin, CDDP), a widely used chemotherapeutic drug (see Figure 7.16). The first NH_3 ligand is added to any of the four positions around the central Pt atom, as all four positions are equivalent. The second NH_3 will be directed cis to the first NH_3 group and cisplatin is obtained. The reason is that the Cl^- ligands have a larger so-called trans effect than NH_3 (Figure 7.17).

> *The **trans effect** or **trans labellising effect** is mainly seen in square planar complexes and describes the ability of some ligands to direct newly added ligands into the trans position. The intensity of the trans effect increases in the following order: F^-, H_2O, $OH^- < NH_3 < py < Cl^- < Br^- < I^-$, SCN^-, $NO_2^- < SO_3^{2-} < CH_3^- < H^-$, NO, CO, CN^-.*

In comparison, if the synthesis is started from $Pt(NH_3)_4^{2+}$, transplatin is obtained. Again, the addition of the first ligand, in this case Cl^-, can occur at any of the four positions. The addition of the second Cl^- ligand will be directed into the trans position by the initial Cl^- ligand as it has a higher trans effect than the NH_3 ligand (Figure 7.18).

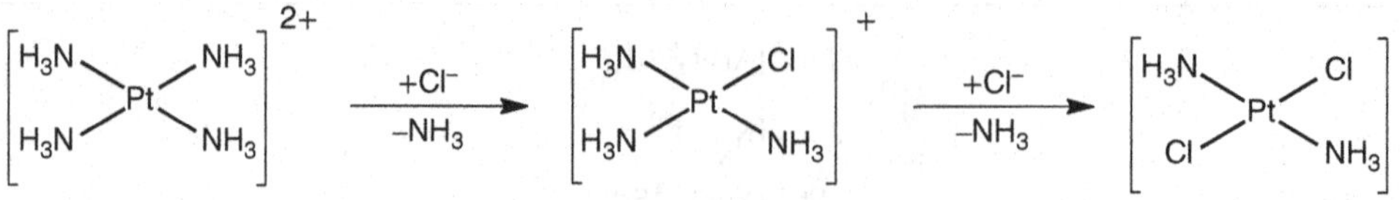

Figure 7.18 *Synthesis of transplatin explaining the translabellising effect [2] (Reproduced with permission from [2]. Copyright © 2009, John Wiley & Sons, Ltd.)*

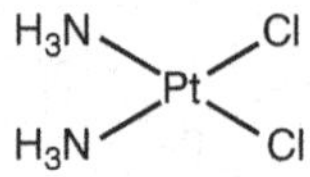

Figure 7.19 *Chemical structure of cisplatin*

7.2.1 *Cisplatin*

CDDP, also referred to as *cisplatinum* or *cisplatin*, is a yellow powder and has found widespread use a chemotherapeutic agent. The platinum complex binds to DNA and causes cross-linking, which triggers the programmed cell death (apoptosis). Cisplatin is specifically used as an effective therapeutic agent against ovarian, testicular, uterus, bladder and head and neck cancers (Figure 7.19).

7.2.1.1 Discovery

Rosenberg, a biophysicist working at Michigan State University, discovered the anticancer activity of cisplatin in 1965 serendipitously. Rosenberg devised an experiment to investigate the effect of electric fields on cell division, in which he passed an alternating electric current through two Pt electrodes immersed in a beaker containing *Escherichia coli* bacteria in a cell growth medium containing ammonium and chloride ions. During the experiment, Rosenberg discovered that the bacteria had grown in size, but not divided as was expected. On carrying out some control experiments, it became soon clear that it was not the electric current that caused this unusual cell growth. Rosenberg realised that a chemical reaction had taken place in the cell medium requiring oxygen, ammonium ions (NH_4^+) and chloride ions (Cl^-) in addition to a small amount of platinum, which was dissolved from the surface of the electrodes. A mixture of platinum salts was accidentally synthesised which contained cisplatin (*cis*-$[Pt(II)Cl_2(NH_3)_2]$). Rosenberg subsequently showed that only *cis*-$[Pt(II)Cl_2(NH_3)_2]$ and not *trans*-$[Pt(II)Cl_2(NH_3)_2]$ could prevent the growth of cancer cells *in vitro*. Typically, cisplatin kills cancer cells at micromolar doses.

Nevertheless, there is a lot of work necessary in between the discovery of a cytotoxic agent and the licensing of an anticancer drug. At the time Rosenberg discovered the potential of cisplatin, only organic compounds were seen to be appropriate for medicinal use in humans and certainly a heavy metal compound was seen as being too toxic for a therapeutic approach. Rosenberg convinced research institutes such as the National Cancer Institute to carry out several tests and trials. In 1979, he finally managed to file a patent on the use of cisplatin as anticancer agent. The synthesis itself had been reported 100 years ago, and cisplatin certainly was not a novel compound anymore, which could be patented. Bristol-Myers became interested in the compound, and the FDA licensed cisplatin as an anticancer drug. This discovery led to a whole new area of drug discovery, as from this point drug development was not only limited to organic compounds anymore.

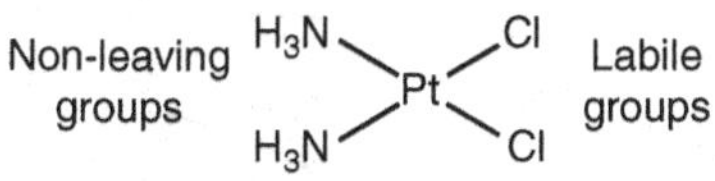

Non-leaving groups
Labile groups

Figure 7.20 *Chemical structure of cisplatin showing the labile and the nonleaving groups*

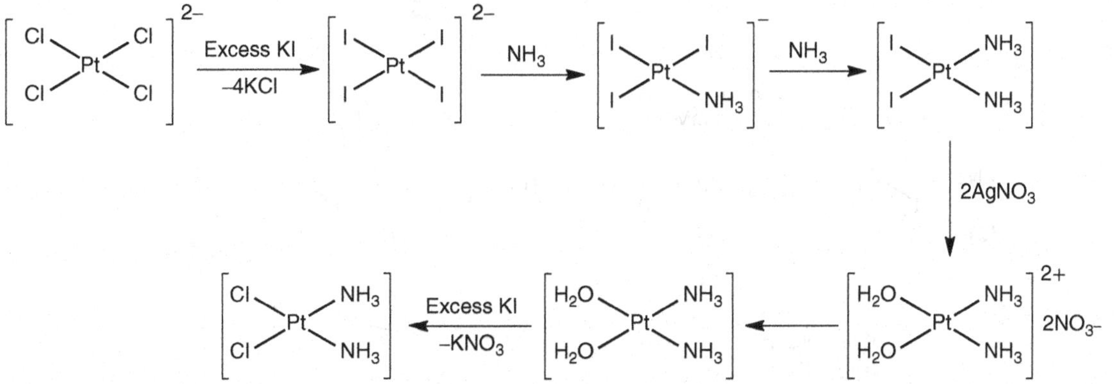

Excess KI, −4KCl; NH_3; NH_3; 2$AgNO_3$; Excess KI, −KNO_3

Figure 7.21 *Synthesis of cisplatin*

7.2.1.2 Mode of action

Cisplatin is a neutral, purely inorganic compound, first synthesised in 1844, containing a platinum(II) centre and two ammonia ligands and two chloride ligands. The ammonia ligands represent the nonleaving groups, whereas the chloride ligands are labile and can be exchanged by nucleophiles (Figure 7.20).

The synthesis of cisplatin starts with $K_2[PtCl_4]$, but it has undergone several improvements since it was published more than 100 years ago. The main problem is the occurrence of impurities and the synthesis of the by-product transplatin. Nowadays, the synthetic routes are mostly based on a method published in the 1970s by Dhara. In the initial step, $K_2[PtCl_4]$ is reacted with KI, and the platinum complex is converted into the tetraiodo analogue. Subsequently, NH_3 is added and *cis*-$[PtI_2(NH_3)_2]$ is obtained. *cis*-$[PtI_2(NH_3)_2]$ precipitates from the solution once $AgNO_3$ is added, and the insoluble AgI can be filtered off. KCl is added to the solution and cisplatin is formed as a yellow solid. The success of the synthesis relies on the strong translabellising effect of the iodo ligands as discussed earlier [3] (Figure 7.21).

Since the success of using cisplatin as a chemotherapeutic agent, significant research has been undertaken to establish the exact mode of action. Some areas still remain unclear, but it is clear that its anticancer ability mainly stems from its ability to form adducts with DNA. It has been generally accepted that the cytotoxic activity of cisplatin is due to the interaction between the metal complex and the genetic DNA, which is located in the cell nucleus [4].

Cisplatin is administered intravenously as the neutral complex, and transported via the blood stream to the cancer cell. The blood stream and the extracellular fluids have a high chloride concentration (>100 mM) and therefore, the platinum complex will not be hydrolysed. There is still much debate about the cellular uptake. It is believed that the neutral complex enters the cancer cell by passive and/or active transport. Apart from the passive diffusion, carrier-mediated proteins have been identified, such as the plasma membrane copper transporter, organic cation transporters and others [5] (Figure 7.22).

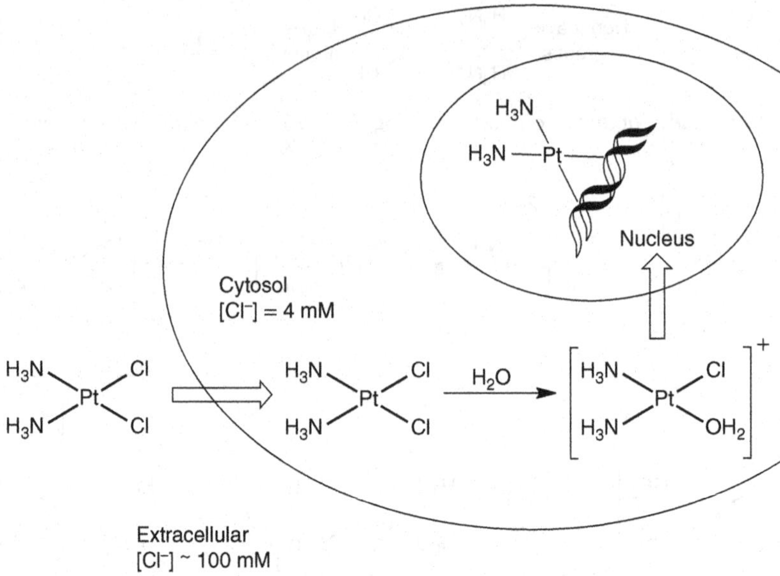

Figure 7.22 Schematic showing the cytotoxic pathway for cisplatin (Reproduced with permission from [4]. Copyright © 2012, Royal Society of Chemistry.)

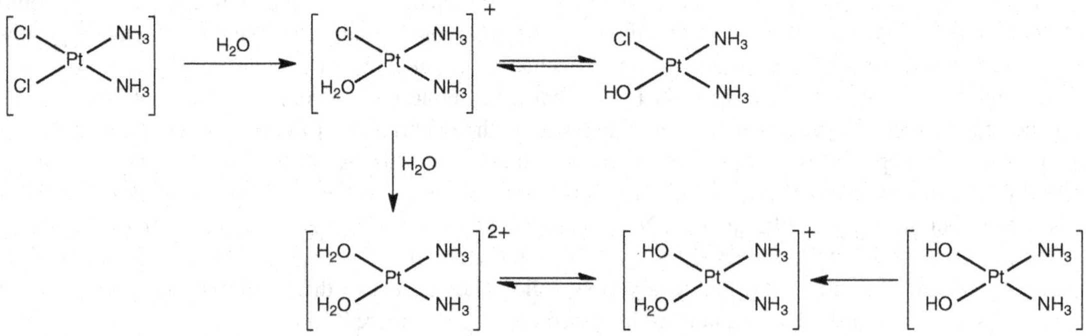

Figure 7.23 Hydrolysis of cisplatin

The mode of action inside the cell begins with the hydrolysis of the platinum–chloride bonds. This hydrolysis is facilitated by the significantly lower chloride concentration inside the cell (4 mM) compared to the high chloride concentration in the blood plasma, which prevents the hydrolysis of cisplatin during the transport in the blood stream. Upon entering the cell, it is proposed that cisplatin loses its chloride ligands and forms the mono and diaqua species. The hydrolysed species are good electrophiles and can bind to a variety of nucleophiles in the cell, such as nucleic acid and thiol-containing proteins (Figure 7.23).

The anticancer activity of cisplatin is based on the interaction of the platinum complex with DNA located in the nucleus. Interaction with the mitochondrial DNA is believed to be less important for the antitumour activity of cisplatin. Cisplatin binds to DNA primarily by coordination to the nitrogen (N7) atom of guanine, whereas it also can bind (to a lesser degree) to N7 and N1 of adenine and N3 of cytosine [2, 6] (Figure 7.24).

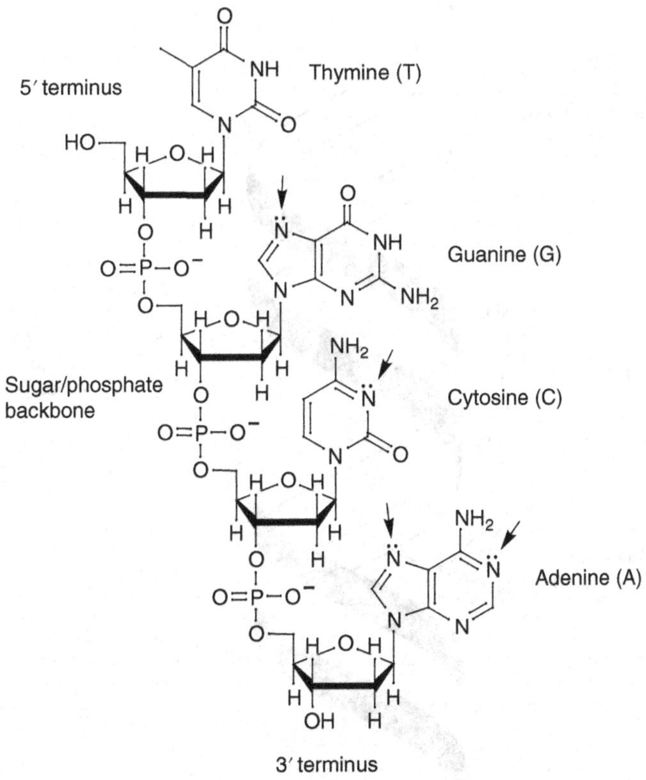

Figure 7.24 *A schematic representation of a DNA segment showing sites available for the platinum binding. Guanine N7 is the preferred binding site for cisplatin [2] (Reproduced with permission from [2]. Copyright © 2009, John Wiley & Sons, Ltd.)*

It was found that the majority of DNA–cisplatin complexes are formed by intrastrand 1,2-(GG) cross-links, which means cisplatin coordinates to two guanine bases within the same DNA strand. This form of DNA-adduct formation makes up around 65% of all DNA–cisplatin complexes. Note that the Arabic numbers refer to two adjacent nucleotides within the DNA sequence and 'intra' means that the cross-link occurs within the same DNA strand. Other binding modes include intrastrand 1,2-(AG) cross-links (25%) as well as inter-strand cross-links (between two DNA strands) and mono-functional binding to DNA [4] (Figure 7.25).

As a result of the formation of DNA–cisplatin adducts, the secondary structure of DNA is affected. In particular, the major intrastrand cross-linking of cisplatin leads to conformational alteration in the DNA. The platinum core binds to N7 of guanine bases located in the major groove and, as a result of its square planar geometry, forces the two bases to tilt towards each other and away from the parallel stacked form of DNA. This leads to a distortion of the helix axis, bending it towards the major groove. In turn, this exposes the minor groove on the opposite site and makes it accessible for other compounds. In essence, the formation of these adducts results in the DNA helix to become kinked and the DNA translation to be interrupted [2, 6, 7] (Figure 7.26).

7.2.1.3 *Resistance and cytotoxicity*

One of the major problems of chemotherapy with cisplatin is that after repeated use the cancer cells often become resistant to the treatment. Unfortunately, cisplatin has a narrow therapeutic window, which means

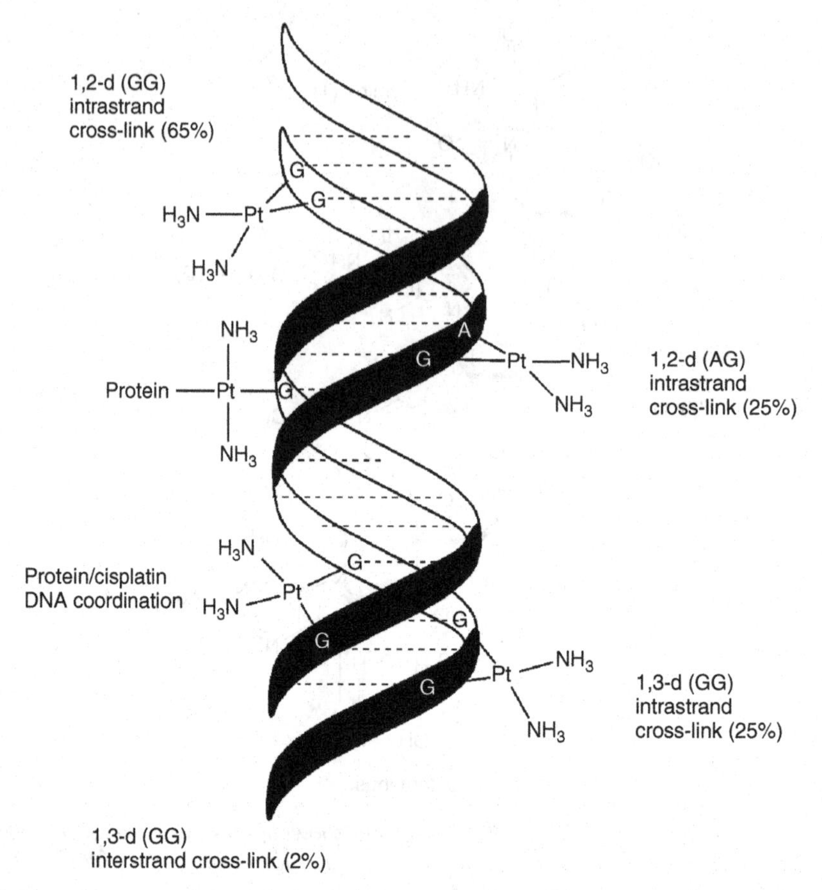

Figure 7.25 Schematic representation of a DNA segment coordinated by cisplatin showing the different coordination modes

that the difference between the therapeutically active and the toxic dose is fairly small. Therefore, it is not simply possibly to increase the dose once resistance is observed.

It has been observed that in platinum-resistant cells, less of the metal is bound. This can be due to the cells being more effective blocking entry to the cell, being able to move cisplatin actively out of the cell or, in the case where the agent made its way to its target, that DNA can be repaired and the metal complex is removed.

It has been shown that cancer cells that develop resistance to cisplatin have a higher concentration of the sulfur-containing proteins such as glutathione (GSH) and metallothionein (MT). Both GSH and MT can form very stable complexes with Pt^{2+} via a S—Pt bond, which will reduce the ability of the platinum complex to interact with other targets in the cell. It is actually assumed that most of the platinum drugs bind to sulfur before it reaches the DNA, as this interaction is the kinetically favoured process. There is evidence that some of the Pt–thioester complexes can be broken in the presence of DNA. Nevertheless, the Pt–thiol complexes are very stable and will therefore inactivate the platinum drug [4]. There is also evidence that the Pt(II)–S adducts could be exported out of the cell via a glutathione S-conjugate [2]. It is interesting to note that only about 1% of the administered cisplatin will complex with DNA [5].

DNA repair mechanisms are important processes within the cells and there are several different ways in which DNA can be repaired. If the DNA has been chemically modified by drugs, UV light or radicals, Nature

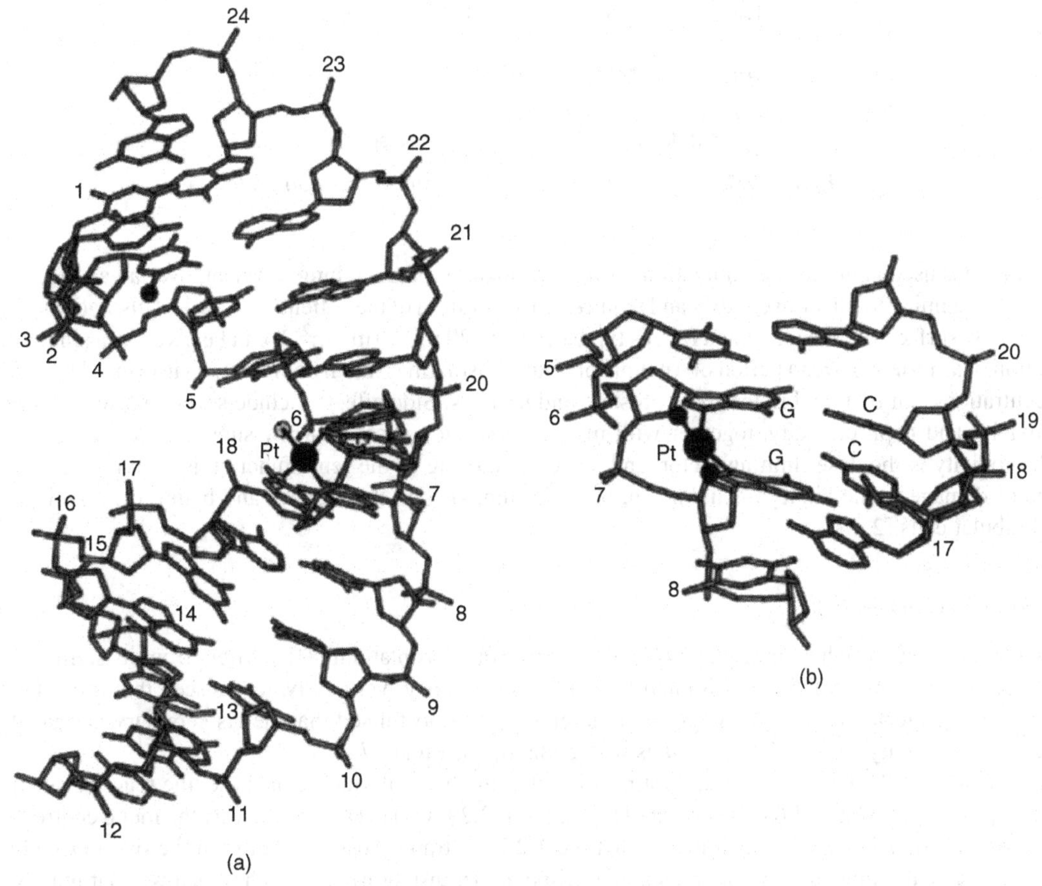

Figure 7.26 *Crystal structure of cisplatin bound to DNA (a) and to N7 of the guanine base (b) (Reprinted by permission from Macmillan Publishers Ltd: Nature [9], Copyright (1995).)*

has established a sophisticated system to check and repair the DNA and ultimately ensure the survival of the organism. One system is called *nucleotide excision repair* (*NER*) and it uses enzymes to remove the single-stranded DNA segment that has been modified. It also replaces this segment by reading the opposite DNA strand. As previously discussed, the 1,2-intrastrand cross-link represents the major Pt–DNA adduct, where cisplatin coordinates to two neighbouring guanine bases within the same DNA strand. In recent animal experiments, it has been shown that the NER system is indeed important to the removal of platinum from DNA. Mice with a compromised NER system were not able to remove the platinum-coordinated segments from their DNA in the kidney cells researched [2]. There are also other mechanisms of resistance under discussion, which are not further explained in this book.

7.2.1.4 *Formulation and administration*

Cisplatin is commercially available either as a powder for reconstitution or as a ready-to-use solution containing the drug in a saline concentration (NaCl solution) at a pH around 3.5–4.5. Under this condition, the majority of platinum is present as $[PtCl_2(NH_3)_2]$. Additionally, the low pH avoids the formation of any multinuclear platinum compounds.

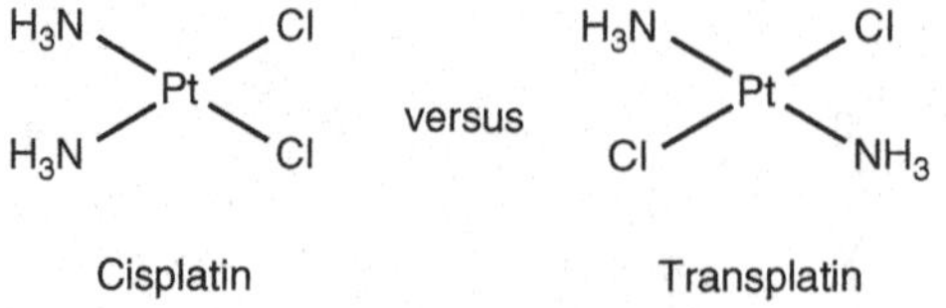

Figure 7.27 Chemical structure of cisplatin versus transplatin

Cisplatin is used alone or in combination for the treatment of bladder, lung, cervical, ovarian and testicular cancer. It is administered intravenously and requires pre-hydration of the patient. The dose of cisplatin depends on the body surface of the patient and ranges typically from 20 to 140 mg/m^2. It can be given as a single-dose injection or as infusion over a period of a couple of hours. Cisplatin is mainly eliminated via urine. The highest concentrations can be found in the liver, prostate and kidneys. Side effects include severe nausea, vomiting, ototoxicity and nephrotoxicity, together with myelosuppression (bone marrow suppression). In particular, nephrotoxicity is the dose-limiting factor, and close monitoring of the renal function is required. The exact mode of damage to the kidneys is unknown, but it accumulates in the kidneys and believed to damage the renal tubular cells [2, 8].

7.2.1.5 *Transplatin*

Soon the question arose how transplatin (*trans*-diamminedichloroplatinum (II)) differs from its stereoisomer cisplatin and how this explains the dramatic difference in efficacy. Very early, it was seen that transplatin is not active when tested in animal models, and therefore it was postulated that the cis geometry is crucial for the cytotoxic activity for which two reasons where identified (Figure 7.27).

A close look at the structure of transplatin shows that the two chloride ligands (i.e. the reactive sites) are further apart in transplatin (4.64 Å) compared to cisplatin (3.29 Å). This affects the way the metal centre Pt(II) can cross-link sites on DNA, which means that the 1,2 intrastrand cross-link between the two purine bases is stereochemically hindered for the transplatinum species (transplatin). Studies have shown that mainly the mono-functional 1,3 adducts and inter-strand adducts are formed, which are easily recognised by the NER repair system [7].

Furthermore, it has been shown that transplatin is kinetically more reactive than cisplatin and therefore more of the compound is inactivated before reaching its target, which contributes to the weaker activity of transplatin. Therefore, current research focusses on the synthesis of novel transplatin compounds in which the hydrolysis rate of the compound can be altered [2].

Recently, a range of transplatinum compounds were proved to have anticancer activity. The most notable compounds are the *trans*-Pt(IV) complexes with the general formula *trans*-$Pt(IV)Cl_2X_2LL'$ or the *trans*-Pt(II) complexes with the general formula *trans*-$Pt(II)Cl_2LL'$. The ligands (L and L′) are mainly imino, aromatic amine and aliphatic amine ligands, whereas the Pt(IV) complexes contain also hydroxyl groups. Interestingly, some of those compounds are also active on cisplatin-resistant cell lines, thereby opening a new area of research for the treatment of these cancer types [9, 10] (Figure 7.28).

7.2.2 Platinum anticancer agents

Despite the success of cisplatin, there is a need to develop new platinum anticancer drugs. Cisplatin is a very toxic compound and can have severe side effects such as nephrotoxicity (kidney poisoning), ototoxicity (loss of high-frequency hearing) and peripheral neuropathy (damage to nerves of the peripheral nervous system), although it is possible to control some of these effects. Very typically, cancer cells can become resistant to

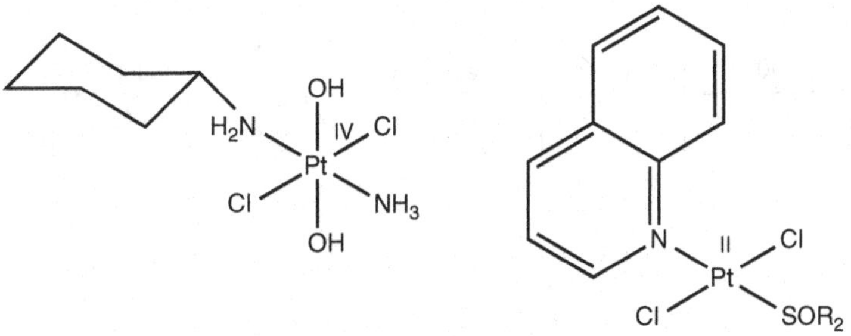

Figure 7.28 Example of a trans-Pt(IV) complex (a) and a trans-Pt(II) complex (b)

cisplatin after repeated administration. This is a fairly common problem experienced at the repeat treatment with cisplatin. Furthermore, compounds active against a variety of cancer types are required to combat cancer.

Two second-generation platinum drugs are so far successfully registered worldwide – carboplatin and oxaliplatin. There are others such as nedaplatin, which is registered in Japan for the treatment of head and neck, testicular, lung, cervical, ovarian and nonsmall-cell lung cancer. In South Korea, heptaplatin is used against gastric cancer, whereas lobaplatin is licensed in China for the treatment of cancers including metastatic breast cancer, small-cell lung cancer and myelogenous leukaemia [10].

Nevertheless, the development of new platinum-based drugs has been less successful than expected. The majority of compounds are not used in a clinical setting because their efficacy is too low, toxicity is too high or the compounds showed a poor aqueous solubility, a fairly common problem for transition-metal-based compounds.

7.2.2.1 *Carboplatin*

Carboplatin, *cis*-diammine(1,1-cyclobutanedicarboxylato)platinum(II), is a second-generation platinum drug. Its structure is based on cisplatin with the difference that the chloride ligands are exchanged for a bidentate chelating ligand. A consequence is that carboplatin is less reactive than cisplatin and therefore is less nephrotoxic and orthotoxic than the parent compound. Unfortunately, it is more myelosuppressive than cisplatin, which reduces the patients' white blood cell count and makes them susceptible to infections [8]. Carboplatin was licensed by the FDA in 1989 under the brand name Paraplatin and has since then gained worldwide recognition. Carboplatin on its own or in combination with other anticancer agents is used in the treatment of a variety of cancer types including head and neck, ovarian, small-cell lung, testicular cancer and others [2] (Figure 7.29).

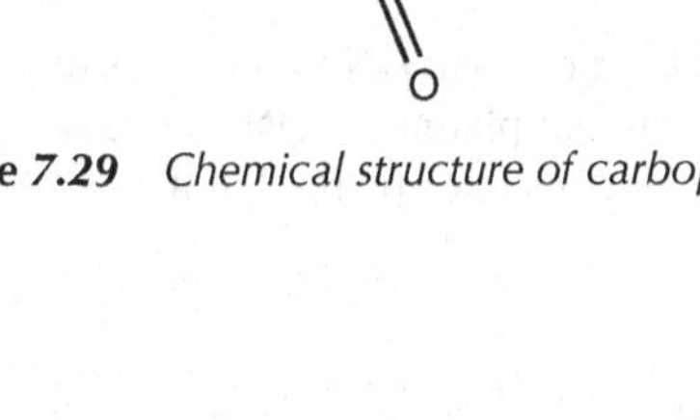

Figure 7.29 Chemical structure of carboplatin

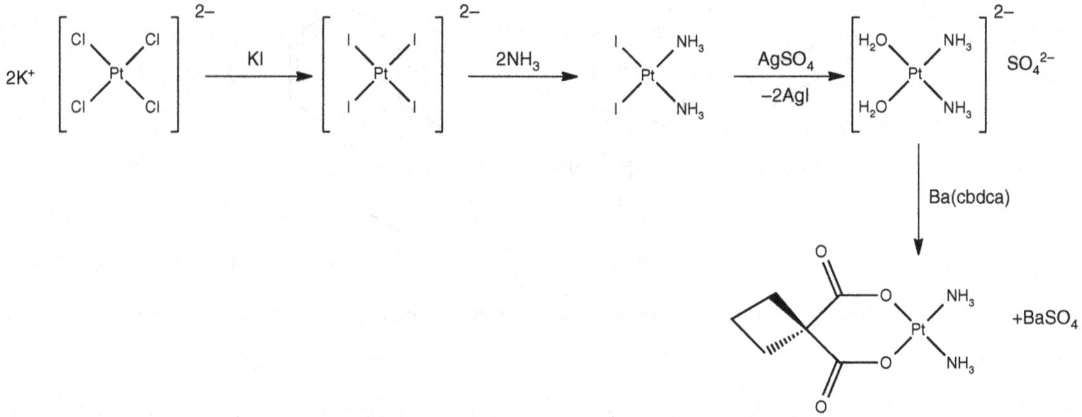

Figure 7.30 Synthesis of carboplatin

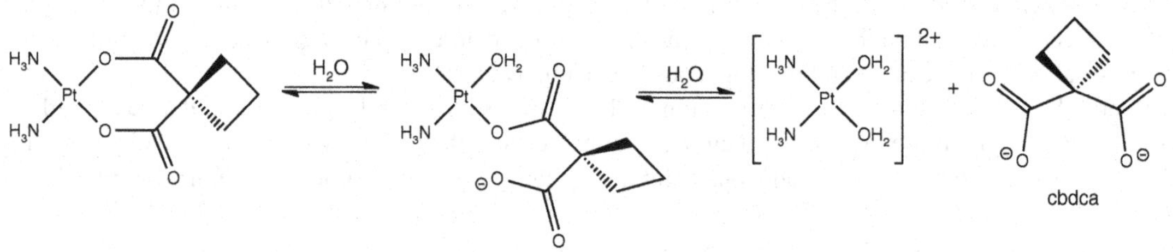

Figure 7.31 Hydrolysis of carboplatin [2]

Carboplatin is a pale-white solid showing good aqueous solubility. The synthesis starts with potassium tetrachloroplatinate, which is reacted to the orange $[PtI_4]^{2-}$ anion. Analogous to the synthesis of cisplatin, in the following steps the anion is reacted with ammonia (due to the translabellizing effect of iodide, the ammonia ligands are directed into the cis position) and converted to *cis*-$[Pt(NH_3)(H_2O)_2]SO_4$. In the final step, the complex is reacted with the chelating agent Ba(cbda) (cbdca, cyclobutane-1,1-dicarboxylate) and carboplatin is formed [2] (Figure 7.30).

Carboplatin is administered by intravenous (IV) injection. A typical solution contains the drug in a high concentration in water or in a mannitol or dextrose solution. The dose is determined either on the basis of the body surface area of the patient or according to their renal function. The doses are typically three to six times higher than cisplatin, which reflects the lower chemical reactivity and toxicity of this drug. Carboplatin can be given on an outpatient basis, as it is better tolerated than cisplatin.

The mode of action relies on the ring-opening of the cbdca chelate ring and interaction of the platinum centre with DNA. Carboplatin is seen as a prodrug, which itself is not very reactive within the human body but once activated shows its full potential. Hydrolysis of carboplatin and removal of the chelate ligand or at least the opening of the ring makes this compound much more cytotoxic than the parent compound itself (Figure 7.31).

In vitro studies have shown that the drug binds to DNA and forms initially a mono-functional adduct, which over time is converted into the di-functional platinum–DNA adduct. There are indications that carboplatin forms DNA intrastrand cross-links analogous to cisplatin, but its reactivity towards DNA is reduced [2].

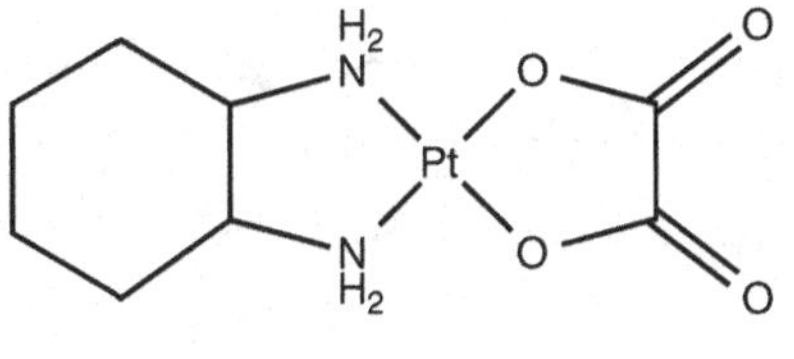

Figure 7.32 Chemical structure of oxaliplatin

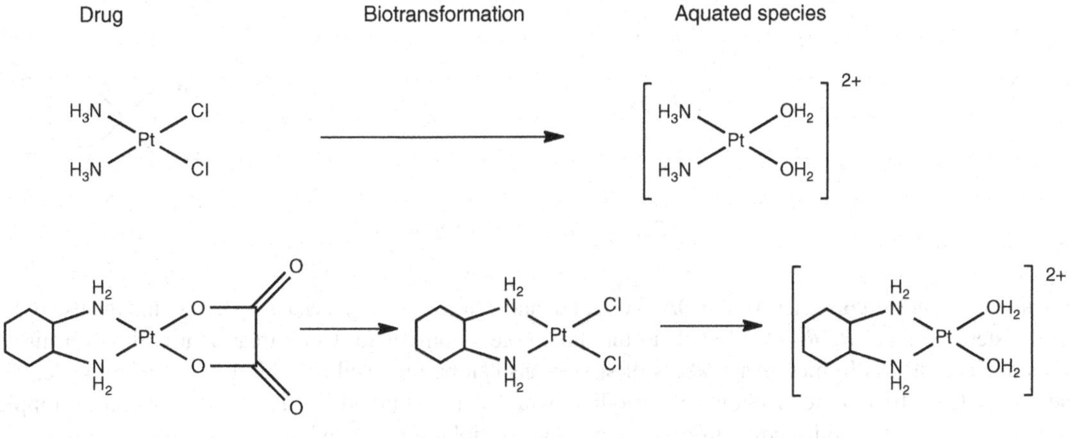

Figure 7.33 Hydrolysis products of cisplatin and oxaliplatin

7.2.2.2 *Oxaliplatin*

Oxaliplatin (*cis*-[oxalato] *trans*-1,2-diaminocyclohexane platinum(II)), for example, marketed under the trade name Eloxatin, is considered as a third-generation platinum-based anticancer drug. Its structure differs from previously synthesised platinum compounds by the configuration of its amino substituents. Its platinum centre is coordinated by two chelating ligands, namely an oxalate ligand and a so-called DACH (1,2-diaminocyclohexane) ligand. In comparison to cisplatin, the two chlorine leaving groups are replaced by an oxalato leaving group. The simple amino groups are replaced by the DACH ligand, which is the nonleaving group (Figure 7.32).

Cisplatin and carboplatin are hydrolysed to a common diamino-platinum species, whereas the hydrolysis product of oxaliplatin contains the bulky DACH group, which sterically hinders the DNA repair mechanism. These mismatch repair enzymes are particularly active in colon cancer and, not surprisingly, oxaliplatin shows excellent activity in the treatment of colon and rectal cancers [11] (Figure 7.33).

The clinical use of oxaliplatin was approved by the European Union in 1999 and by the FDA in 2002. It is most effective in combination with 5-fluorouracil and leucovorin (5-FU/LV) in the treatment of metastatic carcinomas of the colon or rectum [11]. Oxaliplatin induces less side effects than cisplatin; for example, it is less nephrotoxic and ototoxic and leads to less myelosuppression. Unfortunately, treatment with oxaliplatin can lead to nerve damage, which may not be reversible in the case of chronic exposure of the patient to the drug. Oxaliplatin is usually administered intravenously as infusion over a period of 2–6 h in doses similar to cisplatin. The neurotoxic side effects are dose-limiting [8].

The synthesis of oxaliplatin starts with $K_2[PtCl_4]$, the same starting material used for the synthesis of carboplatin. This is reacted with water and 1 equiv of the nonleaving ligand, 1*R*,2*R*-DACH ligand. Note that

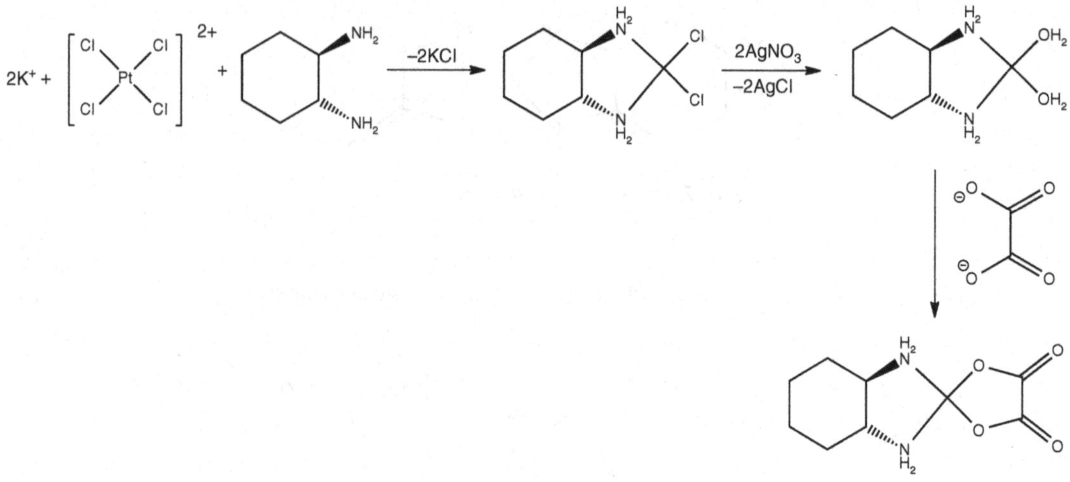

Figure 7.34 Synthesis of oxaliplatin [2]

there are different stereoisomers of the DACH ligand, and cytotoxicity studies have shown that the use of this specific stereoisomer 1*R*,2*R*-DACH leads to the most potent compound. Upon treatment with silver nitrate, the diaqua complex is formed. Any excess of silver ions can be removed by adding potassium iodide. This leads to the formation of the insoluble silver iodide, which can be filtered off. The diaquo platinum complex is subsequently treated with 1 equiv of oxalic acid, and oxaliplatin is formed as a solid (Figure 7.34).

It is believed that DNA is the major cellular target of oxaliplatin, as researchers have shown that it forms intrastrand cross-links similar to cisplatin. The oxaliplatin–DNA adduct also leads to a bending of the DNA similar to the cisplatin–DNA adduct. Nevertheless, there are significant differences to the cisplatin–DNA adduct. The oxaliplatin–DNA adduct forces a narrow minor groove bend (helix bend of 31°), whereas the equivalent cisplatin–DNA adduct leads to a wide minor groove (60–80°). Also, it has been observed in the solid state structure of the oxaliplatin–DNA adduct that there is a hydrogen bond formed between the NH of the DACH group and the oxygen atom of guanine base, which interacts with the platinum centre [2].

7.2.2.3 *Other platinum drug candidates*

There are numerous platinum compounds under research for their potential use as anticancer agents. Only a few of them have found their way into the clinic so far, with cisplatin, carboplatin and oxaliplatin being the most successful ones.

One example is nedaplatin, *cis*-diammineglycolatoplatinum(II), which is structurally similar to carboplatin. The chemical structure consists of a central platinum(II) atom with two *cis*-ammonia groups as nonleaving groups and – in contrast to carboplatin – the dianionic form of glycolic acid as the leaving group. Nedaplatin has been approved for the clinical use in the Japanese market for the treatment of head and neck, testicular, ovarian, lung and cervical cancer. It is typically administered by IV injection and its dose-limiting side effect is myelosuppression (Figure 7.35).

Lobaplatin and heptaplatin are further examples of platinum-based agents being used in China and South Korea, respectively. Lobaplatin is used in the treatment of nonsmall-cell lung cancer and breast cancer. Heptaplatin is used in South Korea to treat gastric cancer. Both drugs show the typical side effects such as myelosuppression and mild hepatotoxicity. Their success is limited and has not led to approval in the EU or by the FDA (Figure 7.36).

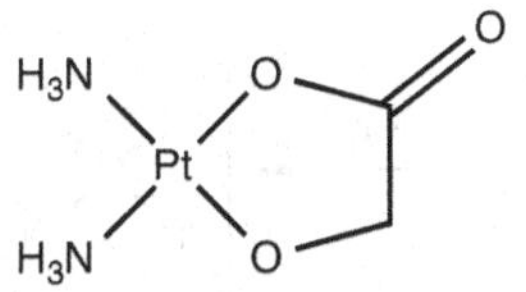

Figure 7.35 *Chemical structure of nedaplatin*

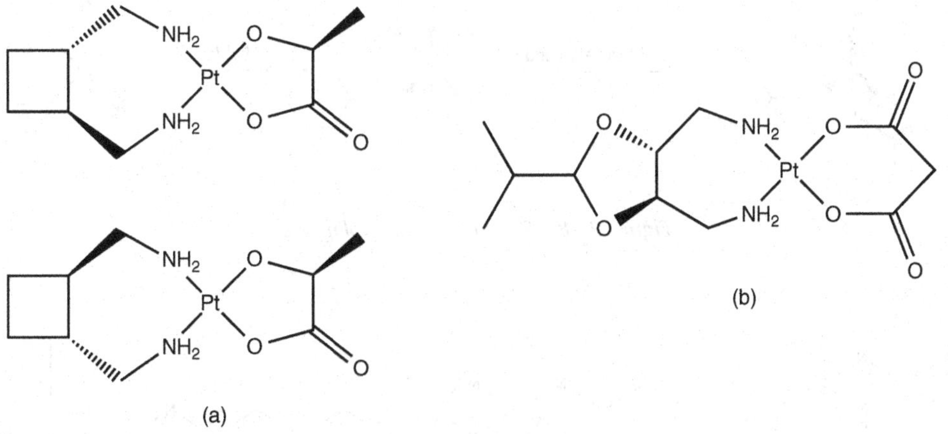

Figure 7.36 *Chemical structures of (a) lobaplatin and (b) heptaplatin*

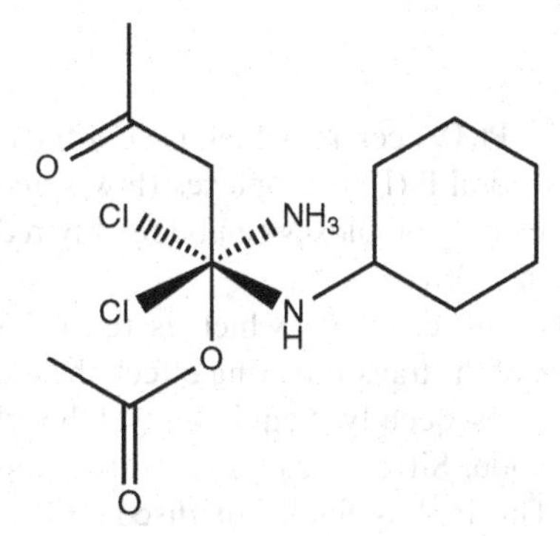

Figure 7.37 *Chemical structure of satraplatin*

Satraplatin (JM216, *cis,trans,cis*-[$PtCl_2(OAc)_2(NH_3)(C_6H_5NH_2)$]) is a Pt(IV) or Pt^{4+} complex, which is active by oral administration, as it is more hydrophobic than cisplatin. This form of administration is very attractive because of the convenience and freedom it provides to the patient. Satraplatin also has a milder toxicity profile and is shows no cross-resistance with cisplatin. Satraplatin in combination with prednisone has completed phase III clinical trials against hormone-refractory prostate cancer. The results were very encouraging, but the overall survival rate did not improve significantly enough. As a result, the fast-track approval of the FDA was not granted [12] (Figure 7.37).

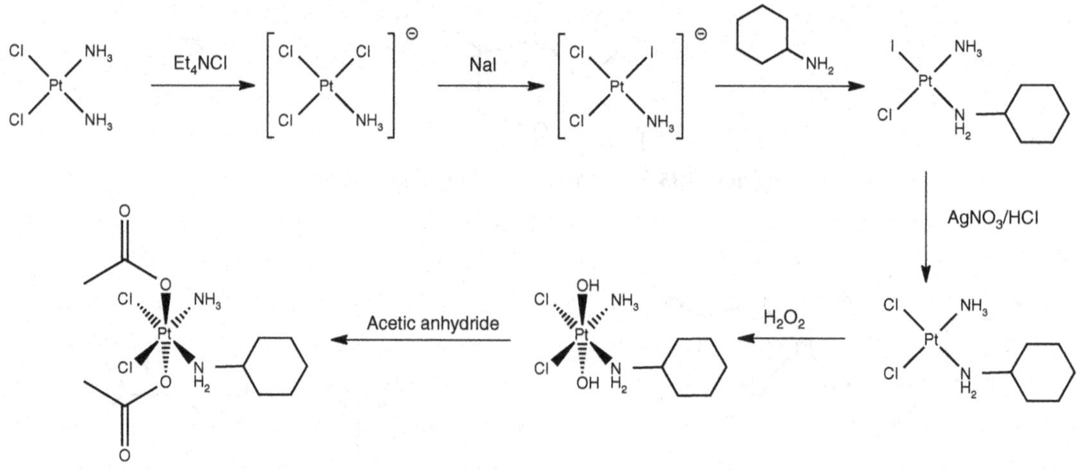

Figure 7.38 Synthesis of satraplatin [2]

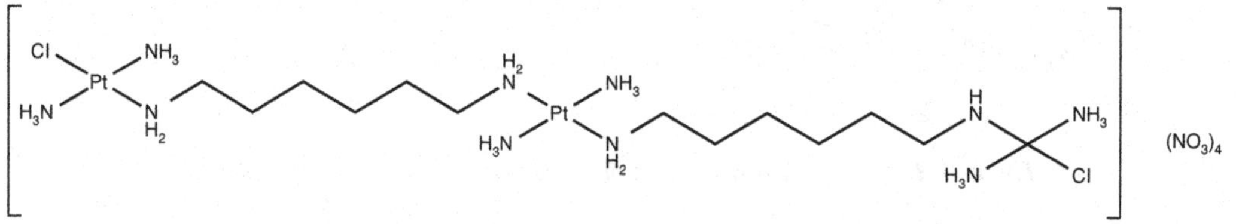

Figure 7.39 Chemical structure of BBR3464

Structurally, satraplatin consists of a Pt(IV) centre, which is coordinated by six ligands forming a close to octahedral geometry. In general, octahedral Pt(IV) complexes (low-spin d^6) are much more kinetically inert than square planar Pt(II) complexes. Pt(IV) complexes can be readily reduced *in vivo* to Pt(II) by reductants such as ascorbate or thiols (e.g. cysteine, GSH).

The synthesis of satraplatin starts with cisplatin, which is reacted with tetraethylammonium chloride (Et_4NCl) – a source of Cl^-. As a result of the trans-directing effect, the iodide ligand is introduced in a second step adjacent to the ammonia group. Subsequently, 1 equiv of cyclohexylamine is added, which coordinates to the platinum centre trans to the iodide. Silver nitrate is used to remove the iodide ligand, as no further 'trans-directing' action is required. The Pt^{2+} is finally oxidised to Pt^{4+}, which expands the coordination sphere from 4 to 6 – octahedral geometry. In the last step, the acetate ligands are introduced (Figure 7.38).

Satraplatin is the only orally administered platinum-based drug that has entered clinical trials so far. The difficulty for this administration route lies in the aggressive conditions that are present in the stomach. In general, metal complexes do not survive the acidic conditions in the stomach and therefore will not reach the gastrointestinal (GI) tract unchanged. The advantage of satraplatin is that the complex is relatively inert to any exchange reactions and therefore has an increased chance of reaching the pH-neutral GI tract unchanged. From here, the drug enters the blood stream. *In vitro* studies with fresh human blood have shown that within minutes the reduction of the platinum centre to Pt^{2+} takes place in the red blood cells. This may be facilitated by haemoglobin, cytochrome c and NADH, and leads to a square planar Pt^{2+} complex containing the chloride and ammonia ligands [2].

Further research has led to the development of multi-platinum complexes. This is against the 'rules' set out for platinum-based anticancer drugs, which state that a successful drug candidate should consist of only one platinum centre with amine-based nonleaving groups in cis position and two leaving groups, also in cis position. Clearly, polynuclear platinum compounds fall outside these rules, but researchers have synthesised the unusual trinuclear complex BBR3464, which was very successful in *in vitro* studies and even reached clinical trials against melanoma and metastatic lung and pancreatic cancer [2] (Figure 7.39).

7.3 Iron and ruthenium

Group 8 of the periodic table of elements consists of the nonradioactive members iron (Fe), ruthenium (Ru) and osmium (Os), as well as the radioactive element Hassium (Hs) (Figure 7.40).

All three nonradioactive elements are silvery white, hard transition metals with a high melting point. Iron has been classified as the most common element within the entire earth, as most of the earth's core is iron. In contrast, ruthenium and osmium are two of the rarest elements on earth. The radioactive element hassium has not been isolated in pure form yet and therefore its exact properties have not been established. It has only been produced in nuclear reactors and never has been isolated.

The electronic configuration of group 8 metals is shown in Figure 7.41.

H																		He
Li	Be												B	C	N	O	F	Ne
Na	Mg												Al	Si	P	S	Cl	Ar
K	Ca	Sc		Ti	V	Cr	Mn	Fe	Co	Ni	Cu	Zn	Ga	Ge	As	Se	Br	Kr
Rb	Sr	Y		Zr	Nb	Mo	Tc	Ru	Rh	Pd	Ag	Cd	In	Sn	Sb	Te	I	Xe
Cs	Ba	La	Ce - Lu	Hf	Ta	W	Re	Os	Ir	Pt	Au	Hg	Tl	Pb	Bi	Po	At	Rn
Fr	Ra	Ac	Th -Lr	Rf	Db	Sg	Bh	Hs	Mt	Ds	Rg	Uub						

Figure 7.40 Periodic table of elements; group 8 elements are highlighted

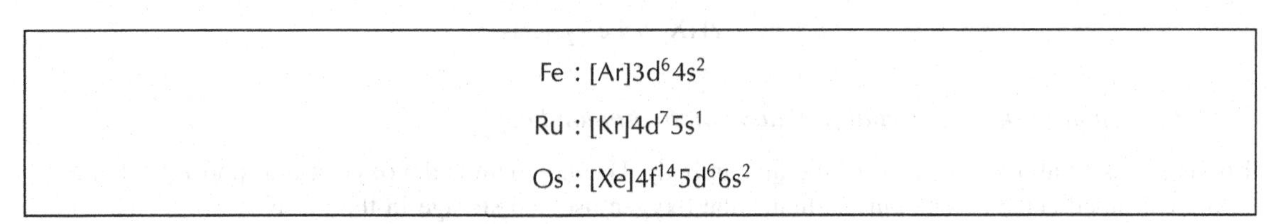

Figure 7.41 Electronic configuration of group 8 metals

7.3.1 Iron

Iron is the chemical element with the symbol Fe (Latin: ferrum) and atomic number 26. It is one of the most used metals because of the relatively low production costs and its high strength. Iron can be found in many everyday items, from food containers to screw drivers or any type of machinery. Steel is a form of iron, which is alloyed with carbon and a variety of other metals.

Iron ions are a necessary trace element used by almost all living organisms with the only exceptions being a few prokaryotic organisms that live in iron-poor conditions. As an example, the lactobacilli in iron-poor milk use manganese for their catalysis processes. Iron-containing enzymes, usually containing haeme prosthetic groups, participate in the catalysis of oxidation reactions in biology and in the transport of a number of soluble gases.

More than other metals, metallic iron has long been associated with health. Chalybeate springs, the iron-containing waters, have been well known for centuries for their healing properties. In the nineteenth century, the 'veritable pills of Blaud', which contain ferrous sulfate and K_2CO_3, were used to 'cure everything'. In the 1930s, the relationship between iron-deficiency anaemia and the lack of dietary iron was established. Nowadays, iron deficiency is the most frequent nutritional deficiency in the world.

7.3.1.1 Chemistry of iron

Iron is a metal extracted from iron ore, and is almost never found in the free elemental state. In order to obtain elemental iron, the impurities must be removed by chemical reduction. Iron is the main component of steel, and it is used in the production of alloys or solid solutions of various metals.

Finely divided iron powder is pyrophoric in air. There is a difference between when elemental iron is heated in dry air or in the presence of humidity. Fe heated in dry air will oxidise, whilst when heated in moist air it will rust characterised by the formation of $Fe_2O_3 \cdot xH_2O$. The formation of rust is an electrochemical process occurring in the presence of oxygen, water and an electrolyte such as NaCl.

$$2Fe \rightarrow 2Fe^{2+} + 4e^-$$

$$O_2 + 2H_2O + 4e^- \rightarrow 4[OH]^-$$

$$Fe(OH)_2 \text{ oxidizes to } Fe_2O_3 \cdot xH_2O$$

The highest oxidation states of iron are Fe(VI) and Fe(IV), whilst Fe(V) is very rare. Examples of compounds where Fe occupies this oxidation state include $[FeO_4]^{2-}$, $[FeO_4]^{3-}$, $[FeO_4]^{4-}$ and $[FeO_3]^{2-}$. Oxidation states +2 and +3 are the most commonly occurring oxidation states for Fe. The old name for Fe(III) is ferric and for Fe(II) is ferrous – this is still reflected in many drug names. Fe reacts with halogens under heat with the formation of FeF_3, $FeCl_3$, $FeBr_3$ and FeI_2. FeF_3 is a white solid, whilst $FeCl_3$ is a dark-green hygroscopic solid. $FeCl_3$ is an important precursor for any Fe(III) chemistry.

Fe(II) halogens are typically synthesised by reacting Fe with the relevant acid, HX, with the exception of FeI_2, which can be synthesised from the elements directly. FeF_2 is a white solid, sparingly soluble in water, whilst $FeCl_2$ forms white, water-soluble, hygroscopic crystals.

$$Fe + 2HX \rightarrow FeX_2 + H_2$$

7.3.1.2 Iron performs many vital functions in the human body

Iron is an essential trace element for the human body. Haemoglobin is the oxygen-transport metalloprotein in the red blood cells; myoglobin facilitates the oxygen use and storage in the muscles; and cytochromes transport electrons. Iron is also an integral part of enzymes in various tissues. The average 70-kg adult body

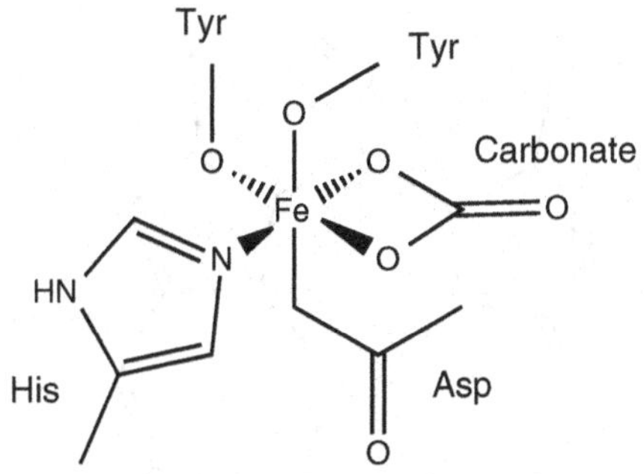

Figure 7.42 Structure of transferrin showing the coordination of Fe^{3+} [2]

contains around 4200 mg of iron ions. The majority (65%) can be found as haemoglobin or myoglobin, which is classified as the functional iron [13].

Iron will pass the stomach and is absorbed predominantly in the duodenum and upper jejunum. Beyond this point, intestinal bicarbonate elevates the pH, rendering iron insoluble. Free iron is, as most metal ions, highly toxic to the human body. Therefore, nature has created a sophisticated transport and storage system which ensures that no free iron ions are present in the blood stream. Iron ions absorbed from the GI tract are transported via transferrins, which are Fe^{3+}-containing metalloproteins, to the storage vessels or until it is incorporated into haemoglobin. In the human body, Fe^{3+} is stored mainly in the liver and spleen in form of ferritin, which is a water-soluble metalloprotein and stores Fe^{3+}.

Transferrins include the so-called serum transferrins, for example, ovotransferrin (present in egg white) and lactoferrin (present in milk), which can transport ~40 mg of iron ions per day in humans. The glycoprotein protein is folded in such a way that there are two pockets suitable for the coordination of Fe^{3+}. Figure 7.42 shows how Fe^{3+} is coordinated within these pockets. Note a CO_3^{2-} ligand is essential for the binding mechanism.

Transferrins *are a group of proteins that are abundant in blood and their primary role is to transport Fe^{3+}, as free Fe^{3+} would be toxic to most organisms. Transferrin consists of a single polypeptide chain independently binding a maximum of two Fe^{3+} ions. Fe^{3+} is bound in a hexa-coordinated high-spin complex within the protein. The metal is coordinated by the N atom from the imidazole residue, two deprotonated phenol groups, a carboxylate and a carbonato ligand.*

Haemoglobin and myoglobin are so-called haeme-iron proteins characterised by the presence of a haeme group (a protoporphyrin group). The main function of haemoglobin is the transport of oxygen to the tissue in need. Myoglobin is the oxygen storage protein present in the tissue.

Myoglobin consists of a monomeric protein chain containing one protoporphyrin group as the functional unit. Within myoglobin, the iron centre is coordinated by the four nitrogen groups of the porphyrin in addition to the coordination of a fifth nitrogen centre from a histidine (His) group. The functional unit containing the Fe(II) centre is called a *haeme group* and is a square-based pyramidal complex. During the oxygen binding mechanism, O_2 will enter trans to the His group to give an octahedrally coordinated iron species (Figure 7.43).

Haemoglobin, the oxygen-transport protein in the red blood cells, is a tetramer and each of the four chains contains a haeme group. It is interesting to note that the four haeme groups in haemoglobin do not operate independently. The release (and binding) of oxygen is a cooperative process, which means that the loss (uptake) of the first oxygen molecule triggers the release of the remaining three.

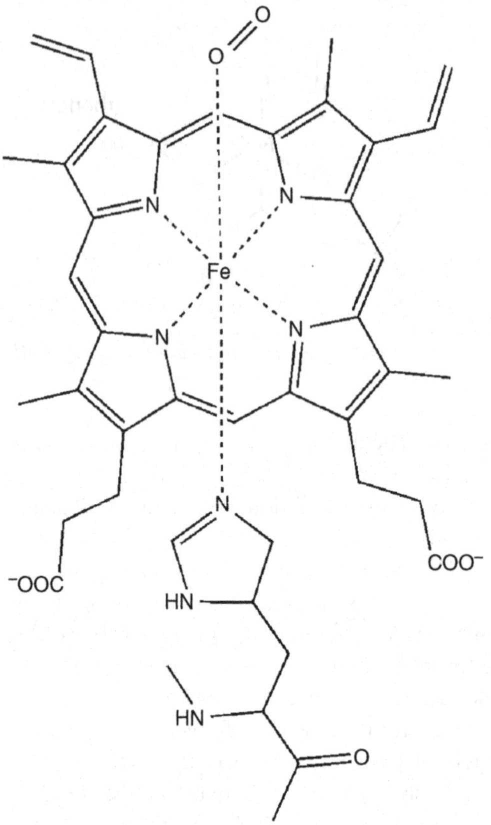

Figure 7.43 Structure of haemoglobin

The current model for oxygen binding in haemoglobin and myoglobin can be explained in the following way. The deoxy form contains a high-spin Fe(II) centre, which, because of its size, does not form a plane with its four nitrogen donor atoms. Instead, it is located slightly above the plane, drawn towards the His residue. Once oxygen enters trans to the His residue, the iron centre is oxidised to a low-spin Fe^{3+} centre and O_2 is reduced to $[O_2]^-$. Both species contain an unpaired electron. The low-spin Fe^{3+} moves into the plane and pulls the His residue down. This affects the remaining protein chain and triggers the uptake/release of oxygen in the other three haeme groups [13] (Figures 7.44 and 7.45).

Cytochromes are part of the mitochondrial electron-transfer chain but also found in the chloroplasts of plants (involved in photosynthesis). There are many different cytochromes; all are involved in reduction–oxidation processes and are grouped into families – cytochromes a, cytochromes b and cytochromes c. They contain a haeme group with an iron centre, which has the ability to change reversibly from Fe(III) to Fe(II), and vice versa. In contrast to the oxygen-coordinated iron centre in haemoglobin, the iron in cytochromes is always six-coordinated. Cytochrome c, for example, is involved in the mitochondrial electron-transfer chain and accepts an electron from cytochrome c1 and transfers it to cytochrome c oxidase. This electron is subsequently used in the reduction of oxygen, where four electrons are needed. This means that actually four cytochrome c transfer an electron to cytochrome c oxidase, where one molecule of O_2 is converted to two molecules of water. Note that Fe(III) forms the core of the oxidised cytochromes whereas Fe(II) is present in the reduced form [13] (Figure 7.46).

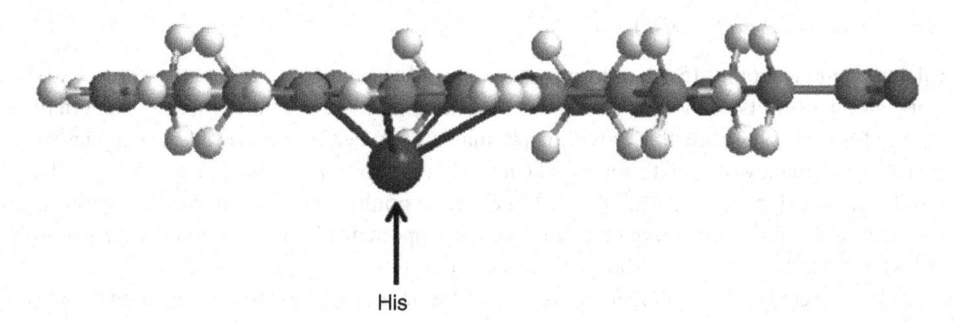

Figure 7.44 *Structure of deoxyhaemoglobin, with the iron atom lowered from the haeme plane*

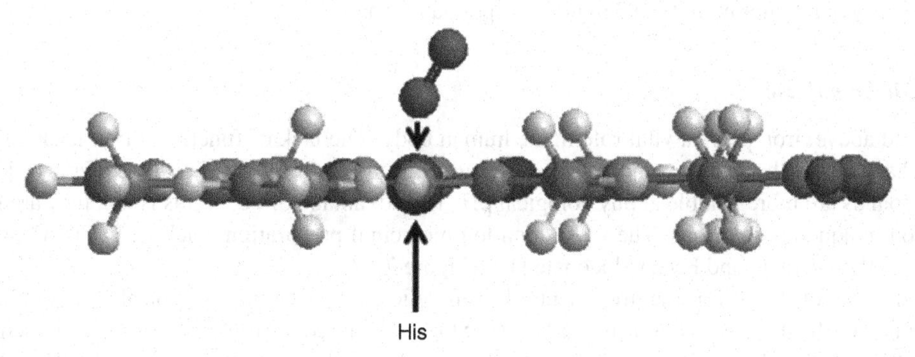

Figure 7.45 *Structure of oxyhaemoglobin, with the iron atom located within the haeme plane and coordinated to the O_2*

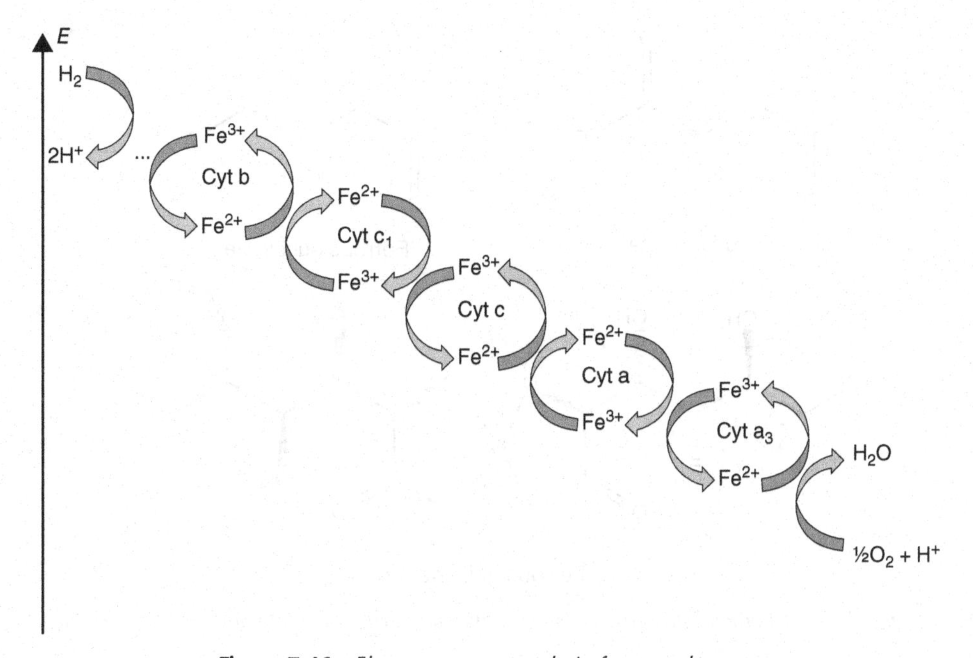

Figure 7.46 *Electron transport chain for cytochromes*

7.3.1.3 Iron uptake and metabolism

The normal diet contains around 15 mg of iron per day and only around 10% is actually absorbed. The absorption is dependent on a variety of factors, such as its bioavailability, the amount of iron present in the food and the body's need for iron. There are phases within a human's life where the body requires an increased amount of iron, such as in pregnancy or growth spurts. The absorption rate can be adjusted accordingly. The bioavailability of iron is highly dependent on the food source. Meat, poultry and fish are rich in easy-to-absorb iron. Absorption can also decrease depending on other dietary components. Some compounds, such as polyphenols (in vegetables) or tannins (in tea), can form chelates with iron.

Iron is actually excreted only in small amounts if there are no major blood losses. The human body regulates the uptake of iron precisely, only replacing the amount of iron lost in order to prevent an iron overload. Typical examples for iron loss would be through superficial GI tract bleeding or menstruation bleeding in women. Nevertheless, typically not more than 3 mg of iron per day is lost.

7.3.1.4 Medicinal use

As discussed above, iron plays a vital role in the human body. The lack of functional iron leads to anaemia, which is characterised by lethargy and weakness. Usually, iron is administered orally as Fe^{2+} or Fe^{3+} salts. Fe^{2+} compounds are more soluble at physiological pH. The advantage of Fe^{3+} salts is that they are not prone to oxidation in aqueous solutions. The most common medicinal preparations include $FeCl_3$, $FeSO_4$, Fe(II) fumarate, Fe(II) succinate and Fe(II) gluconate [13] (Figure 7.47).

The oral dose of Fe^{2+} for the treatment of iron-deficiency anaemia is typically recommended as 100–200 mg/day. In the case of ferrous sulfate ($FeSO_4$), this is equal to 65 mg of Fe^{2+}, which is given three times per day. As a therapeutic response, the haemoglobin concentration should raise about 100–200 mg/100 ml/day. It is recommended continuing the treatment for 3 months once the normal range

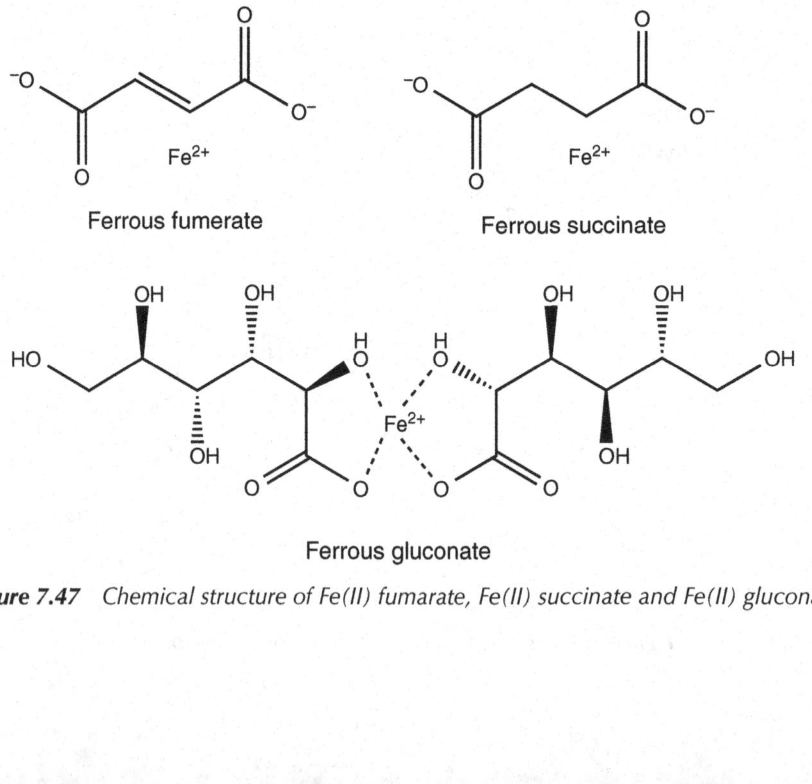

Ferrous fumerate

Ferrous succinate

Ferrous gluconate

Figure 7.47 Chemical structure of Fe(II) fumarate, Fe(II) succinate and Fe(II) gluconate

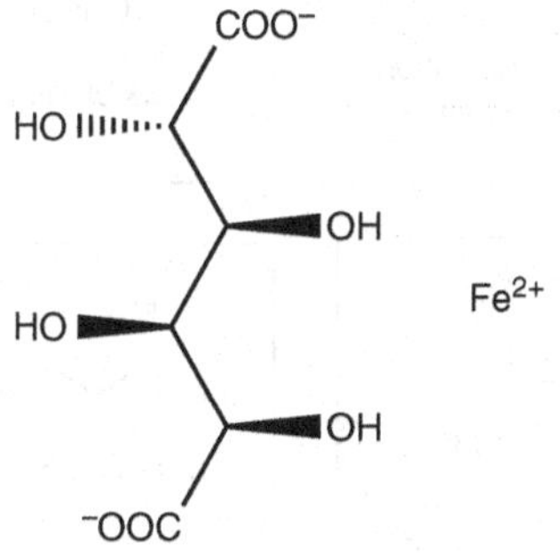

COO⁻ HO OH HO OH ⁻OOC Fe2+

Figure 7.48 Chemical structures of ferrous sucrose

is reached. It is known that oral treatment with iron salts can lead to GI irritations [8]. There is very little difference between the efficiency and absorbance rate of the above-mentioned iron salts. The choice of the preparation is influenced by side effects and cost.

Iron salts such as iron dextran and iron sucrose can be administered by IV infusion or IV injection. This administration route should be chosen only when oral therapy is not successful, as there is a risk of anaphylactic reactions. Patients with chronic renal failure who are on haemodialysis treatment often require IV iron supplementation (Figure 7.48).

7.3.1.5 Bleomycin

The drug Bleomycin (BLM) is successfully used as an anticancer agent, and is known to cause fragmentation of the DNA. The drug is used for the treatment of testicular cancer, non-Hodgkin's lymphoma, Hodgkin's lymphoma and cancers of the head and neck area (Cancer research UK). The name Bleomycin describes a family of water-soluble antibiotics that can be isolated from the bacterium *Streptomyces verticillus*. All family members contain the same core structure, a sulfur-containing polypeptide chain, and are only differentiated by a small side group and the sugar moiety [14] (Figure 7.49).

BLM was discovered 1966 by Umezawa *et al.* when they screened the filtrate of *S. verticillus* for cytotoxic activity. The therapeutically active forms of BLM are BLM A_2 and B_2, which differ only in the side chain [2]. BLM is believed to exhibit its anticancer activity by DNA degradation, a process that is dependent on the presence of molecular oxygen, and the binding of a metal to BLM to form the so-called 'activated BLM complex' [15].

The structure of BLM consists of several biologically important units, each contributing to its anticancer activity. Two structural units of importance to highlight are the metal-binding site and the DNA-binding site. It is believed that the intercalation of DNA by BLM occurs via the C-terminus, which contains two thiazole rings and the positively charged sulfonium salt. The positive charges of the sulfur atom can interact with the negatively charged phosphate backbones of the DNA. The metal-binding site can be found at the N-terminus and contains deprotonated amide and histidine groups. The metal is coordinated in a square planar complex, where a primary amine group occupies the axial position [15]. It can coordinate to a variety of metals such as Cu^{2+}, Co^{2+}, Zn^{2+} and Fe^{2+}, but it shows the highest binding affinity to Fe^{2+}. The metal chelation and subsequent activation of molecular oxygen is crucial to the antiproliferative activity of BLM. The carbohydrate core seems to be less involved in the direct anticancer activity. Nevertheless, it has been suggested that it regulates the cellular uptake and indirectly regulates the anticancer activity [14, 15].

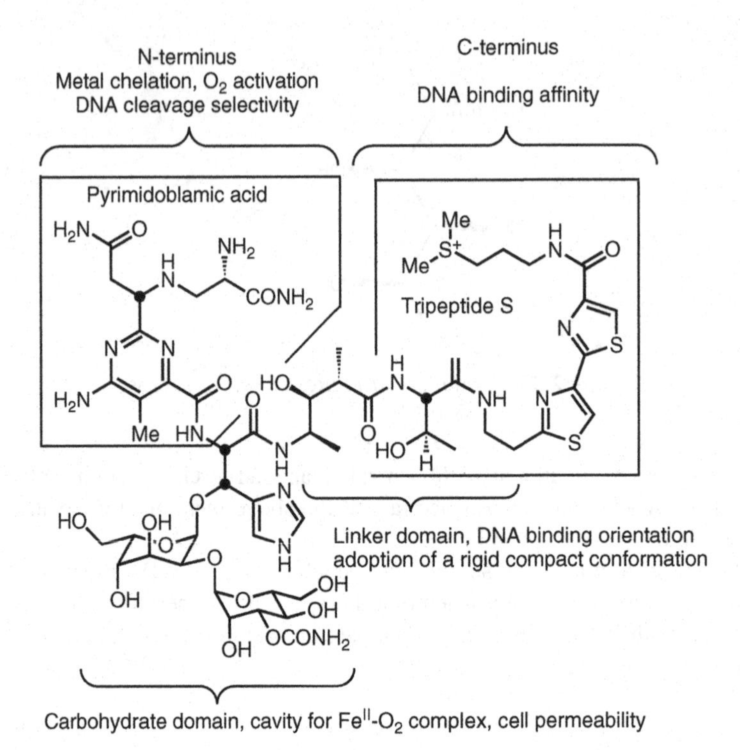

Figure 7.49 *Structure of Bleomycin [15]*

With regard to the mode of action of BLM as an anticancer agent, it is believed to be based on a unique radical mechanism that leads to DNA fragmentation. In the presence of molecular oxygen, the 'activated' BLM complex, HOO–Fe(III)BLM, is formed. This complex is known to cleave double-stranded DNA at the 5′-GC or 5′-GT sites [15].

In the next step, the homolytic splitting of the O—O bond of the HOO–Fe(III)BLM complex produces a radical that is able to degrade DNA. This complex is known to cleave double-stranded DNA at the 5′-GC or 5′-GT sites by abstraction of the C4′-H atom followed by a fragmentation of the deoxyribose backbone [15]. It is believed that the homolytic splitting of the O—O bond and the cleavage of DNA take place in a concerted manner, which means simultaneously. There are also other modes of action known for the DNA degeneration by BLM, but the above-mentioned one is the most prominent one [16].

For a clinical application, usually a mixture of BLM A_2 (~60–70%) and BLM B_2 (~20–30%) and others in combination with other antiproliferative agents are administered J. C. Dabrowiak, Metals in medicine, Wiley-Blackwell, Oxford, 2009. The most commonly used drug Blenoxane (70% BLM A_2) has been developed and marketed by GlaxoSmithKline and is used in the treatment of Hodgkin's lymphoma, testicular cancer and carcinomas of the skin and head and neck [15]. BLM is usually administered intravenously or intramuscularly. It is known to cause very little bone marrow suppression, but has a high dermatological toxicity causing increased pigmentation. The major side effect with the use of BLM is the occurrence of pulmonary fibrosis, which is dose-related. It is therefore important to regularly check the lungs by X-ray and respiratory function in general [8].

7.3.2 Ruthenium

Ruthenium is the chemical element with the symbol Ru and atomic number 44. It occurs as a minor side product in the mining of platinum. Ruthenium is relatively inert to most chemicals. Its main applications are in the area of specialised electrical parts.

The success of cisplatin, together with the occurrence of dose-limiting resistances and severe side effects such as nausea and nephrotoxicity, encouraged the research into other metal-based anticancer agents. Ruthenium is one of those metals under intense research, and first results look very promising, with two candidates – NAMI-A and KP1019 – having entered clinical trials.

7.3.2.1 Ruthenium properties and its biology

Ruthenium has mainly four properties that make it an interesting candidate for drug development: (i) the range of oxidation states, (ii) low toxicity compared to cisplatin, (iii) rate of ligand exchange and (iv) ability to mimic iron binding in biological targets.

Under physiological conditions, ruthenium can be found in several oxidation states such as II, III and IV, and the energy barrier for conversion between these oxidation states is fairly low. Ru(II) and Ru(III) have been extensively used in drug design, and they preferentially form six-coordinated octahedral complexes. Ru(III) species are the most inert biological species compared to the Ru(II) and Ru(IV) compounds. The redox potential of a complex depends on the ligands and can therefore be modified. For pharmaceutical applications, a Ru(III) complex can be so engineered that it easily reduces to the Ru(II) compound. In a biological environment, Ru(III) and Ru(IV) can be reduced by biomolecules such as GSH and ascorbate. In turn, Ru(II) species can be oxidised by molecular oxygen or enzymes such as cytochrome oxidase. Cancer cells are known to have a generally reducing environment. Therefore, ruthenium complexes can be administered as inert Ru(III) compounds, which are then *in situ* reduced to the active Ru(II) species [17]. This would mean that minimal damage is caused to healthy cells whilst cancerous cells become the target of the active ruthenium species; the result is that ruthenium compounds show a significantly lower toxicity than platinum complexes. Nevertheless, this theory, called *activation by reduction* has been questioned in recent years.

Ligand exchange is an important factor for the activity of metal-based drugs, especially anticancer agents. Only very few metal drugs reach their target in the form they have been administered. Most undergo rapid ligand exchange as soon as there are interactions with macromolecules or water. Some interactions are crucial for the activation of metal-based drugs, but often they hinder their activity. Ruthenium complexes have similar ligand exchange kinetics as cisplatin, with Ru(III) compounds being the most inert complexes. In contrast to cisplatin, which forms square planar complexes, ruthenium complexes are octahedral and therefore there is room for two more ligands compared to cisplatin to engineer the potential drug.

Iron and ruthenium are both members of group 8 within the periodic table of elements, and therefore researchers have inquired whether ruthenium can utilise the transport mechanisms normally used by iron. Iron is a key element for many biological processes; nevertheless, it is toxic for biological systems in its isolated form. It is believed that the low toxicity of ruthenium is a result of its iron-mimicking ability. There have been several hypotheses suggesting that ruthenium could use transferrin and albumin as transport mechanisms instead of iron or in a 'piggy-back' mechanism where ruthenium binds to the outside of transferrin when it is loaded with iron.

7.3.2.2 Ruthenium-based anticancer agents

KP1019 and NAMI-A are two ruthenium compounds that have entered clinical trials. They are classical coordination compounds in which the central ruthenium atom is coordinated by Lewis bases in an octahedral

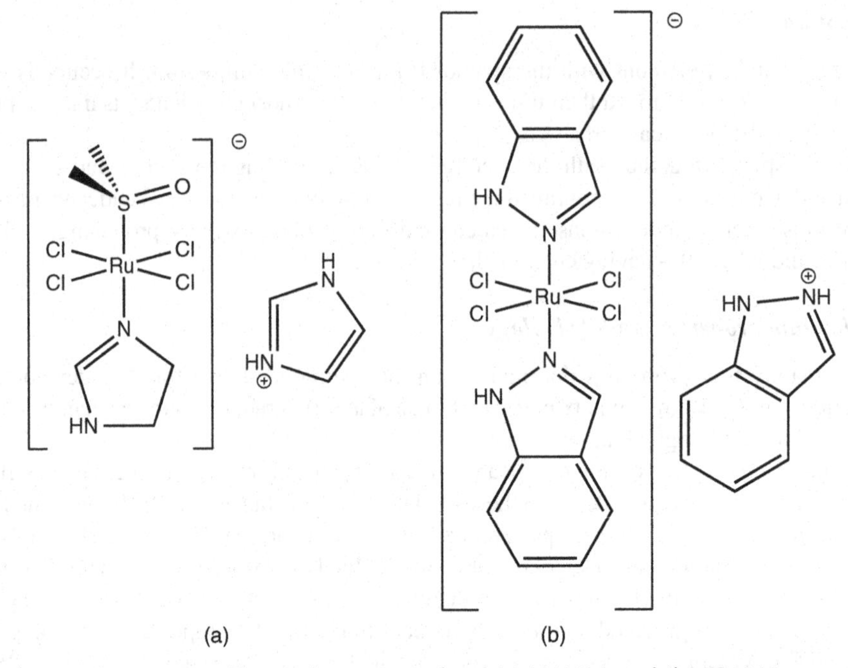

Figure 7.50 Chemical structures of and (a) NAMI-A and (b) KP1019

arrangement. Both complexes are based on a Ru(III) core, and the ligands are typically chloride and organic groups. This means that the complex can easily be hydrolysed and converted into its active form. Despite their similar structures, both ruthenium compounds present a different anticancer activity. KP1019 is active against primary cancers, whereas NAMI-A is active against metastasis, secondary tumours that have moved to other areas. The treatment of metastasis is currently an area where improvement is urgently needed (Figure 7.50).

NAMI-A has shown activity in the treatment of metastatic cancer and has completed phase I clinical trials in the Netherlands. It has been shown that the complex is relatively nontoxic. Significantly higher doses than cisplatin (above 500 mg/m^2/day) lead to side effects such as blisters on the extremities. Within the study, the ruthenium complex was administered intravenously over a period of 3 h in a 0.9% saline solution (pH ~ 4) [2]. The NAMI-A is synthesised by reacting $RuCl_3 \cdot 3H_20$ with HCl and DMSO (dimethylsulfoxide). This reaction results in the *trans* complex Imidazolium *trans*-imidazoledimethyl sulfoxide-tetrachlororuthenate(III) (NAMI-A) [2] (Figure 7.51).

It is interesting to note that Imidazolium *trans*-imidazoledimethyl sulfoxide-tetrachlororuthenate(III) is a paramagnetic compound and the complex is quickly hydrolysed in water. Initially, one chloride ligand is replaced by an aquo ligand, but the DMSO ligand is quickly replaced as well. As previously mentioned, it has been suggested that the complex is activated by bio-reduction of the Ru(III) centre to Ru(II) in the hypoxic environment of cancer cells. There is not much knowledge at present about the biological target for Imidazolium *trans*-imidazoledimethyl sulfoxide-tetrachlororuthenate(III). It is known that it interacts with the imidazoles of proteins and that the interaction with DNA is only weak, questioning DNA as a primary target for the ruthenium drug.

KP1019, *trans*-[tetrachloro-bis(1*H*-indazole)ruthenate(III)], was tested in phase I clinical trials as a possible treatment option against colon cancer. Administered doses ranged from 25 to 600 mg twice weekly, and no significant side effects were noticed. The synthesis of KP1019 also starts with $RuCl_3 \cdot 3H_20$, which is dissolved in ethanolic HCl and reacted with an excess of indazole (Figure 7.52).

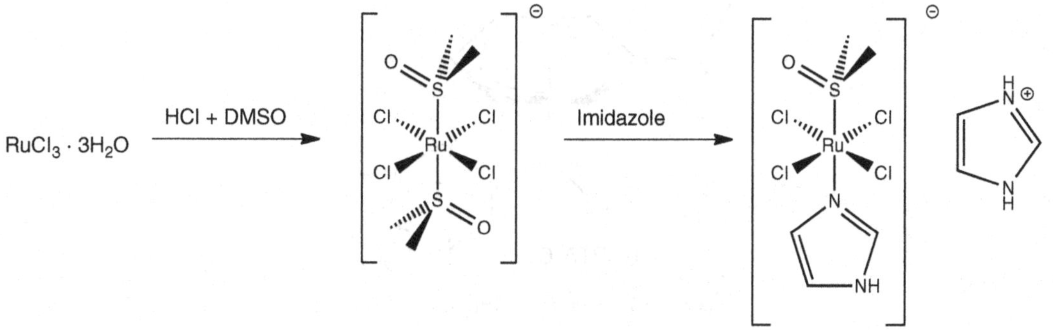

Figure 7.51 *Synthesis of NAMI-A [2]*

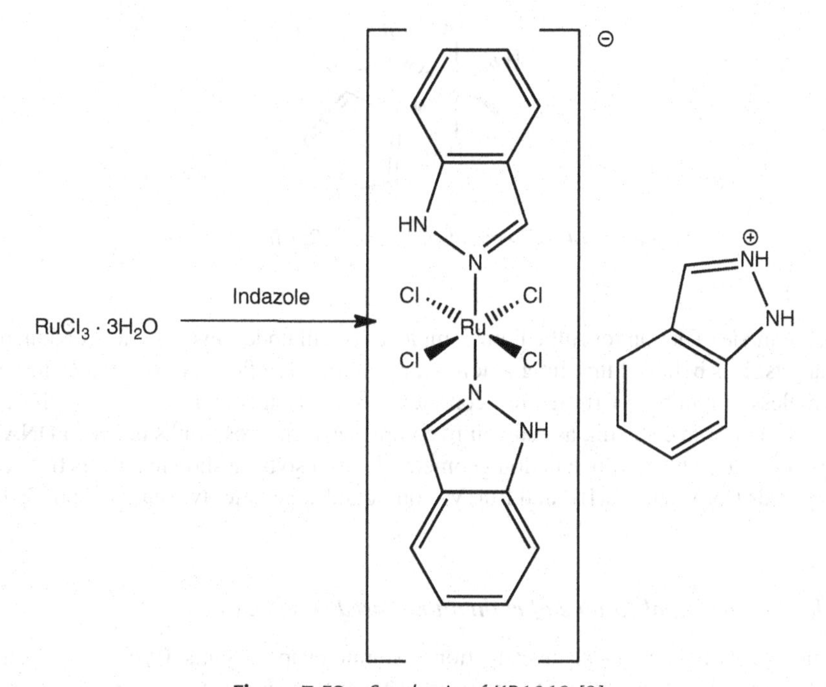

Figure 7.52 *Synthesis of KP1019 [2]*

The ruthenium complex hydrolyses quickly in water and induces apoptosis in cancer cells, potentially by blocking the mitochondrial function. KP1019 is believed to be transported by the protein transferrin to the tumour cells which are known to have a greater number of transferrin receptor on their cell surface.

RAPTA is an organometallic ruthenium complex (see Chapter 8 for the definition of organometallic complex) which is highly water-soluble. The Ru(II) centre is coordinated by chloride, an aromatic ring and 1,3,5-triaza-7-phosphatricyclo-[3.3.1.1]decane which is a form of phosphaadamantane. Interestingly, it shows similar biological activity as Imidazolium *trans*-imidazoledimethyl sulfoxide-tetrachlororuthenate(III) despite the differences in geometry, ligands and oxidation state. This shows that the active species in the cancer cell is different from the administered active pharmaceutical ingredient (API) (Figure 7.53).

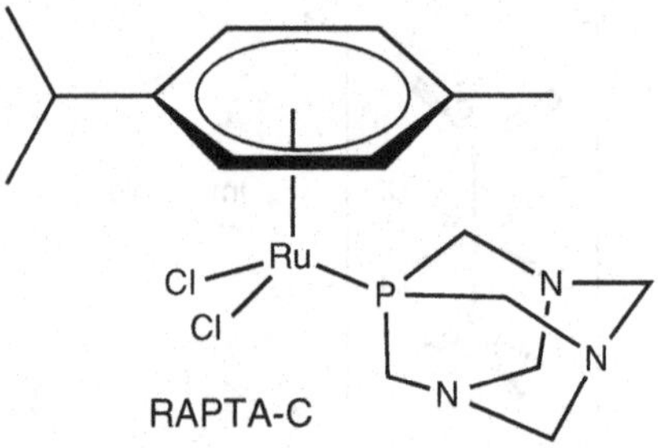

RAPTA-C

Figure 7.53 Chemical structure of RAPTA

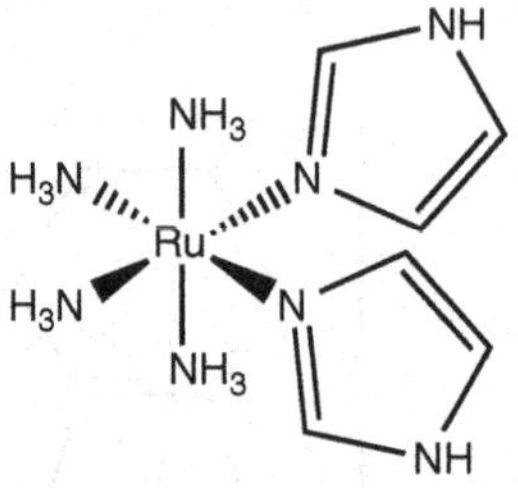

Figure 7.54 Chemical structure of Ru(III) imidazole

In summary, the mode of action for ruthenium complexes is still under investigation, and there are a variety of biological targets. It is believed that interaction with proteins significantly contributes to their anticancer activity. Nevertheless, a number of ruthenium complexes have been shown to bind to DNA. The mode of binding to DNA differs from cisplatin, as ruthenium complexes form cross-links between DNA strands probably due to steric hindrance by their octahedral geometry. It has also been shown that Ru(II) species are much more reactive towards DNA than Ru(III) and Ru(IV) compounds, the latter two can potentially be seen as less toxic prodrugs.

7.3.2.3 Further medical applications of ruthenium-based complexes

Ruthenium complexes have been under investigation as immunosuppressants. Cyclosporin A has severe side effects, such as hypertension, nephrotoxicity and nausea; hence there is a drive to search for new drugs. Ru(III) imidazole $[Ru(NH_3)_4(Im)_2]$ is a fairly stable complex that has been shown to inhibit the T-cell proliferation at nanomolar level [17] (Figure 7.54).

Ruthenium complexes have also shown promising results when initially tested as antimicrobials and antibiotics. Especially in the fight against malaria, new compounds are desperately needed as the Plasmodium parasite has become resistant to many traditional treatment options, mainly chloroquine. Research has shown that the Ru(II) chloroquine complex is significantly more effective, and it is suggested that the uptake of the metal complex follows a different route. Similar results have been seen when ruthenium compounds were tested as antibiotics [18].

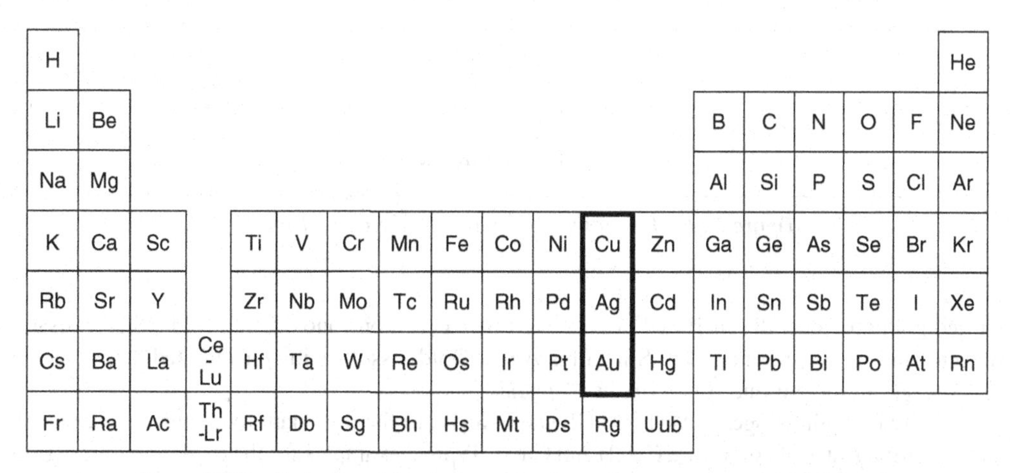

Figure 7.55 *Periodic table of elements; the coinage metals are highlighted*

7.4 The coinage metals

Historically, the three nonradioactive members of group 11 of the periodic table (11th vertical column) are designated as coinage metals, consisting of copper (Cu), silver (Ag) and gold (Au) (Figure 7.55).

Metals used to make coins have to fulfil special requirements, considering that a coin may be in circulation for several decades. Therefore, a coin needs to have anticorrosive properties and should not show any significant signs of wear. Very often, coinage metals are therefore mixed with other metals to form a so-called alloy. These alloys are harder and often more resistant to everyday use than the metals themselves.

Quite often, there is a problem that for low denomination coins the face value is significantly lower than the value of the metal itself. For example, the modern British penny is made out of steel plated with copper, whereas the American penny is made out of zinc with a copper covering.

7.4.1 General chemistry

Group 11 metals, such as Cu, Ag and Au, are known as the *noble metals* and belong to the d-block or transition metals. They are all relatively inert and corrosion-resistant metals and therefore are useful for the production of coins. All three metals are excellent conductors of electricity and heat. The most conductive metal for electricity is Ag, followed by Cu and then Au. Silver is the most thermally conductive element and the most light-reflecting element. Copper is widely used in electrical wiring and circuitry. In precision equipment, where the risk of corrosion needs to be kept as low as possible, gold is quite often used. Silver is widely used in mission-critical applications such as electrical contacts as well as in agriculture, medicine and scientific applications. Probably, one of the best known applications of silver ions and silver is the use in photography. Upon exposure, the silver nitrate in the film reverts to metallic silver itself.

Silver, gold and copper are all quite soft metals and therefore are not very useful as weapons or tools. Nevertheless, because of their malleability, they have been and still are used to make ornaments and jewellery.

The electronic configuration of group 11 metals is shown in Figure 7.56.

Cu : $[Ar]3d^{10}4s^1$

Ag : $[Kr]4d^{10}5s^1$

Au : $[Xe]4f^{14}5d^{10}6s^1$

Figure 7.56 *Electronic configuration of group 11 metals*

Silver and gold are generally inert and not attacked by oxygen or nonoxidising acids. Silver dissolves in salpetric acid, and in the presence of H_2S it forms Ag_2S. Gold dissolves in concentrated hydrochloric acid, forming $[AuCl_4]^-$ in the presence of an oxidising agent.

Both metals react with halogens. Ag(I), Au(III) and Au(I) are the dominant oxidation states, whereas the dominant oxidation states of copper are Cu(II) and Cu(I). Typical examples for the oxidation states III include AuF_3, $AuCl_3$ and $AuBr_3$, whereas there is only AgF_3. There are many examples for silver and gold salts where the metal takes the oxidation state I.

7.4.2 Copper-containing drugs

Copper is a valuable metal and has been mined for more than 2000 years. It has had many uses throughout history. Initially, copper was mainly used to make alloys such as brass and bronze, which are harder and stronger than copper itself. Nowadays, copper is mainly used because it conducts heat and electricity (e.g. wiring) and it is corrosion-resistant (e.g. as roofing material).

Historically, copper was used for the treatment of a variety of diseases, including chronic ulcers, headaches, ear infections, rheumatoid arthritis (RA), and so on. In 1832, copper workers were found to be immune to an outbreak of cholera in Paris, which stimulated further research into the medicinal use of copper. Almost every cell in the human body uses copper, as most contain copper-dependent enzymes. Unfortunately, excessive amounts of copper are toxic for the human body, whereas low amounts of copper also lead to health problems, manifested in Menkes disease.

Copper ions from food sources are processed by the liver, and transported and excreted in a safe manner. Inorganic metallic copper from sources such as drinking water mainly enters the blood directly and can be toxic as it can penetrate the blood–brain barrier. Typically, 50% of the daily copper intake is absorbed in the GI tract and transported to the liver from where it is transported to the peripheral tissue bound to ceruloplasmin, a copper-binding glycoprotein. A smaller amount of copper is also bound to albumin. Excess copper is mainly excreted in bile into the gut and then the faeces.

Copper is an essential trace metal, and copper ions are incorporated into a number of metalloenzymes – so called *cuproenzymes*. In the human body, the majority of copper ions can be found as Cu^{2+}; nevertheless, the oxidation state shifts between the cuprous (Cu^+) and cupric (Cu^{2+}) forms. It is important that copper can accept and donate electrons easily for its role in a variety of cuproenzymes, and examples are outlined in the following. Lysyl oxidase, a cuproenzyme, is responsible for the cross-linking of collagen and elastin, which forms strong and at the same time elastic tissue connections. These tissues are used, for example, for the formation of blood vessels and the heart. Ceruloplasmin (ferroxidase I) and ferroxidase II, two copper-based enzymes, can oxidise ferrous iron to ferric iron. Ferric iron can then be transported with the help of transferrin, for example, to form red blood cells. Furthermore, a variety of copper-dependent enzymes, such as cytochrome c and superoxide dismutase, work as antioxidants and are involved in the reduction of reactive oxygen species (ROS). There are also a variety of cuproenzymes that are involved in the synthesis (dopamine-b-monooxygenase) and metabolism (monoamine oxidase) of neurotransmitters [19].

7.4.2.1 *Wilson disease*

Wilson disease is a genetic disorder in which excessive amounts of copper build up in the human body. The copper is mainly stored in the liver and brain, and therefore causes liver cirrhosis and damage to the brain tissue. The damage to the brain tissue occurs mainly at the lenticular nucleus and a typical brown ring is visible around the iris; therefore Wilson disease is also called *hepatolenticular degeneration.*

Wilson disease was first described in 1912 by Wilson, pointing out the 'progressive lenticular degeneration' accompanied by chronic liver disease, which was leading to cirrhosis. It was Kayser and Fleischer who later made the association between these symptoms and a deposition of copper in the human body. Later on, it was established that Wilson disease is a genetic disorder with an autosomal recessive inheritance pattern. The abnormal gene was identified to be ATP7B, a metal-transporting adenosine triphosphatase (ATPase) mainly expressed in hepatocytes, which has the main function of transporting copper across the membrane. This means that, in the absence of this gene, the excretion of copper via the liver is reduced. This leads to copper accumulation in the liver, which damages the liver and eventually releases copper into the blood stream from where it can further poison the organs. Mainly the brain, kidneys and the cornea are affected by the copper accumulation.

Wilson disease presents itself mostly as a liver disease or a psychological illness and was fatal until treatments were developed. The main route of treatment is chelation therapy (see Chapter 11), with British anti-lewisite (BAL) being the first chelating agent used in 1951. In 1956, the orally available chelating agent D-penicillamine simplified the treatment of Wilson disease. Further treatment options include liver transplant, which has the potential to cure these patients [20]. Zinc supplementation can also be used in patients with Wilson's disease, as it prevents the absorption of copper ions (see Section 7.5). It is important to note that this therapy has a slow onset time and chelating treatments have to be continued for 2–3 weeks after the start with zinc supplementation [8].

7.4.2.2 *Copper and wound healing*

Glycyl-L-histidyl-L-lysine (GHK) is a tripeptide known for its high binding affinity to Cu^{2+} and its complex role in wound healing. The GHK–Cu(II) complex was isolated from human plasma in the 1970s and it was shown to be an activator for wound healing. GHK–Cu(II) has two main functions: as an anti-inflammatory agent to protect the tissue from oxidative damage after the injury, and as an activator for wound healing itself as it activates the tissue remodelling [21].

The structure of GHK is very similar to that of common drugs used to treat ulcers (Figures 7.57–7.59).

After the initial stages of wound healing are activated, such as blood coagulation and neutrophil invasion, a second stage of wound healing begins, which includes the population of GHK at the wound, which has a high affinity to Cu^{2+}. Mast cells, which are located in the skin, secrete GHK, which accumulates Cu^{2+} and forms the copper complex GHK–Cu(II) and therefore increases the metal–tripeptide concentration at the wound. First, GHK–Cu(II) has an anti-inflammatory effect by protecting the tissue from oxidative damage and by suppressing local inflammatory signals (i.e. cytokine interleukin-1 (IL-1)). Second, GHK–Cu(II) is released into the blood stream and encourages the production of wound macrophages that support the wound repair by removing the damaged tissue and secreting a family of several growth factor proteins. GHK–Cu(II) also hinders fibroblast production of TGF-β-1 and therefore suppresses the scar development. The GHK–Cu(II) complex also stimulates the growth of blood vessels, neurons and elastin, and, in general, supports most processes of wound healing [7].

Because of its versatile properties during the wound-healing process, it is not surprising that researchers tried to design commercial products based on GHK–Cu(II). Initial results were promising, but unfortunately the stability of the tripeptide GHK was not sufficient enough resulting in rapid breakdown. In the human body, GHK is permanently reproduced and therefore stability issues are not a major problem [22].

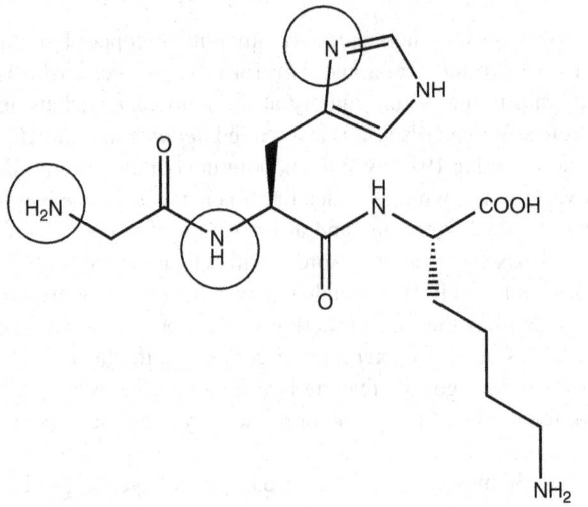

Figure 7.57 *Chemical structure of GHK; potential atoms for coordination to Cu^{2+} are circled*

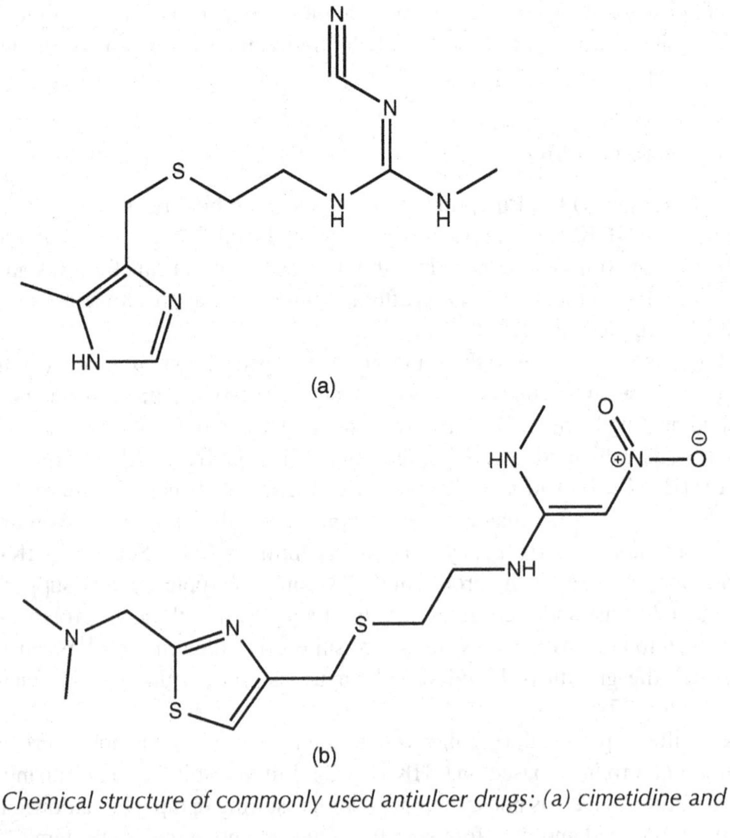

(a)

(b)

Figure 7.58 *Chemical structure of commonly used antiulcer drugs: (a) cimetidine and (b) nizatidine*

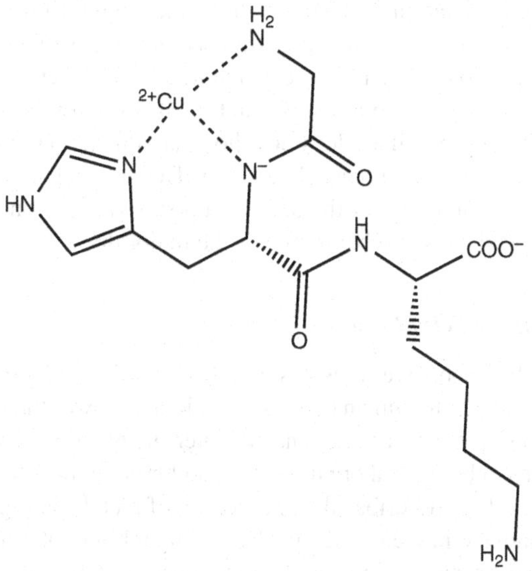

Figure 7.59 Chemical structure of the GHK–Cu(II) complex

7.4.2.3 Copper and cancer

Cancer progression has been linked to increased ceruloplasmin and copper levels in a variety of tissues. Copper deficiency has been considered as an anticancer strategy, but several clinical studies have not been encouraging. The exact role of copper in cancer is not yet fully understood, but it is possible to be involved via oxidation processes and the production of ROS and its involvement in angiogenic processes, as copper is believed to stimulate proliferation of endothelial cells [23].

7.4.3 Silver: the future of antimicrobial agents?

The name silver is derived from the Saxon word 'siloflur', which has been subsequently transformed into the German word 'Silabar' followed by 'Silber' and the English word 'silver'. Romans called the element 'argentum', and this is where the symbol Ag derives from.

Silver is widely distributed in nature. It can be found in its native form and in various ores such as argentite (Ag_2S), which is the most important ore mineral for silver, and horn silver (AgCl). The principal sources of silver are copper, copper–nickel, gold, lead and lead–zinc ores, which can be mainly found in Peru, Mexico, China and Australia.

Silver has no known active biological role in the human body, and the levels of Ag^+ within the body are below detection limits. The metal has been used for thousands of years mainly as ornamental metal or for coins.

Furthermore, silver has been used for medicinal purposes since 1000 BC. It was known that water would keep fresh if it was kept in a silver pitcher; for example, Alexander the Great (356–323 BC) used to transport his water supplies in silver pitchers during the Persian War. A piece of silver was also used, for example, to keep milk fresh, before any household refrigeration was developed. In 1869, Ravelin proved that silver in low doses acts as an antimicrobial. Around the same time, the Swiss botanist von Nägeli showed that already at very low concentration Ag^+ can kill the green algae spirogyra in fresh water. This work inspired the gynaecologist Crede to recommended use of $AgNO_3$ drops on new born children with conjunctivitis.

In 1884, Crede introduced the application of a 1% silver nitrate solution for the prevention of blindness in newborn, and the results were so impressive that this still used nowadays in America [24].

Today, airlines and NASA rely on silver filters to guarantee good water quality on board their aircrafts. A contact time of hours is necessary to see a disinfectant against coliforms and viruses of silver in water. In the concentrations normally applied, silver does not show any impact on the taste, odour or colour of the water. There is also no negative effect on human cells known. The only negative health effect known is called *argyria*, which is an irreversible darkening of the skin as a result of a prolonged application of silver. The existence of silver-resistant organisms is probably one of the major drawbacks for silver-based therapies.

7.4.3.1 Silver ions and their medicinal use

The antibacterial activity of silver and silver ions has long been known and led to many applications. This is mainly due to the fact that its toxicity to human cells is considerably lower than to bacteria. Most commonly, it is used for the prophylactic treatment of burns and in water disinfection. Unfortunately, the mechanisms and the chemistry of silver ions in biological organisms are so far not entirely clear. von Nägeli proposed the so-called oligodynamic effect, which describes the toxic effect of metals on organisms. It has been proposed that silver ions irreversibly damage key enzymes in the cell membrane of bacteria. This would lead to an inactivation of the pathogen. Silver reacts, as many other transition metals, preferentially with the thiol groups and also amino, carboxyl, phosphate and imidazole groups.

Silver nitrate ($AgNO_3$), after salicylic acid, is widely used for the treatment of warts. $AgNO_3$ is a highly water-soluble salt, which readily precipitates as AgCl, black in colour, when in contact with the skin. Warts are caused by a human papillomavirus, and mostly hands, feet and the anogenital areas are affected. The treatment is based on the destruction of the local tissue, and the silver salt is applied via a caustic pen to the affected area. Silver nitrate is highly corrosive and is known to destroy these types of tissue growth. Care has to be taken when this treatment option is used, as the resulting AgCl stains any skin or fabric which it has been in contact with.

7.4.3.2 Silver(I) sulfadiazine

Silver sulfadiazine is indicated for the prophylaxis and treatment of infections in burn wounds. Silver sulfadiazine is highly insoluble in water, and as a result, it does not cause hypochloraemia in burns in contrast to silver nitrate.

The active ingredient silver sulfadiazine is a sulfur-derived topical antibacterial used primarily on second- and third-degree burns. It is known to be active against many Gram-negative and Gram-positive bacteria as well as against yeast. The cream is kept applied to the burned skin at all times for the duration of the healing period or until a graft is applied. It prevents the growth of a wide array of bacteria, as well as yeast, on the damaged skin. Caution has to be given to large-area application as sulfadiazine levels in the plasma may well reach therapeutic levels that can cause side effects (Figure 7.60).

7.4.3.3 Silver dressings

All kinds of dressings containing silver ions have become more and more popular because of their antimicrobial effect. Nevertheless, the effect of silver nanoparticles on wound healing is still under discussion. Also, the use of dressings containing silver sulfadiazine and hydrocolloid dressings containing silver for the treatment of foot ulcers is still under discussion [25].

So far, clinical recommendation suggests that antimicrobial dressings containing silver should be used only when there are clinical signs of an infection present. They should not be routinely used for wound dressing

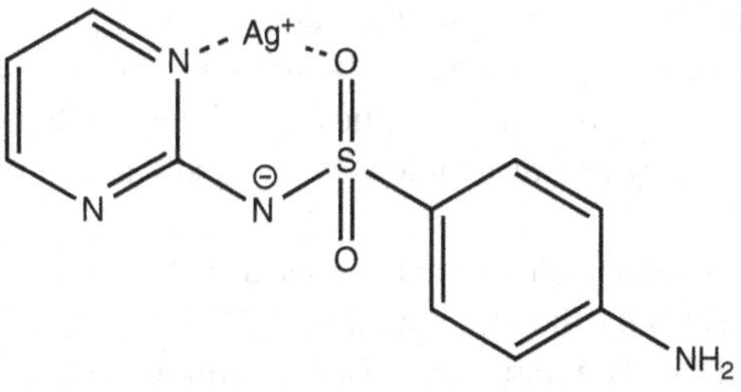

Figure 7.60 Chemical structure of silvadene

or the treatment of ulcers. Silver dressing will work only in the presence of wound secretion, as only then the silver ions will show an antimicrobial effect. There is also some evidence that silver dressings delay the healing process of acute wounds and therefore silver dressings are not recommended for use in those cases [8].

7.4.4 Gold: the fight against rheumatoid arthritis

From ancient times, gold has been regarded as one of the most beautiful metals and ever since treasured by man. The metal was first used to make tools, weapons and jewellery but was soon used for trade and as coins. Gold is a soft yellow metal, which is characterised by its high ductility. Very often, gold is alloyed with other metals to give it more strength. For example, white gold is gold alloyed with palladium.

Gold has also a long-standing tradition in medicine, as it has been used by many nations for thousands of years. From as early as 2500 BC, Arabians, Chinese and Indians used gold compounds for medicinal purposes. In mediaeval times, the elixir 'aurum potabile', which was an alcoholic mixture of herbs with some gold flakes, was sold by medicine men travelling around Europe and this elixir was supposed to cure most diseases. In the nineteenth century, $Na[AuCl_4]$ was reported to treat syphilis, whilst others used it to cure alcoholism. On a more serious note, Koch discovered in 1890 the antibacterial properties of gold cyanide. *In vitro* experiments with the *Mycobacterium tuberculosis* showed that gold cyanide has the potential as a tuberculosis therapy. Gold compounds were also investigated for the treatment of RA, when it was believed that RA was caused by bacteria, and many other health problems [7].

7.4.4.1 Gold therapy for rheumatoid arthritis

RA is a chronic, inflammatory, progressive autoimmune disease that primarily affects the joints. The disease is characterised by swelling of the joints and increasing pain leading to stiffness. Inflammation occurs initially in the synovial membrane surrounding the joints and then spreads to the synovium. An irreversible erosion of the articular cartilage on the bone joints means that bones will directly rub against each other and cause severe pain. The peak period of onset is between 35 and 55 years of age, with premenopausal women more often affected than men (ratio of 3 : 1). There is no obvious inheritance pattern to be found, but a genetic predisposition seems to be underlying. Patients are diagnosed on mainly three factors: painful joints, inflammation and the presence of the so-called rheumatoid factor.

Treatment options include anti-inflammatory drugs and/or disease-modifying antirheumatic drugs (DMARDs). Anti-inflammatory drugs comprise mostly nonsteroidal anti-inflammatory drugs (NSAIDs) or corticosteroids. NSAIDs are the first drugs of choice for the treatment of mild RA, as they possess analgesic and anti-inflammatory properties and also can be given in conjunction with DMARDs. Patients respond and tolerate NSAIDs quite variably, but generally the onset is rather quick. If there is no relief within 2–3 weeks, the treatment options are reconsidered. Corticosteroids are the most potent anti-inflammatory agents and additionally can be used as immunosuppressants. Corticosteroids are very commonly used in RA despite their severe side effects, which really limit their application. Additionally, corticosteroids have very little influence on the disease itself, but they are invaluable when the pain gets intolerable and mobility is severely decreased.

DMARDs are a category of unrelated drugs that are used in RA to slow down the progression of the disease. In contrast, NSAIDs treat only the inflammation, and corticosteroids are also insufficient to slow down the progression of RA. DMARDs are comprised of immune modulators, sulfasalazine and gold compounds amongst others, and they are used when the RA progresses from a mild to a more severe form. The onset time for DMARDs is significantly longer than for NSAIDs and it can take 2–6 months until any effect is seen. Typically, irreversible joint damage is already observed in the 2 years following diagnosis of RA, and therefore it is recommended to consider DMARDs as soon as moderate to severe RA is diagnosed. Some patients respond well to a combination therapy of NSAIDs and DMARDs, but patient response varies from individual to individual. In general, DMARDs have a higher toxicity and therefore come with significantly more side effects than NSAIDs. Gold salts are long known for their therapeutic effects in RA. In general, the application of gold compounds to treat diseases is called *chrysotherapy*, with RA being the main application area. Gold salts are clinically available as oral and intramuscular formulations.

> ***Chrysotherapy*** *is the treatment of certain diseases, especially RA, with gold compounds. Side effects can be quite severe and dose-limiting and include discolouration of skin, diarrhoea, nausea, flushing, vomiting, metallic taste in mouth and even damage of the kidneys and liver* [26].

7.4.4.2 Examples of gold-containing DMARDs

Au(I)thiolates, such as aurothioglucose, disodium aurothiomalate and trisodium bis(thiosulfato)gold, are the first-generation gold-based DMARDs. They feature linear, two-coordinated Au(I) thiolates and are polymeric with the exception of trisodium bis(thiosulfato)gold. The thiolate group stabilises the oxidation state of +1 for the gold atom, and therefore hinders disproportionation to Au(O) and Au(II).

> ***Disproportionation*** *is a redox reaction in which a species is oxidised and reduced at the same time and two different products are formed.*

The first-generation gold drugs are water soluble because of their hydrophilic groups and they are therefore commonly administered by intramuscular injection. The injection has to be done by a healthcare professional, which means the patient requires regular visits to the clinic. These gold-based compounds typically accumulate in the kidneys, where they are nephrotoxic and cause a leakage of proteins at the glomerulus. Typically, chrysotherapy is discontinued if there is no beneficial effect seen after 6 months, but side effects often last longer.

Sodium aurothiomalate is a commonly used gold-based DMARD and is indicated for active progressive RA. It is administered by deep intramuscular injection. Administration is started with a test dose of 10 mg followed by weekly intervals of 50 mg doses. An improvement is expected to be seen once 300–500 mg is administered. Treatment should be discontinued if there is no improvement after administering 1 g or 2 months. Intervals of administration should be gradually increased to 4 weeks in patients in whom an effect

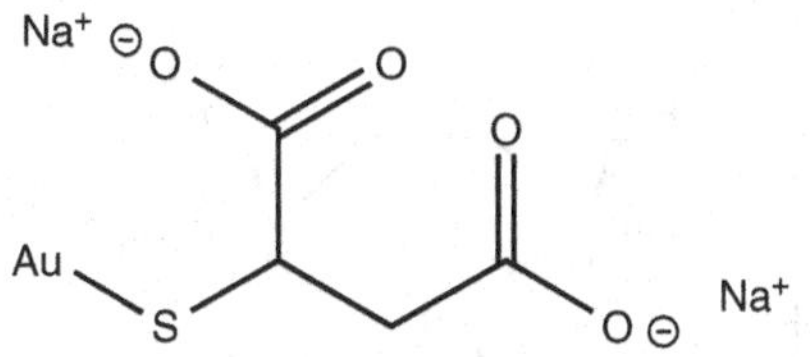

Figure 7.61 Chemical structure of sodium aurothiomalate

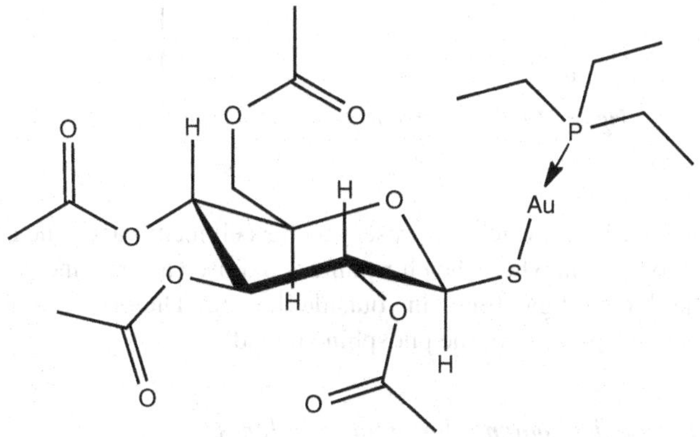

Figure 7.62 Chemical structure of Auranofin

can be seen. If any blood disorders or other side effects such as GI bleedings or proteinuria are observed, sodium aurothiomalate should be discontinued [8, 27] (Figure 7.61).

A second-generation gold-based drug came on the market, called *Auranofin* – ([tetra-*O*-acetyl-β-D-(glucopyranosyl)thio]-triethylphosphine)gold(I) – licensed as an orally available gold drug for the treatment of RA. It features a linear S–Au–P geometry, as shown by X-ray analysis. It is more lipophilic than the first-generation drugs, which makes oral administration possible. Treatment with Auranofin requires less visits to the clinic, but it is believed to be less successful in the treatment of RA compared to gold drugs being administered intramuscularly (Figure 7.62).

7.4.4.3 *Metabolism of gold drugs*

The clinically used gold drugs can be considered as prodrugs, which undergo rapid metabolism *in vivo* to form active pharmacological species. The precise mechanism of action is not known, but most probably it involves a thiol exchange. This means that the thiolate ligand will be replaced by a biological thiol such as albumin, which is a major protein in the serum and is sulfur-rich. Cysteine-34, one of the cysteine residues in albumin, is likely to be deprotonated at physiological pH and is a likely target for the gold compounds. In the case of Auranofin, the triethylphosphine (Et_3P) is oxidised, whilst the disulfide link in the albumin is reduced, and the harmless oxidised species $Et_3P{=}O$ is excreted via the kidneys.

The water-soluble first-generation gold drugs are injected intramuscularly and interact with the cells directly. They do not enter the cells but bind to the cell membranes via thiol groups on the cell surface and interfere with normal cell signalling pathways. Orally available gold drugs, such as Auranofin, enter the cells by a so-called thiol shuttle. Auranofin is transported to the cell, where it reacts with sulfhydryl-dependent

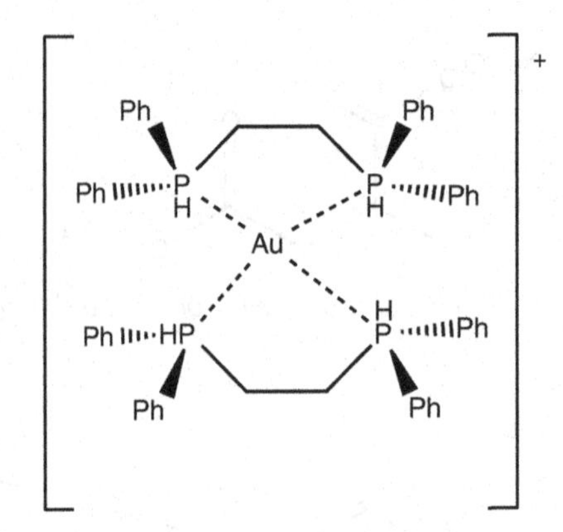

Figure 7.63 Chemical structure of $[Au(dppe)_2]^+$

membrane transport proteins (MSH), which are present on the cell membrane. The Et_3PAu^+ moiety binds to MSH and is transported into the cell, where the phosphine is oxidised as explained earlier and the Au moiety binds to the proteins. The thiolate ligand remains outside the cell. The gold cation can also leave the cell using the same mechanism, independent of the phosphine ligand.

7.4.4.4 Further pharmacological potential of gold complexes

The major clinical use of gold complexes relates to RA. Nevertheless, screening for the *antitumour activity* of antiarthritic gold drugs has encouraged studies towards the use as anticancer drugs. Au(III) species are isoelectronic to cisplatin (the most widely used metal-based anticancer drug).

*The term **isoelectronic** describes two or more molecular entities that have the same number of valence electrons and the same chemical structure independent of the actual elements present.*

The Au(I) complex of bis(diphenylphosphino)ethane (dppe) has an exciting cytotoxic profile when tested *in vitro* and *in vivo*. Experiments have shown that it was more effective when co-administered with cisplatin than each of the compounds individually. Unfortunately, no further studies were undertaken; especially, clinical trials could not proceed, as an acute toxicity to heart, liver and lungs was detected. Furthermore, some *antimicrobial* activity and some *antimalarial* activity of Au(I) complexes have been reported [7] (Figure 7.63).

7.5 Group 12 elements: zinc and its role in biological systems

The three natural occurring elements of group 12 of the periodic table (12th vertical column) are zinc (Zn), cadmium (Cd) and mercury (Hg) (Figure 7.64).

There is no significant abundance of the metals zinc, cadmium and mercury in the earth's crust, but they can be obtained from the respective ores. Zinc blende (ZnS) and sphalerite [(ZnFe)S] are the main sources of zinc, whereas CdS-containing ores are the only ores of importance for cadmium extraction. In order to obtain the pure metal, the relevant ores are roasted and the metal oxides are isolated. The corresponding metal is then extracted under high temperatures in the presence of carbon.

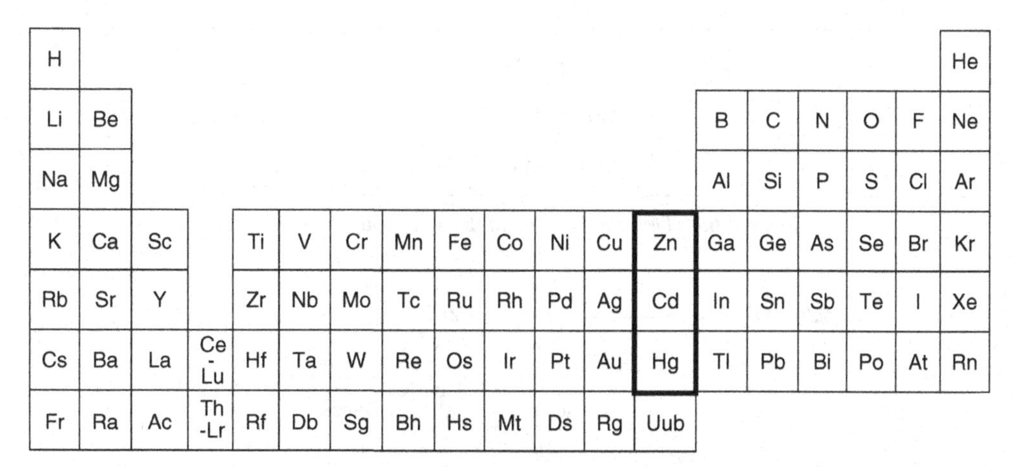

Figure 7.64 *Periodic table of elements; group 12 metals are highlighted*

Mercury is liquid at room temperature, the only metal that shows this behaviour. Therefore, it was and still is often used in thermometers and barometers despite its toxicity. Mercury and cadmium are both highly toxic elements. Cadmium can often be found in batteries. Mercury is also well known for its formation of amalgams with other metals. Amalgams are formed by reacting mercury with other metals (except silver) and have been generally widely used as dental fillings, which mostly contain mercury and variable amounts of silver together with other metals such as tin and copper. Nowadays, there is an increased concern about the safety of those dental amalgams and alternative fillings are more and more used. Mercury(II) nitrate was used in the eighteenth and nineteenth century to cure felts for the production of hats. Hat makers who were exposed for long periods to the mercury compound showed symptoms of mercury poisoning including uncontrollable muscle tremors, confused speech patterns and, in extreme cases, hallucinations. This was the inspiration for the term *mad as a hatter*.

Zinc has many applications (e.g. for galvanisation or in batteries) and is often used in alloys. Zinc is also an essential element for living organisms, plays a vital role in their biochemistry and is often found as the active centre in many enzymes. Cadmium and mercury can compete with zinc at these enzyme-binding sites, which leads to their characteristic toxicity.

7.5.1 General chemistry

The elements of group 12 and their chemical behaviour differ from other d-block metals as they have a completely filled valence shell (d shell) and two electrons in the s shell. The latter two electrons are easily removed, leading to the divalent cation. Group 12 elements do not form ions with oxidation states higher than +2 as a result of the closed full d shell. The electronic configuration of group 12 elements is shown in Figure 7.65.

The chemistry of Zn^{2+} and Cd^{2+} is expected to be fairly similar to that of Mg^{2+} and Be^{2+}. Indeed, there are some overlaps in regard to biological targets, but the chemical behaviour itself differs between members of group 2 and group 12 as a result of their different electronic configurations. It is important to note that mercury has some properties that are unique to this element and cannot be compared with the chemical behaviour of zinc or cadmium. Therefore, the chemistry of mercury will be discussed in less detail in this book.

Zinc and cadmium both dissolve in a variety of acids with the formation of hydrogen gas and the relevant metal cation M^{2+}. In contrast, mercury is inert to reactions with acids. A similar trend is seen for the formation

Zn : $[Ar]3d^{10}4s^2$

Cd : $[Kr]4d^{10}5s^2$

Hg : $[Xe]4f^{14}5d^{10}6s^2$

Figure 7.65 *Electronic configuration of group 12 elements*

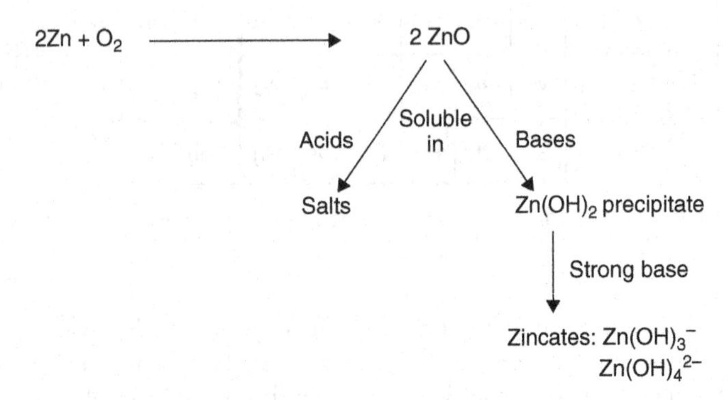

Figure 7.66 *Formation of zinc oxide and its chemical behaviour in acids and bases*

of oxides. Zinc and cadmium form the corresponding oxide when heated under oxygen. Mercury can also form the oxide, but the process is fairly slow. The resulting oxides, ZnO and CdO, are soluble in both acids and bases. ZnO will form salts when dissolved in acids, and the precipitate $Zn(OH)_2$ when dissolved in a base. $Zn(OH)_2$ can be dissolved in a strong base and the so-called zincates [$Zn(OH)_3^-$, $Zn(OH)_4^{2-}$] can be obtained. Cadmium oxides can also be dissolved in acids and bases, but the obtained $Cd(OH)_2$ is insoluble in even strong base solutions, but the hydroxide can be dissolved in ammonia. Halides of the metals (M) zinc and cadmium follow the general formula MX_2 and are either insoluble ($X = F$) in water or show a low aqueous solubility (Figure 7.66).

7.5.2 The role of zinc in biological systems

The average human body contains around 2 g of Zn^{2+}. Therefore, zinc (after iron) is the second most abundant d-block metal in the human body. Zinc occurs in the human body as Zn^{2+} (closed d^{10} shell configuration), which forms diamagnetic and mainly colourless complexes. In biological systems, zinc ions are often found as the active centre of enzymes, which can catalyse metabolism or degradation processes, and are known to be essential for stabilising certain protein structures that are important for a variety of biological processes.

Already from ancient times, Zn^{2+} was known to have important biological properties. Zinc-based ointments were traditionally used for wound healing. Low Zn^{2+} concentrations can lead to a variety of health-related problems especially in connection with biological systems of high Zn^{2+} demand such as the reproductive system. The daily requirement for Zn^{2+} is between 3 and 25 mg, depending on the age and circumstances [28].

The enzymatic function of Zn^{2+} is based on its Lewis acid activity, which are electron-deficient species (see Chapter 4). In the following chapters, examples will be shown to further explain this. Carboanhydrase (CA),

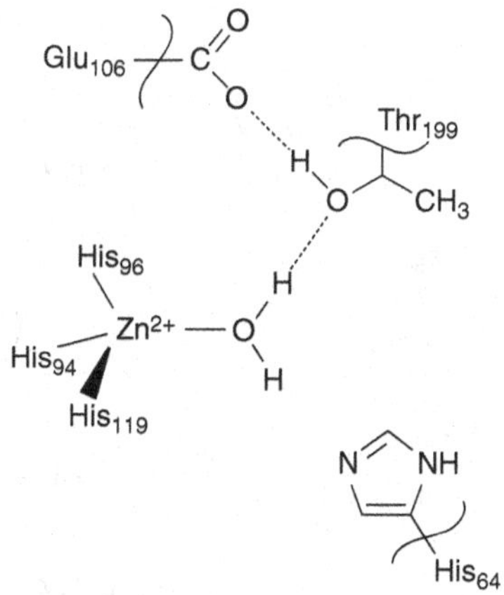

Figure 7.67 Scheme depicting the Zn^{2+} site of CA [28]

carboxypeptidase and superoxide dismutase are some examples for well-studied zinc-containing enzymes. The so-called zinc fingers have been discovered because of the crucial role of Zn^{2+} in the growth of organisms. Within the zinc finger, Zn^{2+} stabilises the protein structure and therefore enables its biological function.

7.5.2.1 Carboanhydrase (CA)

CAs are enzymes that catalyse the hydrolysis of carbon dioxide. These enzymes are involved in many biological processes such as photosynthesis (CO_2 uptake), respiration (CO_2 release) and pH control.

$$H_2O + CO_2 \leftrightarrow HCO_3^- + H^+$$

The human CA, form II(c), consists of 259 amino acids with a molecular weight of around 30 kDa. The catalytic site contains a Zn^{2+} ion which is coordinated by three neutral histidine (His) residues and a water molecule. The water molecule is believed to be important for structural reasons and enzymatic functionality (Figure 7.67).

A hypothetical mechanism for the mode of action for CA is shown in Figure 7.68. In a first step (i), a proton is transferred to His_{64} from the coordinated water molecule. In a second step (ii), a buffer molecule (B) coordinates this proton and transports it away from the active site. The remaining hydroxide ligand reacts quickly and forms a transition state via hydrogen bonding with CO_2 (iii). Following some more transformations, HCO_3^- is released as it is replaced by another molecule of water (iv) and (v) [28].

7.5.2.2 Carboxypeptidase A (CPA)

Carboxypeptidase A (CPA) is an enzyme of the digestive system that is known to cleave amino acids favouring the C-terminal end as well as certain esters. This enzymatic activity depends on the metal at the catalytic site. Zn^{2+} and some Co^{2+}-containing CPAs exhibit peptidase function, whilst esterase function has been seen by CPAs containing a variety of divalent d-block metals. CPA has a size similar to CA, consisting of about 300 amino acids and a molecular mass of 34 kDa. The metal centre is coordinated by two neutral histidine residues and one deprotonated glutamate residue as well as a water molecule.

At physiological conditions, the hydrolysis of proteins and peptides is a fairly sophisticated and slow chemical process. Within the catalysed reaction, the peptide or protein is attacked by an electrophile and

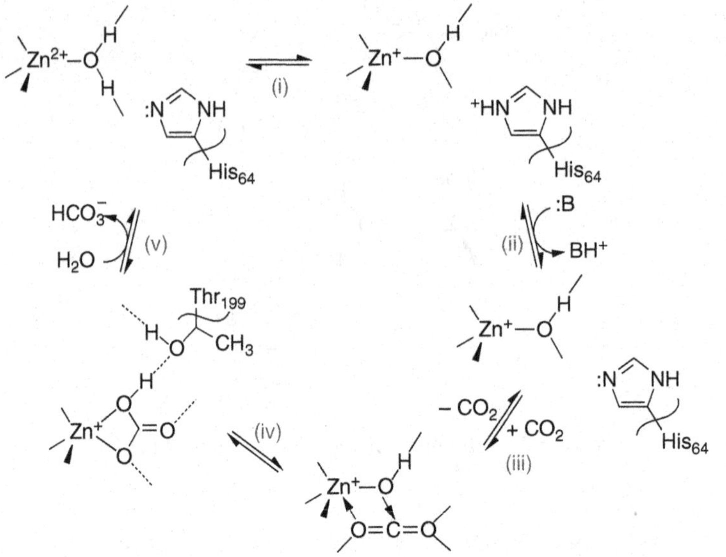

Zn^{2+}–O(H)(H) + :N–NH (His_{64}) ⇌ (i) Zn^{+}–O(H) + $^{+}$HN–NH (His_{64})

(ii) :B → BH^{+}

Zn^{+}–O–H + :N–NH (His_{64})

(iii) – CO_2 / + CO_2

Zn^{+}–O(H) ... O=C=O

(iv)

Zn^{+}(O)(O)C=O ... H–O(Thr_{199})–CH_3

(v) HCO_3^- / H_2O

Figure 7.68 *Scheme showing a hypothetical mechanism for the mode of action of CA [28]*

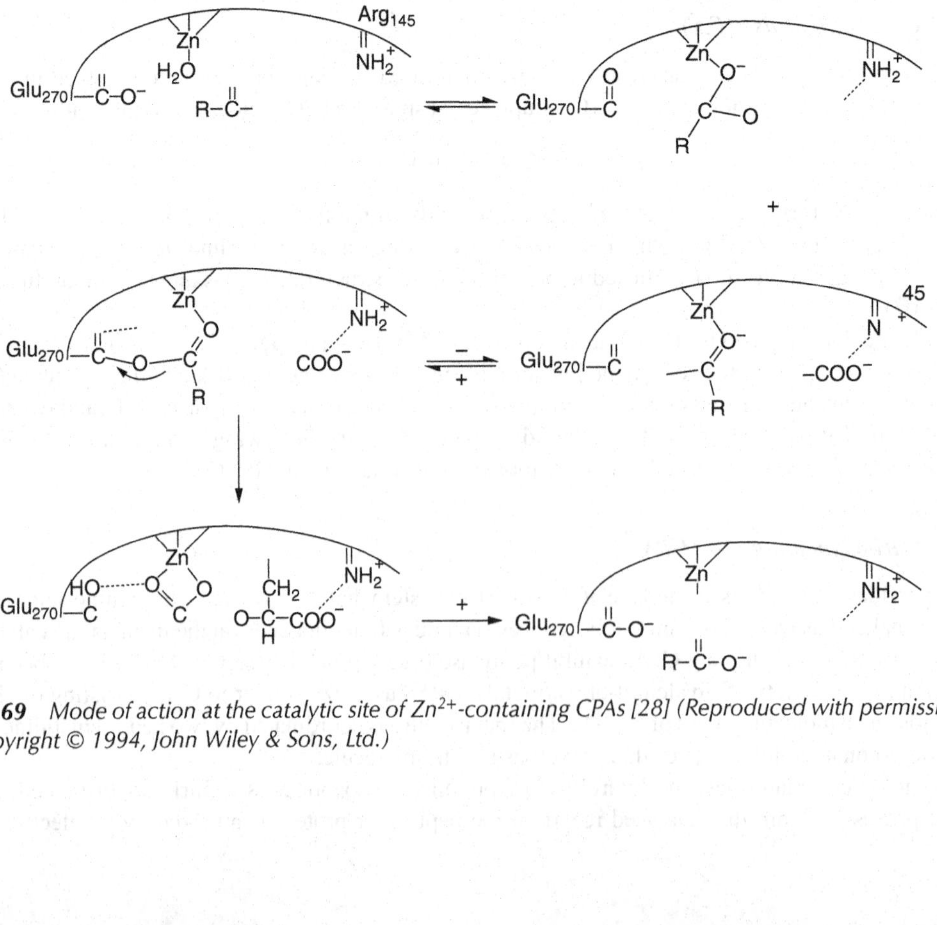

Arg_{145}

Zn

H_2O

NH_2^+

Glu_{270}–C(=O)–O$^-$

R–C(=O)–

⇌

Glu_{270}–C(=O)

Zn–O$^-$

C–O

R

NH_2^+

\+

Zn

Glu_{270}–C(=O)–O–C(=O)–R

NH_2^+

COO$^-$

⇌ (– / +)

Glu_{270}–C(=O)

Zn–O$^-$

–C(=O)–R

45

N$^+$

–COO$^-$

↓

Zn

Glu_{270}–C–HO

O–C(O)–O

CH_2

O–C(H)–COO$^-$

NH_2^+

\+ →

Glu_{270}–C(=O)–O$^-$

Zn

NH_2^+

R–C(=O)–O$^-$

Figure 7.69 *Mode of action at the catalytic site of Zn^{2+}-containing CPAs [28] (Reproduced with permission from [28]. Copyright © 1994, John Wiley & Sons, Ltd.)*

a nucleophile, leading to a fast reaction. Several theories describing the mode of action have been published, and one of those is described below in order to give an idea about the catalytic processes. The electrophile (metal centre) is coordinated to the oxygen of the carbonyl group and therefore allows the glutamate-270 to attack the activated carbonyl group. A mixed anhydride is formed, which subsequently is hydrolysed to form the desired products [28] (Figure 7.69).

Another theory describes the CPA-catalysed hydrolysis of a peptide bond analogous to the CA mechanism. The Zn^{2+} ion is coordinated by one molecule of water, which acts as a nucleophile and can attack the carbonyl group of esters or peptide bonds. A complicated system of hydrogen bonding further facilitates the substrate binding and final steps of this hydrolysis process (Figure 7.70).

Angiotensin converting enzymes (ACEs) are zinc-containing CPAs that convert angiotensin I into angiotensin II. Angiotensin II regulates the reabsorption of water and sodium ions in the kidneys and contracts the blood vessels leading to an increase of the blood pressure. ACE inhibitors are a class of drugs that block the activity of these enzymes. These drugs are used to lower the blood pressure in patients with hypertension and there are a variety of drug examples, such as captopril and lisinopril. The mode of action of these ACE inhibitors is based on their ability to bind to the Zn^{2+} centre and the active site of these CPAs and thereby blocking their enzyme activity. Captopril contains a thiol (SH) group, which coordinates directly to Zn^{2+}, whilst the carbonyl and carboxyl groups interact with the amino acid residues mainly via hydrogen bonding (Figure 7.71).

7.5.2.3 Zinc finger

It is well known that Zn^{2+} is essential for the growth of organisms and transcription of genetic material. It has been shown that there are special proteins that recognise certain DNA segments leading to the activation or regulation of genetic transcription. These proteins contain residues that can coordinate Zn^{2+}. This coordination leads to folding and a specific conformation, and they are called *zinc fingers*. Typically Zn^{2+} is coordinated by two neutral histidine (His) and two deprotonated cysteine (Cys) residues (Figures 7.72 and 7.73).

7.5.3 Zinc: clinical applications and toxicity

Clinical applications of Zn^{2+} range from its use in barrier creams and as a treatment option for Wilson disease to the use of zinc ions for the stabilisation of insulin. Long-acting human or porcine insulin is usually on the market as insulin zinc suspension. It is a sterile solution of usually human or porcine insulin, which is complexed by Zn^{2+}.

Zinc sulfate in the form of either injection or tablets can be used to treat zinc deficiency and as supplementation in conditions with an increased zinc loss. Zinc acetate is one treatment option for Wilson disease, as the zinc supplementation prevents the absorption of copper. It is important to note that zinc treatment has a slow onset time, which is crucial to take into account when switching from another therapy such as chelation therapy. Zinc acetate is usually offered to the market in an oral delivery form, mainly in capsules [8] (Figure 7.74).

Zinc ions are can also be found in barrier creams and lotions. Zinc oxide is present in barrier creams, for example, in creams used against nappy rash, often formulated with paraffin and cod-liver oil. Calamine lotion and creams are indicated for the treatment of pruritus, and both contain zinc oxide. It is interesting to note that the application of zinc oxide may affect the quality of X-ray images and it is therefore recommended not to apply these creams or lotions before X-ray tests [8].

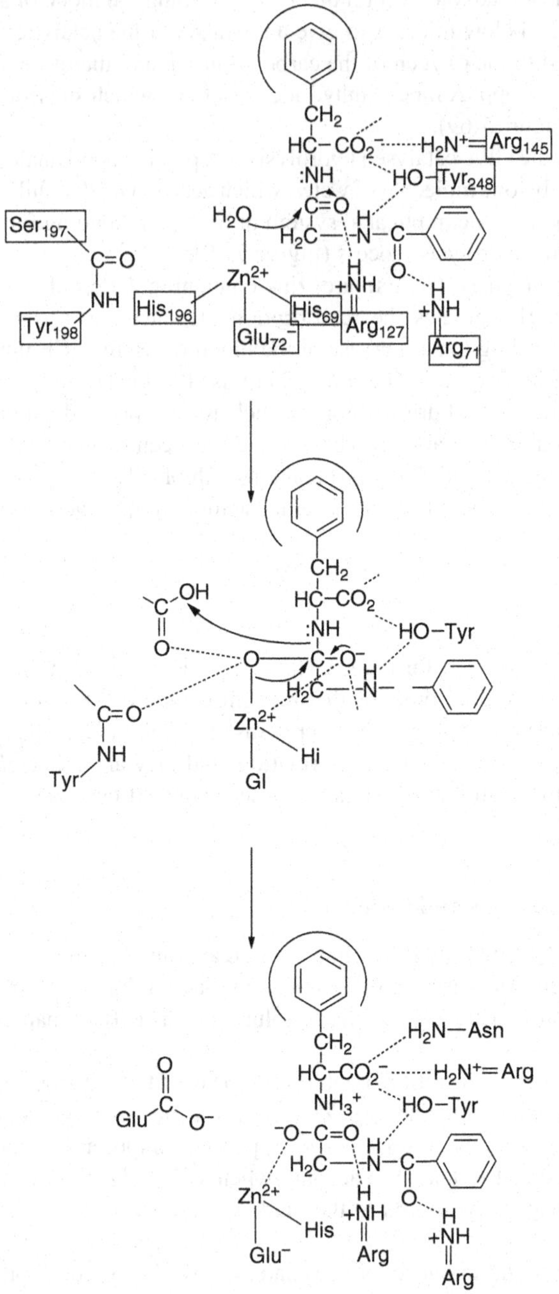

Figure 7.70 *Alternative mode of action at the catalytic site of Zn^{2+}-containing CPAs [28] (Reproduced with permission from [28]. Copyright © 1994, John Wiley & Sons, Ltd.)*

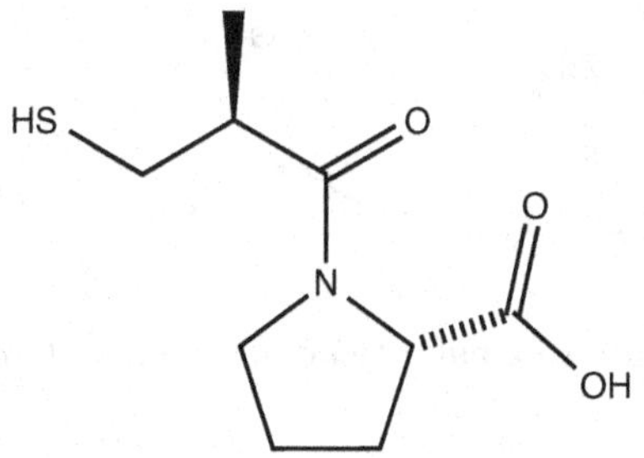

Figure 7.71 *Chemical structure of the antihypertensive drug captopril*

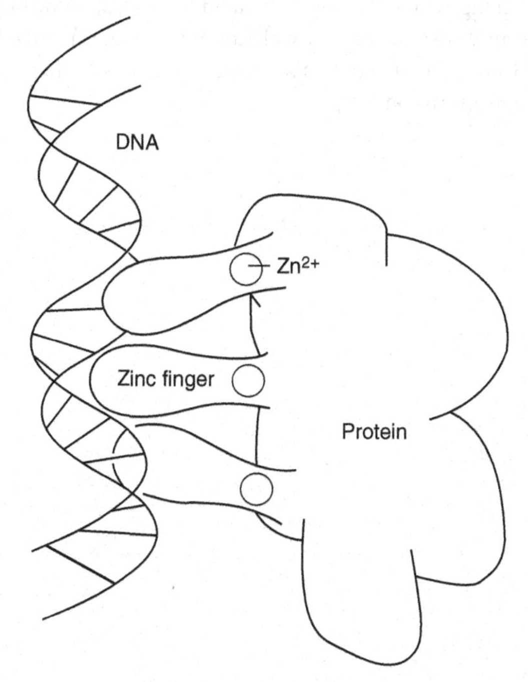

Figure 7.72 *Schematic representation of the interaction between DNA and a zinc finger protein [28] (Reproduced with permission from [28]. Copyright © 1994, John Wiley & Sons, Ltd.)*

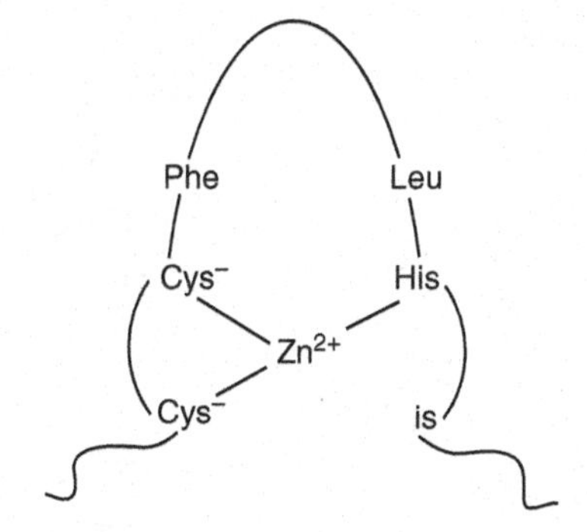

Figure 7.73 *Zinc finger motif from TF IIIA [28] (Reproduced with permission from [28]. Copyright © 1994, John Wiley & Sons, Ltd.)*

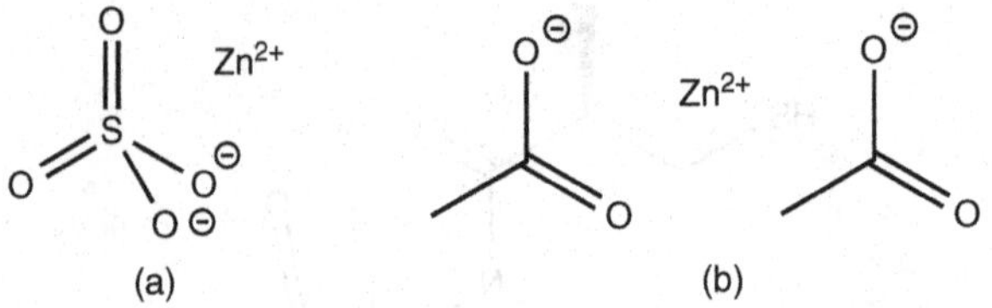

(a) (b)

Figure 7.74 Lewis formula of zinc sulfate (a) and zinc acetate (b)

Zinc is an essential element, but excess zinc can have a negative impact on human health. Zinc toxicity might be seen if the intake exceeds 225 mg. Typical symptoms include nausea, vomiting, diarrhoea and cramps. The so-called zinc-shakes are seen in workers, such as welders, who inhale freshly formed zinc oxide. Ingested metallic zinc dissolves in the stomach acid and zinc chloride ($ZnCl_2$) is formed. Zinc chloride is toxic to most organisms, depending on the concentration [29].

7.6 Exercises

7.6.1 What is the oxidation state of the central metal atom in the following complexes?

(a) $[Au(CN)_4]^-$
(b) $[V(CO)_6]$
(c) $[Fe(H_2O)_6]^{2+}$
(d) $[PdCl_6]^{2-}$

7.6.2 What is the charge of a complex formed by

(a) Co^{2+} and $3C_2O_4{}^{2-}$
(b) Pt^{4+} and $3H_2O$ and $3Br^-$
(c) Au^+ and $2CN^-$

7.6.3 Draw the energy diagrams displaying the d-orbital splitting for the low and high-spin complexes of the following examples assuming the formation of an octahedral complex:

(a) Fe^{3+}
(b) Zn^{2+}
(c) Cu^{2+}
(d) Cr^{3+}
(e) Cr^{2+}

7.6.4 Predict the geometry of the following complexes:

(a) $Zn(NH_3)_4$
(b) $[Mn(OH_2)_6]^{2+}$
(c) $Ag(NH_3)_2{}^+$

7.7 Case studies

7.7.1 Silver nitrate solution

Your pharmaceutical analysis company has been contacted by an important client and asked to analyse a batch containing silver nitrate ($AgNO_3$) solution for topical use. The description of your brief states that you are supposed to analyse the API in these solutions following standard quality assurance guidelines.

Typical analysis methods used for quality purposes are based on titration reactions. A certain amount of solution is diluted and treated with nitric acid and ammonium iron(III) sulfate. The solution is then titrated with ammonium thiocyanate [30].

(a) Research the type of titration described. Describe the chemical structure of any reagents involved in the analysis.
(b) Formulate the relevant chemical equations. What is the role of ammonium iron(III) sulfate?
(c) The package states that each the solution contains 10% silver nitrate. A volume containing 50 mg of silver nitrate is diluted with water and 2 ml of nitric acid and 2 ml of ammonium iron(III) sulfate solution are added. The solution is then titrated with 0.02 M ammonium thiocyanate solution until a red colour appears [30].
(d) For each titration, the following volume of ammonium thiocyanate solution has been used:

14.5 ml	14.8 ml	15.0 ml

Calculate the amount of $AgNO_3$ present in your sample. Express your answer in grams and moles.
(e) Critically discuss your result in context with the stated value for the API.
(f) Research the typically accepted error margins.

7.7.2 Ferrous sulfate tablets

Your pharmaceutical analysis company has been contacted by an important client and asked to analyse a batch of tablets containing dried ferrous sulfate ($FeSO_4$) tablets. The description of your brief states that you are supposed to analyse the API in these tablets following standard quality assurance guidelines.

Typical analysis methods used for quality purposes are based on titration reactions. A certain amount of ferrous sulfate tablets are crushed and dissolved. The solution is then titrated with cerium(IV)sulfate under acidic conditions using ferroin as the indicator British Pharmacopoeia.

(a) Research the type of titration described. Describe the chemical structure of any reagents involved in the analysis.
(b) Formulate the relevant chemical equations.
(c) The package states that each tablet contains 200 mg ferrous sulfate. The weight of 20 tablets has been determined to be 6.24 g. A certain amount of powder containing 0.5 g of dried ferrous sulfate is dissolved in water and sulfuric acid, and ferroin is added as an indicator. The solution is then titrated with 0.1 M ammonium cerium(IV)sulfate solution [30].

For each titration, the following volume of ammonium cerium(IV)sulfate solution has been used:

32.5 ml	33.3 ml	32.9 ml

Calculate the amount of $FeSO_4$ present in your sample. Express your answer in grams and moles.

(d) Critically discuss your result in context with the stated value for the API.
(e) Research the typically accepted error margins.

7.7.3 Zinc sulfate eye drops

Zinc sulfate solution has been traditionally used as an astringent and a mild antiseptic. Nowadays, its use is limited. Your pharmaceutical analysis company has been contacted by an important client and asked to analyse a batch of eye drops containing zinc sulfate ($ZnSO_4$). The description of your brief states that you are supposed to analyse the API in these tablets following standard quality assurance guidelines.

Typical analysis methods used for quality purposes are based on titration reactions. A certain volume of the eye drops is diluted and typically titrated with ammonium edetate using morbant black as indicator [30].

(a) Research the type of titration described. Describe the chemical structure of any reagents involved in the analysis. You might want to familiarise yourself with the concept of chelation (see Chapter 11).
(b) Formulate the relevant chemical equations.
(c) The package states that each sterile solution contains 0.25% w/v of zinc sulfate heptahydrate. For the titration, 5 ml of the zinc sulfate solution is diluted and an ammonia buffer is added to achieve pH 10.9. This solution is titrated with 0.01 M disodium edetate and mordant black as indicator [30].

For each titration, the following volume of edetate solution has been used:

5.0 ml	4.6 ml	4.3 ml

Calculate the amount of $ZnSO_4$ present in your sample. Express your answer in grams and moles.

(d) Critically discuss your result in context with the stated value for the API.
(e) Research the typically accepted error margins.

References

1. N. G. Connelly, *Nomenclature of inorganic chemistry: IUPAC recommendations 2005*, Royal Society of Chemistry, Cambridge, **2005**.
2. J. C. Dabrowiak, *Metals in medicine*, Wiley-Blackwell, Oxford, **2009**.
3. R. A. Alderden, M. D. Hall, T. W. Hambley, *J. Chem. Educ.* **2006**, *83*, 728–734.
4. Z. D. Bugarcic, J. Bogojeski, B. Petrovic, S. Hochreuther, R. van Eldik, *Dalton Trans.* **2012**, *41*, 12329–12345.
5. F. Arnesano, G. Natile, *Coord. Chem. Rev.* **2009**, *253*, 2070–2081.
6. P. M. Takahara, A. C. Rosenzweig, C. A. Frederick, S. J. Lippard, *Nature* **1995**, *377*, 649–652.
7. E. R. Tiekink, M. Gielen, *Metallotherapeutic drugs and metal-based diagnostic agents: the use of metals in medicine*, Wiley, Chichester, **2005**.
8. *British national formulary*, British Medical Association and Pharmaceutical Society of Great Britain, London.
9. M. Coluccia, G. Natile, *Anti-Cancer Agent Med.* **2007**, *7*, 111–123.
10. I. Kostova, *Recent Pat. Anti-Cancer Drug Discovery* **2006**, *1*, 1–22.
11. (a) H. Hochster, A. Chachoua, J. Speyer, J. Escalon, A. Zeleniuch-Jacquotte, F. Muggia, *J. Clin. Oncol.* **2003**, *21*, 2703–2707; (b) S. Goel, A. Bulgaru, H. Hochster, S. Wadler, W. Zamboni, M. Egorin, P. Ivy, L. Leibes, F. Muggia, G. Lockwood, E. Harvey, G. Renshaw, S. Mani, *Ann. Oncol.* **2003**, *14*, 1682–1687.
12. H. Choy, C. Park, M. Yao, *Clin. Cancer Res.* **2008**, *14*, 1633–1638.
13. C. E. Housecroft, A. G. Sharpe, *Inorganic chemistry*, 3rd ed., Pearson Prentice Hall, Harlow, **2008**.
14. R. H. Blum, S. K. Carter, K. Agre, *Cancer* **1973**, *31*, 903–914.
15. D. L. Boger, H. Cai, *Angew. Chem. Int. Ed.* **1999**, *38*, 448–476.
16. E. Alessio, *Bioinorganic medicinal chemistry*, Wiley, Weinheim, **2011**.
17. C. S. Allardyce, P. J. Dyson, *Platinum Met. Rev.* **2001**, *45*, 62–69.
18. Y. K. Yan, M. Melchart, A. Habtemariam, P. J. Sadler, *Chem. Commun.* **2005**, 4764–4776.
19. (a) M. C. Linder, M. HazeghAzam, *Am. J. Clin. Nutr.* **1996**, *63*, S797–S811; (b) R. Uauy, M. Olivares, M. Gonzalez, *Am. J. Clin. Nutr.* **1998**, *67*, 952S–959S.
20. (a) A. K. Wernimont, D. L. Huffman, A. L. Lamb, T. V. O'Halloran, A. C. Rosenzweig, *Nat. Struct. Biol.* **2000**, *7*, 766–771; (b) E. A. Roberts, M. L. Schilsky, *Hepatology* **2003**, *38*, 536.
21. C. K. Sen, S. Khanna, M. Venojarvi, P. Trikha, E. C. Ellison, T. K. Hunt, S. Roy, *Am. J. Physiol. Heart C* **2002**, *282*, H1821–H1827.
22. (a) S. F. Swaim, D. M. Vaughn, S. A. Kincaid, N. E. Morrison, S. S. Murray, M. A. Woodhead, C. E. Hoffman, J. C. Wright, J. R. Kammerman, *Am. J. Vet. Res.* **1996**, *57*, 394–399; (b) F. X. Maquart, G. Bellon, B. Chaqour, J. Wegrowski, L. M. Patt, R. E. Trachy, J. C. Monboisse, F. Chastang, P. Birembaut, P. Gillery, J. P. Borel, *J. Clin. Invest.* **1993**, *92*, 2368–2376; (c) I. Tenaud, I. Sainte-Marie, O. Jumbou, P. Litoux, B. Dreno, *Br. J. Dermatol.* **1999**, *140*, 26–34.
23. (a) A. Gupte, R. J. Mumper, *Cancer Treat. Rev.* **2009**, *35*, 32–46; (b) V. L. Goodman, G. J. Brewer, S. D. Merajver, *Curr. Cancer Drug Targets* **2005**, *5*, 543–549; (c) F. Tisato, C. Marzano, M. Porchia, M. Pellei, C. Santini, *Med. Res. Rev.* **2010**, *30*, 708–749.
24. N. Cioffi, M. Rai, *SpringerLink (online Service)*, Springer, Berlin; Heidelberg, **2012**, p. xvi, 556 p.
25. Z. Aziz, S. F. Abu, N. J. Chong, *Burns* **2012**, *38*, 307–318.
26. A. J. Griffin, T. Gibson, G. Huston, A. Taylor, *Ann. Rheum. Dis.* **1981**, *40*, 250–253.
27. J Barber, C. Rostron, *Pharmaceutical chemistry*.
28. W. Kaim, B. Schwederski, *Bioinorganic chemistry: inorganic elements in the chemistry of life: an introduction and guide*, Wiley, Chichester, **1994**.
29. G. J. Fosmire, *Am. J. Clin. Nutr.* **1990**, *51*, 225–227.
30. *British pharmacopoeia*, Published for the General Medical Council by Constable & Co, London.

Further Reading

F. A. Cotton, G. Wilkinson, C. A. Murillo, M. Bochmann, *Advanced inorganic chemistry*, 6th ed., Wiley, New York; Chichester, **1999**.

H.-B. Kraatz, N. Metzler-Nolte, *Concepts and models in bioinorganic chemistry*, Wiley-VCH, [Chichester: John Wiley, distributor], Weinheim, **2006**.

J. D. Lee, *Concise inorganic chemistry*, 5th ed., Chapman & Hall, London, **1996**.

G. A. McKay, M. R. Walters, J. L. Reid, *Lecture notes. Clinical pharmacology and therapeutics*, 8th ed., Wiley-Blackwell, Chichester, **2010**.

R. M. Roat-Malone, *Bioinorganic chemistry: a short course*, Wiley, Hoboken, N.J. [Great Britain], **2002**.

G. J. Tortora, B. Derrickson, *Principles of anatomy and physiology*, 12th ed., international student/Gerard J. Tortora, Bryan Derrickson. ed., Wiley [Chichester: John Wiley, distributor], Hoboken, N.J., **2009**.

8

Organometallic Chemistry

8.1 What is organometallic chemistry?

Organometallic chemistry is the area of chemistry that deals with compounds containing a metal–carbon bond. As such, this area combines aspects from both organic and inorganic chemistry.

*An **organometallic compound** is characterised by the presence of one or more carbon–metal bonds.*

It is important to note that the metal can either be a member of the s or p block on one hand or a d-block metal (transition metal) on the other. There are no real direct pharmaceutical applications known for group 1 and 2 organometallic compounds, as they are very reactive reagents, except that they are commonly involved in the synthesis of modern medicines. Examples of such synthetic reagents include sodium cyclopentadienide (NaCp, NaC_5H_5) and butyl lithium (BuLi) compounds.

***Cyclopentadienyl** or **Cp^-** ($C_5H_5^-$) is a commonly used ligand in organometallic chemistry, which is versatile in the number of bonds it can form to a metal centre (M). There are different ways of representing these interactions (see also* Figure 8.1).

NaCp is an organometallic agent that is mainly used to introduce a cyclopentadienyl anion ($C_5H_5^-$) to a metal centre in order to form a so-called metallocene (see Section 8.2 for a definition of metallocenes). NaCp can be synthesised by the reaction of either sodium with cyclopentadiene or from dicyclopentadiene under heating. Sodium hydride (NaH) can also be used as a base, instead of sodium, to deprotonate the acidic CH_2 group of the cyclopentadiene (Figure 8.2).

It is interesting to look at the bonding in this Cp ligand. First of all, it is important to understand the formation of the cyclopentadienyl anion, the Cp^- ligand. Cyclopentadiene is surprisingly acidic, which is

Essentials of Inorganic Chemistry: For Students of Pharmacy, Pharmaceutical Sciences and Medicinal Chemistry,
First Edition. Katja A. Strohfeldt.

Companion website: www.wiley.com/go/strohfeldt/essentials

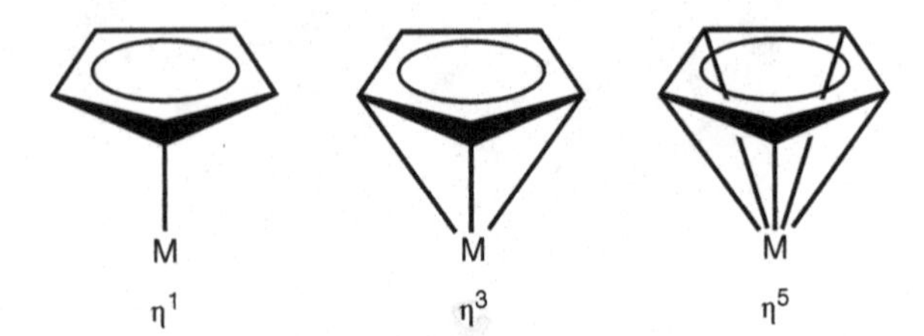

Figure 8.1 Illustration showing different binding modes of Cp^- ligand to metal centre (M)

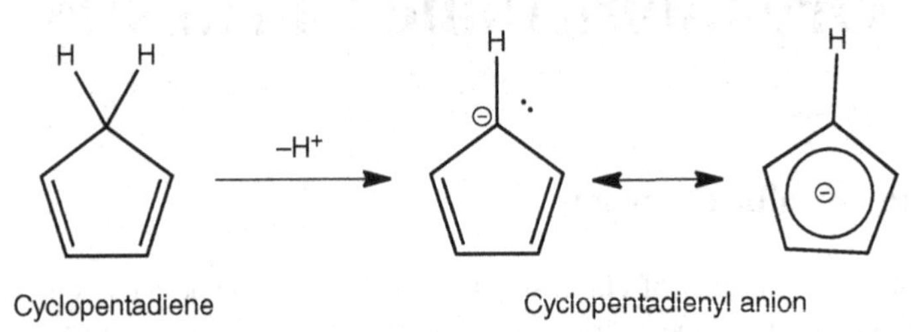

Figure 8.2 Deprotonation of cyclopentadiene

a result of the resonance stabilisation of the resulting cyclopentadienyl anion (Cp^-). The cyclopentadienyl anion follows Huckel's rule ($4n+2$, $n=0, 1, 2$, etc.), which means that the Cp^- anion is a planar carbocycle of aromatic nature.

> *An* ***aromatic molecule*** *has to be a cyclic and planar molecule with an uninterrupted network of π electrons. Furthermore, it has to fulfil the Huckel rule, which states it has $(4n+2)$ π electrons.*

Organolithium compounds (Li—C bond) are probably the best known organometallic agents. In organic synthesis and drug design, they can be used as either an extremely potent base or as nucleophile; in the latter case, the organic moiety will be introduced to the target molecule. They are typically synthesised by reacting an organohalide (RX) with elemental lithium. The best known examples are *n*-butyllithium (*n*-BuLi), *sec*-butyllithium (*s*-BuLi) and *tert*-butyllithium (*tert*-BuLi), with *tert*-BuLi being the most reactive (Figure 8.3).

Alkali metal organometallics are more or less pyrophoric, which means they combust spontaneously on contact with air. They have to be handled under the exclusion of air, humidity and oxygen and are mostly stored in hydrocarbons. The solvent plays an important role and can be responsible for potential decomposition

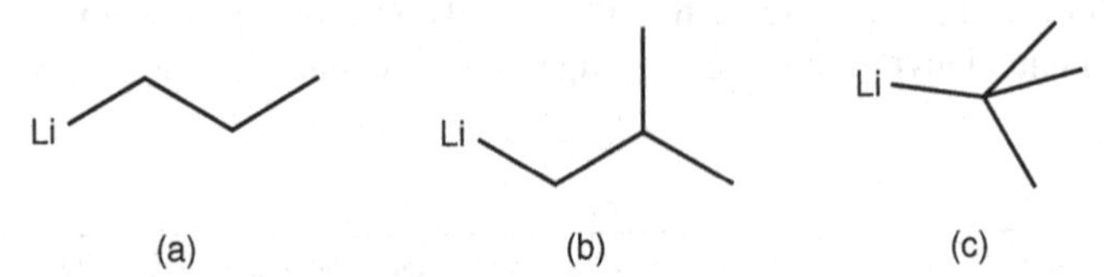

Figure 8.3 Chemical structures of (a) n-BuLi, (b) s-BuLi and (c) t-BuLi. Note that these compounds can form complex structures in the solid state and in solution

processes or an increased or reduced reactivity. The heteroatoms of solvents can potentially also coordinate to the alkali metal and therefore influence potential cluster formation of the organometallic compound. Alkali metal organometallics are known to form relatively complex clusters in solution and in the solid state, which influences their reactivity.

Organometallic compounds, containing d-block metals, are currently under intense research within the pharmaceutical chemistry area in order to find new treatment options for cancer and diabetes, amongst others. d-Block organometallics are generally fairly stable complexes, which in contrast to alkali metal organometallics can be handled in the presence of air. Whilst s- and p-block organometallics form σ and π bonds between the metal and the organic group, in d-block organometallics the number of bonds, which is called *hapticity* (see Section 7.1), can be further increased.

The most common ligands for d-block organometallics include carbon monoxide (CO) in the form of the carbonyl group, phosphanes (PR_2H) and derivatives of the cyclopentadienyl (Cp^-) ligand. A characteristic example is the bonding of the metal with the carbonyl ligand, which can be described as one M—CO interaction. A vacant (hybridised) orbital of the metal centre forms a σ bond with the CO ligand, which means that electronic charge is donated from the CO ligand to the metal centre. As CO is also a π-acceptor ligand, a back donation of electronic charge from the metal centre can occur. This donation/back donation interplay results in a strengthening of the metal–carbon bond and a weakening of the carbon–oxygen bond. CO is classified as a σ-donor and π-acceptor molecule.

The chemistry of d-block organometallic chemistry covers a vast amount of material and therefore we will concentrate on the area of so-called metallocenes, which are complexes containing typically a d-block metal and two Cp^- ligands. This area encompasses the most promising drug-like candidates so far.

8.2 What are metallocenes?

Metallocenes are a special class of the so-called sandwich complexes in which the central metal lies between two cyclopentadienyl (Cp^-) ligands. Their chemical formula is typically of the type $(\eta^5\text{-Cp})_2M$, which means each Cp^- ligand forms five bonds with the metal centre (Figure 8.4).

> ***Sandwich complexes** are defined as organometallic compounds containing a central metal atom, which is bound by haptic covalent bonds to two arene ligands. The metal is typically placed between the two ring systems, which gives the complex the term sandwich.*

Typically, sandwich structures are classified as parallel or bent. Within a parallel sandwich structure, the two arene ligands are arranged parallel to each other. In a bent sandwich structure, the two ligands are at an angle to each other.

One of the best known metallocenes is ferrocene $(\eta^5\text{-Cp})_2Fe$, which is an iron (Fe^{2+})-based parallel sandwich complex. The metal centre is sandwiched between two cyclopentadienyl ligands, which are coparallel. There are different ways of representing the structure of ferrocene, as shown in Figure 8.5. The structure on (a) gives the best understanding of the bonds present in ferrocene, whilst the structure on (b) is the most commonly used one. The structure on (c) shows the electronic properties of the individual components in ferrocene.

Ferrocene follows the so-called 18-electron rule, which most low-oxidation-state d-block metal organometallic complexes seem to follow. The number 18 is the effective number of electrons that a transition metal can accommodate in its nine valence orbitals (comprising five d orbitals, three p orbitals and one s orbital).

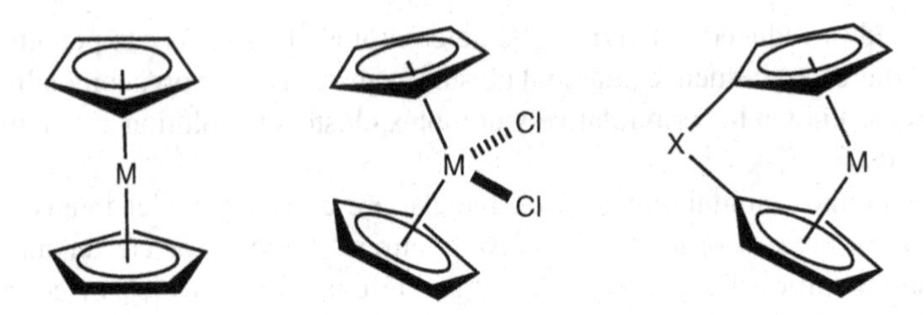

Figure 8.4 Examples for the generic structure of a metallocenes (M = metal)

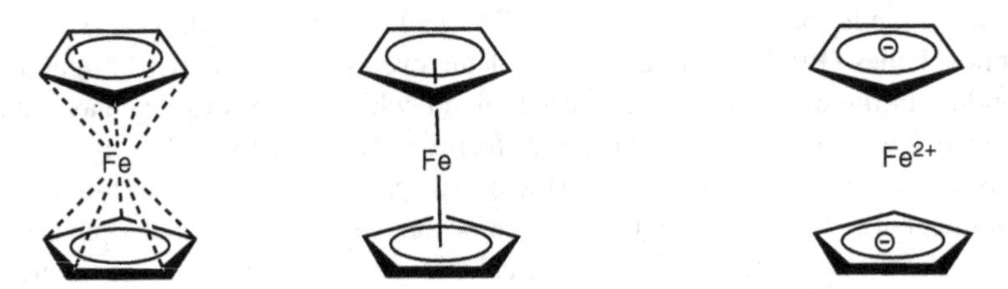

Figure 8.5 (a–c) Different notations for the chemical structure of ferrocene

> *The **18-electron rule** is mainly used to predict the formula for stable transition-metal complexes. The rule is based on the principle that each transition metal has usually nine valence orbitals, one s-type, three p-type and five d-type orbitals. These interact with the ligands and accommodate up to 18 electrons. Transition-metal complexes containing 18 valence electrons (VEs) have the same electronic configuration as the noble gas in this period, similar to the octet rule described for main group metals.*
>
> *Note that there are many examples known that do not follow the 18-electron rule.*

Ferrocene is a good example to explain this rule, and there are two ways of counting the electrons in ferrocene. On one hand, it can be assumed that all partners are neutral, which means that ferrocene consists of an Fe(0) centre and two Cp ligands. That means the Fe(0) centre contributes eight electrons (group 8, eight VEs), whilst each Cp ligand contributes five electrons. This adds up to a total of 18 electrons. On the other hand, the second method of counting is based on the fact that each partner carries a charge, which means an Fe^{2+} centre and two Cp^- ligands. Therefore the iron centre contributes six electrons and each Cp^- ligand also contributes six electrons, which sums up to also 18 electrons (Figure 8.6).

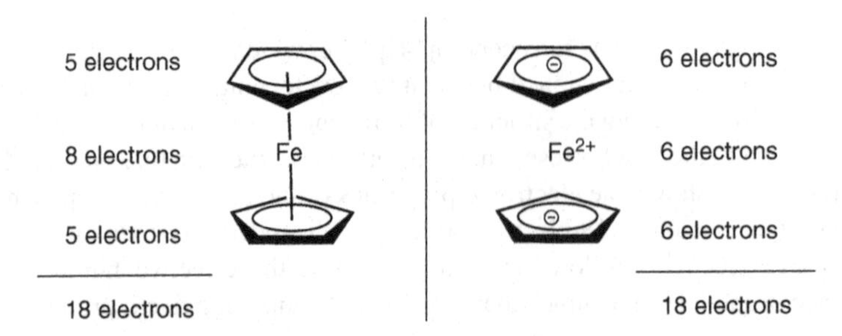

Figure 8.6 The 18-electron rule exemplified on ferrocene

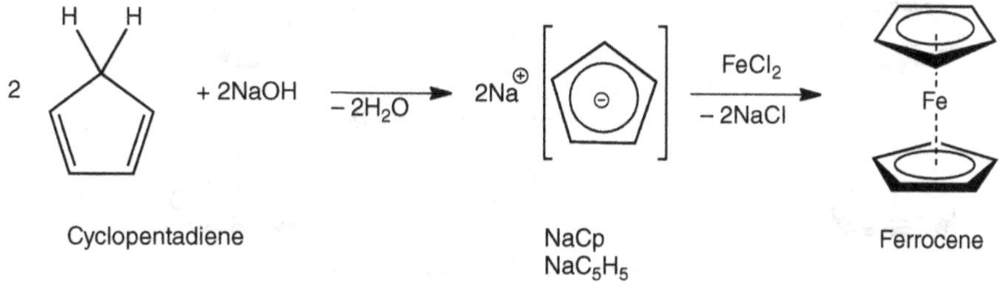

Figure 8.7 Synthesis of ferrocene

It is important to note that this rule can be used to predict stable metal complexes, but it is only a formality. There are many examples of stable transition-metal complexes that do not follow the 18-electron rule, such as Pd(0) or Pt(0) complexes, or indeed some of the titanocenes and vanadocenes presented in the following.

8.3 Ferrocene

Ferrocene (or bis(η^5-cyclopentadienyl)iron, $(C_5H_5)_2Fe$) is an orange powder and is probably one of the best studied metallocenes. As previously mentioned, its structure follows the 18-electron rule and it is a very stable complex. Its Cp^- ligands can be easily derivatised to introduce functional groups. Functionalised ferrocene derivatives are currently used as biosensors in blood glucose measuring equipment and they are also under intense research as potential anticancer agents.

Ferrocene was discovered by Paulson and Kealy in 1951. Cyclopentadienyl magnesium bromide and ferric chloride were reacted in a so-called Grignard reaction (reaction involving R-MgBr) in order to create a fulvalene. Instead, they created ferrocene. At that time, it was difficult to identify the correct structure of ferrocene, but Wilkinson, Rosenblum, Whitting and Woodward managed to do this soon after its discovery [1].

$$2C_5H_5MgBr + FeCl_2 \rightarrow Fe(C_5H_5)_2 + MgBr_2 + MgCl_2$$

Nowadays, ferrocene is synthesised via a so-called transmetallation reaction. Typically, commercially available sodium cyclopentadienide is deprotonated with KOH or NaOH, and the obtained anion is reacted with anhydrous ferrous chloride ($FeCl_2$). Instead of purchased sodium cyclopentadienide, freshly cracked cyclopentadiene is often used (Figure 8.7).

Ferrocene is a very stable complex and can be easily functionalised by derivatising its Cp^- ligands. The Cp^- ligands are aromatic, as previously mentioned, and therefore show a chemical behaviour similar to benzene. This means that reactions known for benzene chemistry can be used with ferrocene, such as the Friedel–Crafts acylation reaction. Ferrocene can be acylated by reacting it with the corresponding aluminium halide (AlX_3). Indeed, this chemical behaviour of ferrocene helped in identifying its real structure (Figure 8.8) [2].

Ferrocene and its derivatives are under intense screening for medicinal purposes. Research has shown that especially ferrocene derivatives exhibit very promising effects for a variety of clinical applications, such as antimalarial and anticancer agents as detailed below. Interestingly enough, ferrocene itself is not a particularly toxic compound, as it can be administered orally, injected or inhaled with no serious health concerns. It is believed to be degraded in the liver by cytochrome P_{450}, similar to benzene. Its degradation process involves the enzymatic hydroxylation of the cyclopentadienyl ligand. Animal studies on beagles have shown that treatment with up to 1 g/kg ferrocene did not result in acute toxicity or death, although it did lead to a severe iron overload, which was reversible [2].

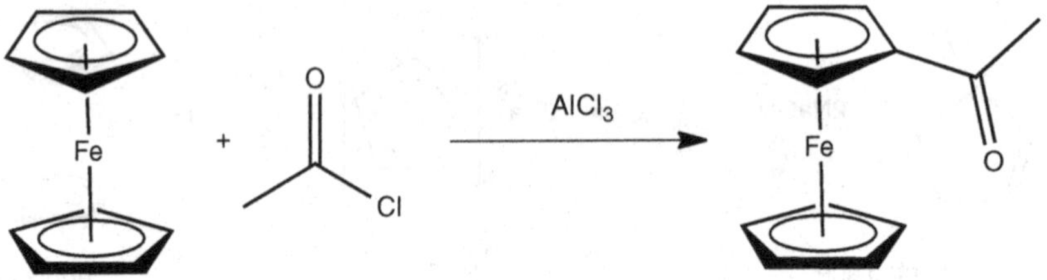

Figure 8.8 Acylation of ferrocene

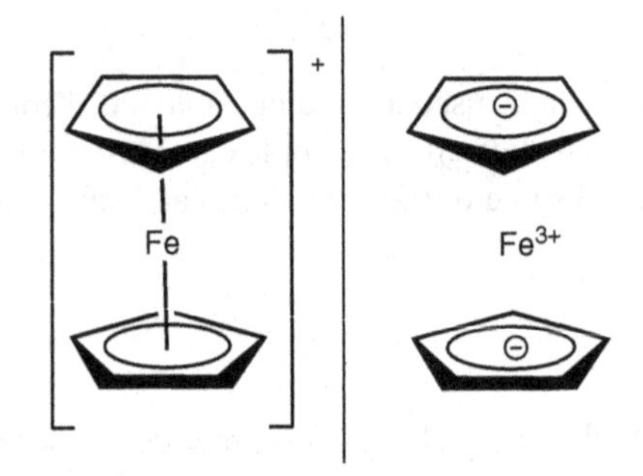

Figure 8.9 Chemical structure of the ferrocenium cation

Ferrocene can easily undergo oxidation to the ferrocenium cation in a one-electron oxidation process. The formed cation is fairly stable, and the whole process is reversible. This redox potential, together with a change in lipophilicity, is the main characteristic that makes ferrocene-based compounds interesting for a variety of potential clinical applications, especially the ones outlined in the following (Figure 8.9) [2].

8.3.1 Ferrocene and its derivatives as biosensors

Diabetes is a major health problem with hundreds of millions sufferers worldwide. As part of the illness, diabetic patients have increased glucose levels in their blood due to a lack of insulin or cells not reacting to insulin. Insulin promotes the uptake of glucose into the cells. There are several options to manage diabetes, but it is extremely crucial for the welfare of the patients that the blood glucose levels are closely monitored. In order to facilitate these regular measurements, a significant amount of research has gone into the development of portable and easy-to-use devices. Modern blood glucose monitors benefit from the technical advances of the so-called biosensor research, an area where the majority of the biosensors are used.

Biosensors are based on enzymes that contain redox-active groups. This means that the redox group can change its redox state as a result of a biochemical reaction. In nature, the enzymes glucose oxidase (GOx) or glucose dehydrogenase (GDH) are used as biosensors for blood glucose monitors. Typically, these enzymes accept electrons from the substrate, glucose in this case, and oxidise it. The enzyme changes to its reduced state, which normally deactivates the enzyme. In order to activate the enzyme again, electrons are transferred and the enzyme is oxidised. GOx and GDH in their reduced form transfer electrons to molecular oxygen, and hydrogen peroxide (H_2O_2) is produced. Oxygen or peroxide electrodes can then be used to measure any change of the substrate, which directly relates to the glucose levels present in the sample. Unfortunately, this method has problems, as, for example, molecular oxygen can be a limiting factor and a lack of oxygen can lead to wrong readings (Figure 8.10).

$$\text{Glucose} + O_2 \xrightarrow{\text{Glucose oxidase}} \text{Gluconolactone} + H_2O_2$$

Figure 8.10 *Enzymatic oxidation of glucose*

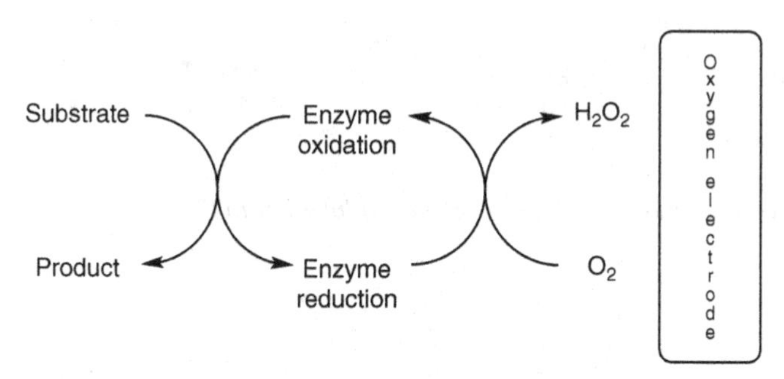

Figure 8.11 *Scheme of a biosensor (Adapted from [3], with permission from Elsevier.)*

Enzymes such as GOx are very specific to the substrate they accept electrons from, that is, the substrates they oxidise, but they are more flexible to the substrate they donate electrons to. Therefore, a variety of inorganic redox-active compounds have been tested as so-called mediators. Mediators function by accepting electrons from the enzyme and thus oxidising the enzyme to its active form. They shuttle electrons from the enzyme to the electrode and are also called *electron sinks*. Electrodes can measure any changes in the redox potential of these mediators, and these changes can directly be related to the amount of glucose present in the sample. This technology excludes the need for molecular oxygen and problems connected to that (Figure 8.11) [3].

In 1984, the first ferrocene-based mediator in conjunction with GOx was used as a biosensor for glucose. Derivatives of ferrocene are still the most important examples for mediators in biosensors, mainly due to their wide range of redox potential, which is independent of any pH changes. Furthermore, the chemistry involved in synthesising these ferrocene derivatives is well explored and fairly straight forward. Additionally, the mediator must successfully compete with the natural mediator (molecular oxygen) in order to ensure accurate readings. From the point of its application as biosensor for blood glucose measurement, it is clear that ferrocene-based mediators can be used only once. This is due to the fact that, whilst ferrocene is relatively insoluble, the reduced form, the ferrocenium ion, is fairly soluble. Mediators should be insoluble in order to lead to reproducible results or, as in this case, can only be used once (Figure 8.12) [3, 4].

$$[Cp_2Fe] \rightarrow [Cp_2Fe]^+ + e^- \qquad (8.1)$$

Equation 8.1: Oxidation of ferrocene to the paramagnetic ferrocenium ion.

Initially, ferrocene itself was used as mediator, but later on a variety of its derivatives were tested for their redox potential in biosensors. Some of these examples are shown below. This research has led to development of the modern blood glucose analysers, which are only the size of a pen and highly mobile devices. These devices use disposable strips and are simple to use (Figure 8.13) [4].

8.3.2 Ferrocene derivatives as potential antimalarial agent

A variety of compounds containing the ferrocene group have been synthesised and tested for their clinical properties, especially as antimalarial and anticancer agents. In this context, especially the ferrocene-based

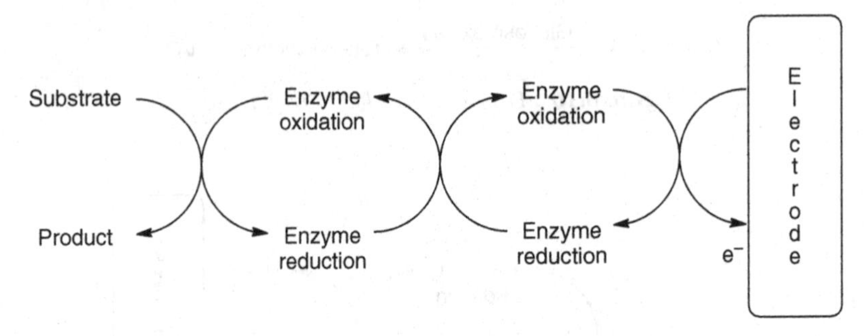

Figure 8.12 *Scheme for mediated biosensor (Reproduced from [3], with permission from Elsevier.)*

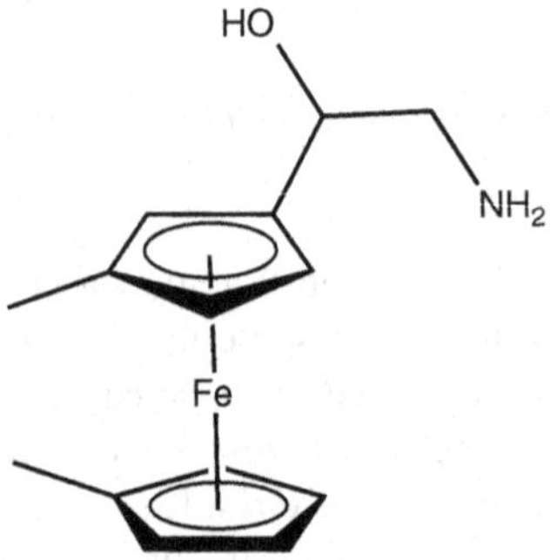

Figure 8.13 *Chemical structure of a simple example of a ferrocene-based biosensor*

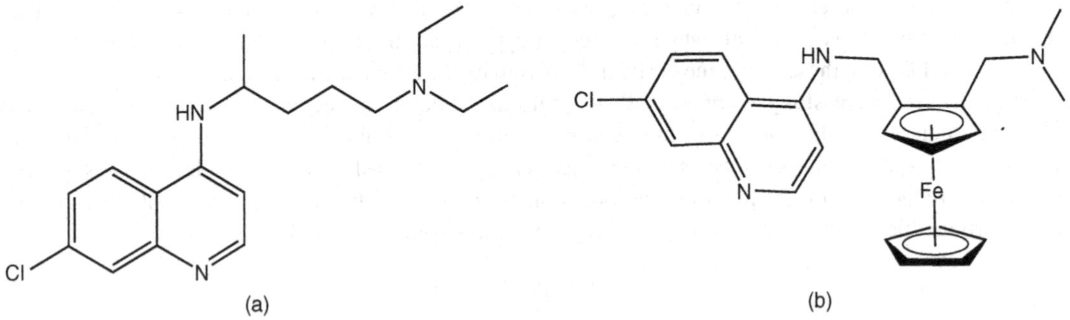

Figure 8.14 *Chemical structures of chloroquinone (a) and ferroquine (b)*

analogue of chloroquine, ferroquine, has shown significant promise. It successfully passed phase II clinical trials and is awaiting results from field testing. Chloroquine is a well-known drug used in the treatment of malaria caused by the parasite *Plasmodium falciparum.* Ferroquine is active against this parasite as well. Even more exciting is the fact that it is also active against the chloroquine-resistant strain of *P. falciparum.* The changed biological activity might be due to the changed lipophilicity and/or the redox action that is present after the introduction of the ferrocene group (Figure 8.14) [2].

8.3.3 Ferrocifen – a new promising agent against breast cancer?

Ferrocene and its derivatives were, and still are, under intense scrutiny as potential anticancer agents. Initially, a range of ferrocenium salts was tested for their cytotoxic activity. The mode of action is still unclear, but DNA, cell membrane and enzymes have been proposed as potential targets. Ferrocenium salts are believed to generate hydroxyl radicals under physiological conditions. These may damage the DNA, possibly by oxidising the DNA. Furthermore, it is believed that the cell membrane might be a target. Research has shown that the counter-ion is crucial for the cytotoxic activity as well as their aqueous solubility. Ferrocenium salts such as the picrate and trichloroacetate derivates display good aqueous solubility and high cytotoxic activity. As part of this research, ferrocene was also successfully bound to polymers in order to improve their water solubility and therefore cytotoxic activity [2].

Jaouen and coworkers substituted phenyl rings in existing drugs and natural products by ferrocene groups in order to introduce a redox-active metal group into these molecules and to change their lipophilicity. A breakthrough was achieved when a phenyl group in tamoxifen, a selective oestrogen receptor modulator (SERM) used as front-line treatment of hormone-dependent breast cancer, was replaced by ferrocene. The active metabolite of tamoxifen is actually the hydroxylated form 4-hydroxytamoxifen, which is highly active in the fight against oestrogen-dependent breast cancer. Breast cancer can be divided into hormone-dependent (also called *oestrogen-dependent*, ER(+)), which is characterised by the presence of an oestrogen receptor, and hormone-independent (also called *oestrogen-independent*, ER(−)) [2, 5].

> ***Selective oestrogen receptor modulators*** *are defined as a class of compounds* that *interact with the oestrogen receptor. This interaction can happen in various tissues leading to different actions.*

The combination of tamoxifen derivatives with ferrocene was a very successful approach, and has led to the development of a class of compounds called *ferrocifens*. Whilst around two-thirds of patients are diagnosed with ER(+) breast cancer and can be treated with hormone therapy such as with SERMs, there is still an urgent need to develop drugs to be used against ER(−) breast cancer. Preclinical studies have shown that ferrocifen is active against the latter type of breast cancer which is not susceptible to the treatment with tamoxifen (Figure 8.15) [2, 5].

The cytotoxic effect of tamoxifen results from the competitive binding to the oestrogen receptor and repressing DNA transcription, which is mediated by oestradiol. It is believed that ferrocifen follows the same mode of action. Research has shown that the replacement of the phenyl group in tamoxifen by ferrocene results in a reduced binding affinity to the receptor (RBA, receptor binding affinity). Variation, and especially increase, of the chain length has a negative effect on the RBA and also on the bioavailability. The optimum chain length seems to be when $n=4$. It is also important to note that the *Z*-isomer binds more strongly to the receptor. Very surprisingly, ferrocifen (with $n=4$) showed also an antiproliferative effect when tested on the oestrogen-independent cell line MDA-MB231, which does not have oestrogen receptors and is not accessible for treatment with tamoxifen. This means that there must be an additional mode of action that is independent of the oestrogen receptor.

Replacing the ferrocenyl group by a ruthenium group resulted in a drop of cytotoxic activity, indicating that the iron group is important. It has been proposed that the additional mode of action of ferrocifen could rely on the redox activation of the ferrocenyl group and the presence of reactive oxygen species (ROS) [2].

These extremely promising results stimulated further research in this area. Tamoxifen was coupled to a variety of known metal-based compounds with potential anticancer activity, such as oxaliplatin, titanocene dichloride and others. Oxaliplatin contains the so-called DACH–Pt group (DACH, 1,2-diaminocyclohexane),

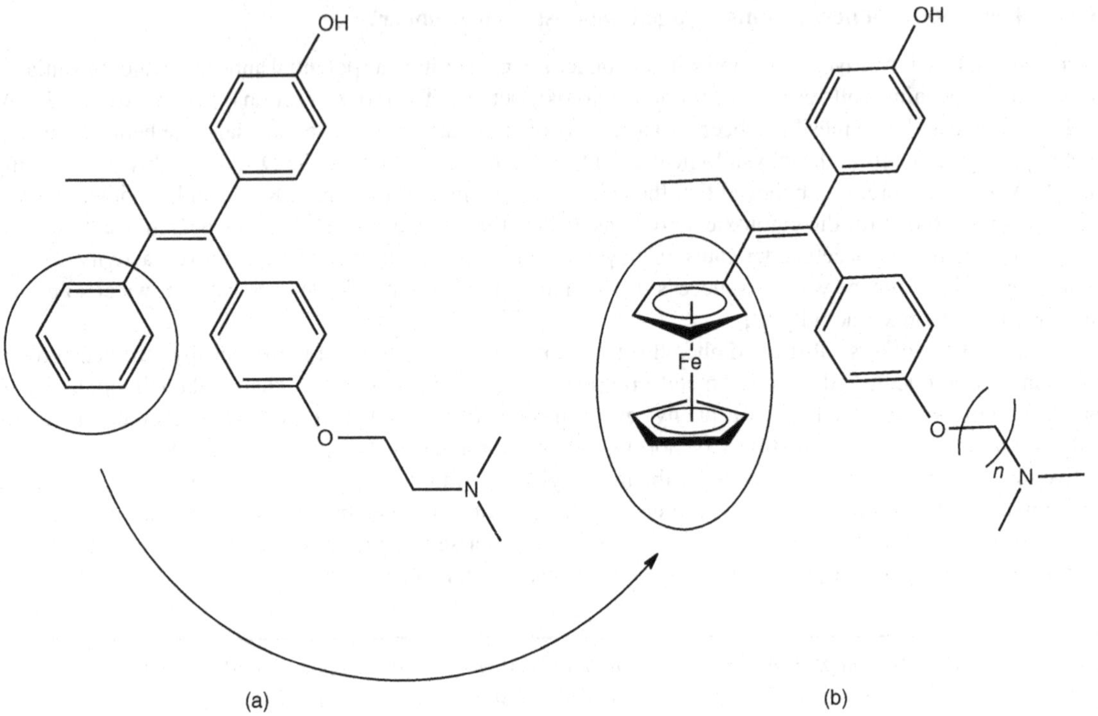

OH OH Fe O N O n N

(a) (b)

Figure 8.15 Chemical structure of 4-hydroxytamoxifen (a) and ferrocifen (b)

which has been used as a basis for a new DACH–Pt–tamoxifen derivative. Oxaliplatin showed a cytotoxic effect of 6.3 μM when tested on the oestrogen-dependent human breast cancer cell line MCF-7, whilst the tamoxifen-vectorised derivatives (see Figure 8.16; R = H, 14 μM and R = OH, 4 μM) also presented an antiproliferative effect at a similar magnitude. Looking in more detail, research shows that the derivative that contains the hydroxyl group displays a higher RBA and also a better IC_{50} value. This shows that the hydroxyl group (also present in the active metabolite of tamoxifen) is important for the recognition by the oestrogen receptor. Nevertheless, the vectorisation of DACH–Pt does not really result in a significant improvement in comparison to oxaliplatin itself, and therefore this combination is not really beneficial as an SERM for the fight against breast cancer [5].

The titanocene-tamoxifen derivative has an RBA value of 8.5%, which means it should recognise the oestrogen receptor well. The results of the cytotoxicity tests were very unexpected, where a proliferative effect was observed. The estrogenic effect was comparable to that of oestrogen itself. It is believed that this estrogenic effect is due to the titanium moiety and/or its hydrolysis products (Figure 8.17) [5].

In order to bring ferrocifen into clinical studies, it was important to find a suitable formulation. This is an area notoriously difficult for metal-based drugs, and OH-ferrocifen finally entered phase II clinical trials. A variety of pharmaceutical approaches have been researched, including the use of nanoparticles, cyclodextrins and lipid nanocapsules [2].

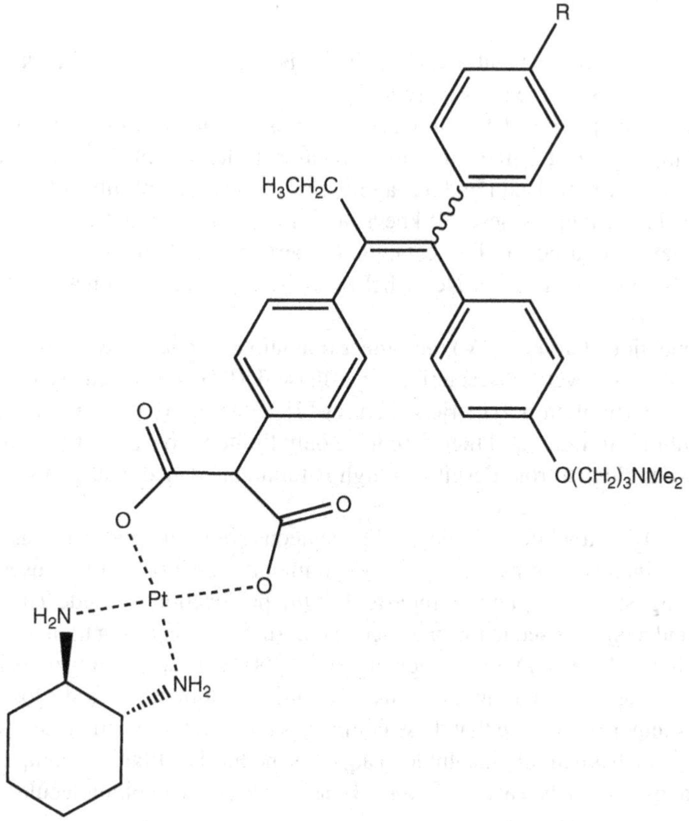

Figure 8.16 *Chemical structure of the DACH–Pt–tamoxifen derivative (R = H, OH)*

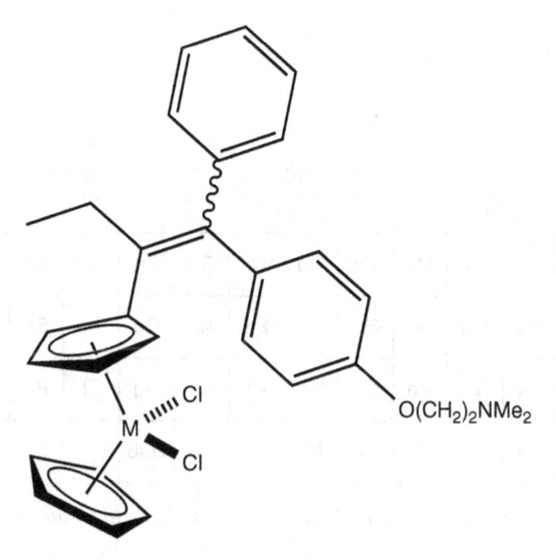

Figure 8.17 *Chemical structure of titanocifen*

8.4 Titanocenes

Titanium, with chemical symbol Ti and atomic number 22, is a member of the d-block metals and belongs to group 4 of the periodic table of elements (Figure 8.18).

Titanium is a metal, which puzzles the bioinorganic scientists, as it has no known native role in the human body (or any organism) despite it is the ninth most abundant element in the earth's crust [6]. The metal titanium is extremely corrosion-resistant and has a remarkable biocompatibility, which has led to its clinical applications. These include endoprostheses of knees and hips, dental implants, heart pacemakers and many more. Titanium complexes are used for their catalytic properties, and there are complexes that interact with biomolecules (this will be shown below). Nevertheless, there is no known function of titanium as a native essential metal [6, 7].

The most stable oxidation state is Ti(IV), and most titanium compounds with lower oxidation states are easily oxidised to Ti(IV). It is very characteristic that these Ti(IV) species easily undergo hydrolysis and form Ti—O bonds and fairly often polymeric structures [7]. Titanium oxides and many Ti(IV) compounds do not show high solubility in water and therefore have only limited bioavailability. This might explain why titanium has no known biological role despite its high natural abundance, and is not an essential metal for any life form [6].

Titanium dioxide (TiO_2) is a white solid with a low aqueous solubility and is the natural occurring oxide of titanium present in minerals such as rutile (composed mainly of TiO_2). TiO_2 is used as a white pigment in paints, food colouring, sunscreens and toothpaste. For the purification of crude TiO_2, the raw material is reduced with carbon and then oxidised in the presence of chlorine. The resulting titanium tetrachloride ($TiCl_4$) is purified by distillation and reacted with oxygen at around 2000 K to form TiO_2 in its pure form.

Current research into the medicinal applications of titanium complexes mainly focusses on their use as anticancer agents. It is important to note that these complexes are different from the above-described titanium alloys/metals or TiO_2, which are highly insoluble in aqueous media. The titanium complexes described below contain functional groups that can be easily exchanged and/or interact with biomolecules in order to fulfil their medicinal role.

H																	He
Li	Be											B	C	N	O	F	Ne
Na	Mg											Al	Si	P	S	Cl	Ar
K	Ca	Sc	**Ti**	V	Cr	Mn	Fe	Co	Ni	Cu	Zn	Ga	Ge	As	Se	Br	Kr
Rb	Sr	Y	Zr	Nb	Mo	Tc	Ru	Rh	Pd	Ag	Cd	In	Sn	Sb	Te	I	Xe
Cs	Ba	La-Lu	Hf	Ta	W	Re	Os	Ir	Pt	Au	Hg	Tl	Pb	Bi	Po	At	Rn
Fr	Ra	Ac-Lr	Rf	Db	Sg	Bh	Hs	Mt	Ds	Rg	Uub						

Figure 8.18 Periodic table of elements; the element titanium is highlighted

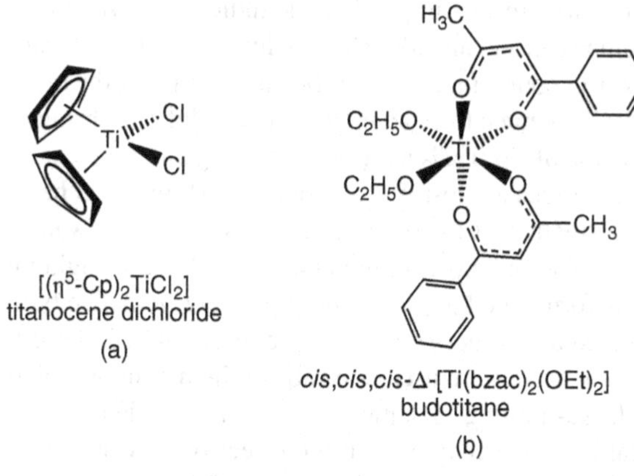

$[(\eta^5\text{-Cp})_2TiCl_2]$
titanocene dichloride
(a)

cis,cis,cis-Δ-$[Ti(bzac)_2(OEt)_2]$
budotitane
(b)

Figure 8.19 *Chemical structures of Cp_2TiCl_2 (a) and budotitane (b) [9]*

8.4.1 History of titanium-based anticancer agents: titanocene dichloride and budotitane

The success of cisplatin stimulated the search for further metal-based anticancer drugs. Two titanium-based complexes, namely titanocene dichloride (Cp_2TiCl_2) and budotitane ($Ti(IV)(bzac)_2(OEt)_2$, bzac, 1-phenylbutane-1,3-dionate), were identified as promising drug candidates and entered clinical trials [7]. Both compounds contain labile groups in the cis position analogous to cisplatin. Unfortunately, hydrolysis in titanium compounds is faster than in cisplatin, and in these cases can lead to the formation of hydroxo-bridge species and potentially to TiO_2, which is insoluble in water. This represents the main challenge for the clinical application of this class of compounds (Figure 8.19) [8].

Cp_2TiCl_2 was first synthesised in 1954 and is a red crystalline solid in its pure form. The Ti(IV) centre is coordinated by two η^5-Cp ligands and two chlorine ligands. Cp_2TiCl_2 can be synthesised by reacting titanium tetrachloride ($TiCl_4$) with NaCp [1]. Compared to those of ferrocene, the planar cyclopentadienyl ligands are coordinated to the titanium centre in a bent sandwich configuration [7]. Furthermore, Cp_2TiCl_2 is an exception to the 18 VEs rule, as it has only 16 VEs, but it is still a stable complex. Each Cp^- ligand contributes six electrons, whilst each Cl^- ligand contributes two electrons. Ti(IV) has no VEs left, which makes a total of 16 VEs.

$$2NaC_5H_5 + TiCl_4 \rightarrow (C_5H_5)_2TiCl_2 + 2NaCl)$$

Compared to cisplatin, much less is known on the mode of action of titanium-based drug candidates. In fact, the active species responsible for the anticancer activity of Cp_2TiCl_2 *in vivo* has yet not been identified, nor has the coordination mechanism to DNA or eventual repair processes. Studies have shown that titanium will accumulate near phosphorus-rich areas, indicating a titanium–DNA interaction takes place. Crystallographic studies suggest that, within Cp_2TiCl_2, the Cp groups sterically prevent the cross-linking of DNA. A direct comparison between Cp_2TiCl_2 and cisplatin shows that their spectrum of activity differs, leading to the conclusion that their mode of action must also differ [7].

Current studies have confirmed the uptake of Ti(IV) by transferrin and transport to the cancer cells. Within the cell, it is believed that Ti(IV) may bind, in contrast to cisplatin, to the DNA backbone, which means coordinating to the negatively charged phosphates. Generally, it is believed that this coordination is not very strong [7].

Cp_2TiCl_2 showed very promising results in preclinical studies. *In vitro* studies have been carried out by Köpf-Maier on a variety of human cancers including human lung and renal cell cancer carcinomas xenografted into mice. The results were very encouraging, as a significant response of the tumour to the chemotherapeutic agent was observed and Cp_2TiCl_2 seemed to be an interesting candidate to bring into the clinic [7]. The major challenge for a clinical application of Cp_2TiCl_2 as an anticancer agent is presented by its hydrolytic instability at pH below 5.5, where both chlorides are lost. Even at neutral pH, research has shown that the Cp rings are lost quickly, resulting in the formation of hydrolysed products which are water insoluble [6]. Therefore, it was necessary to develop a formulation before this compound could be brought into clinical trials. For its use in clinical trials, Cp_2TiCl_2 was formulated as a lyophilised powder in malate buffer (pH = 3.2) or malic acid (pH = 3.5). Cp_2TiCl_2, formulated as a water-soluble powder, has entered phase I and phase II clinical trials [7]. Results from phase I clinical trials were used to establish the maximum tolerable dose and to define the pharmacokinetic properties. Dose-limiting toxicities have been identified, as reversible damage to the liver and kidneys. Phase II clinical trials concentrated on the effect of Cp_2TiCl_2 treatment of renal cell cancer and metastatic breast cancer. Unfortunately, the encouraging effect seen in preclinical studies could not be repeated in the clinical studies. The tumours did not show a significant response to Cp_2TiCl_2, and therefore clinical trials were discontinued [6].

Budotitane is an octahedral Ti(IV) complex that was developed as a potential anticancer agent around the same time as the anticancer properties of Cp_2TiCl_2 were discovered. Budotitane can be synthesised by reacting $TiCl_4$ with the appropriate diketonate in an anhydrous solvent. Important structural features are the unsubstituted aromatic rings, which positively contribute to the cytotoxic activity, and the hydrolysable group. The latter group seems to be less important in regard to the cytotoxicity of budotitane but crucial for formulation purposes as it seems to determine the aqueous stability. Research has shown that the aqueous stability increases with the hydrolysable group in the order: $I^- < Br^- < Cl^- < F^- < OR^-$ (Figure 8.20) [7].

Another major drawback for the clinical use of budotitane is the presence of isomers. There is the possibility of a cis or trans arrangement of the —OC_2H_5 group, with the cis arrangement believed to be the more stable one. Furthermore, the β-diketonato ligand is not symmetrical. This means that there are eight possible arrangements in this octahedral scheme. In solution, there is a mixture of isomers and it is difficult to isolate the isomers. So far, it is not known which isomer exhibits the anticancer activity (Figure 8.21) [7].

Preclinical studies using human xenografts in nude mice highlighted the potential of budotitane as novel chemotherapeutic agent in 1984. Budotitane showed cytotoxic activity comparable to cisplatin. Keppler *et al.* intensively studied its antitumour activity in a variety of animal models and found specific activity against Ehrlich ascetic tumour and colon cancer. Budotitane, similar to Cp_2TiCl_2, is prone to hydrolysis in conjunction with a low aqueous solubility. Therefore, significant research went into the development of an appropriate formulation [7]. Several clinical studies were performed using cremophor EL as a nonionic surfactant, which has been used as a vehicle for the solubilisation of a wide variety of hydrophobic drugs (cremophor). Furthermore, 1,2-polyethyleneglycol has been added to encourage the coprecipitation of the drug. All three ingredients (budotitane, cremophor EL and 1,2-polyethyleneglycol) are dissolved in anhydrous ethanol and mixed, and the solvent is then evaporated off. Using this procedure, micelles containing 100–200 mg/100 ml of active pharmaceutical ingredient are formed, which are stable for a few hours. Budotitane went through several phase I and phase II clinical trials. Budotitane was administered as intravenous (IV) infusion twice a week to patients with solid tumours refractory to previous treatments. Cardiac arrhythmia was identified as the dose-limiting

$$2\text{Hbzac} + \text{TiCl}_4 \rightarrow [\text{Ti(bzac)}_2(\text{OEt})_2] + 4\text{HCl}$$

Figure 8.20 *Synthesis of budotitane*

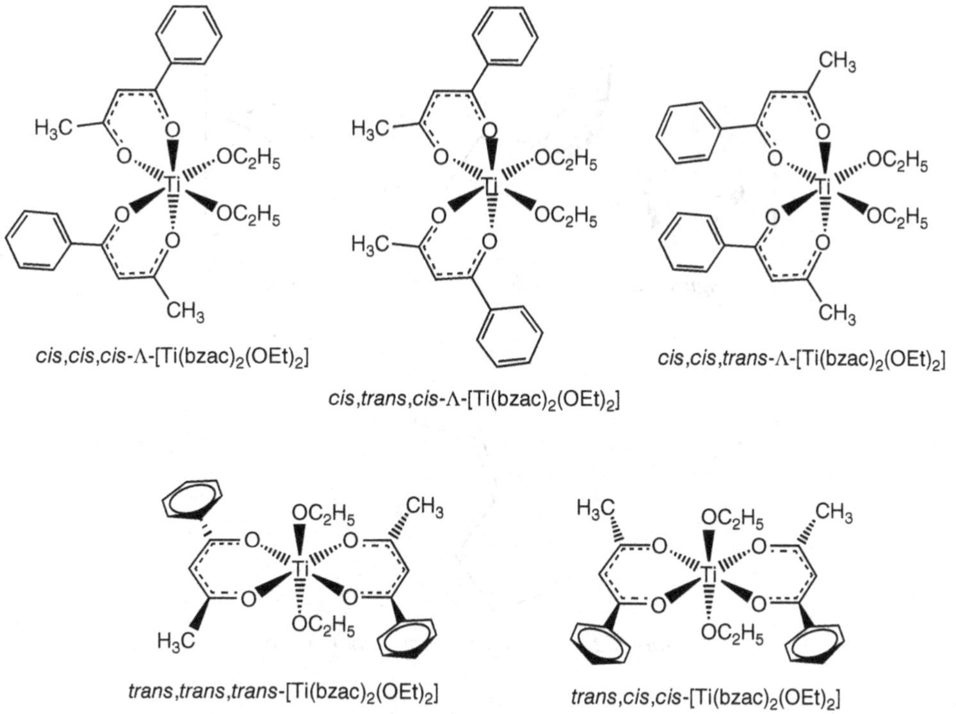

cis,cis,cis-Λ-[Ti(bzac)$_2$(OEt)$_2$]

cis,trans,cis-Λ-[Ti(bzac)$_2$(OEt)$_2$]

cis,cis,trans-Λ-[Ti(bzac)$_2$(OEt)$_2$]

trans,trans,trans-[Ti(bzac)$_2$(OEt)$_2$]

trans,cis,cis-[Ti(bzac)$_2$(OEt)$_2$]

Figure 8.21 *Isomers of budotitane [9]*

side effect. Unfortunately, problems with the formulation, such as the existence of isomers and the difficulty in analysing and characterising the loaded micelles, led to a discontinuation of the clinical trials [7].

Poor water solubility and fast hydrolysis are the main problems for the use of titanocene or titanium-based compounds as anticancer agents. Nevertheless, this intensive research in the 1980s and 1990s has encouraged further research as described in the following.

8.4.2 Further developments of titanocenes as potential anticancer agents

As mentioned, poor aqueous solubility and instability in water are the main problems that restrict titanocenes in clinical applications. McGowan and coworkers renewed the interest in research in this area with their elegant synthesis of ring-substituted cationic titanocene dichloride derivatives. The idea was based on the introduction of charges in order to improve the aqueous solubility. Indeed, the resulting cationic titanocenes are more water-soluble and show a significant cytotoxic activity especially against cisplatin-resistant ovarian cancer (Figure 8.22) [10].

Following on from these results, Tacke and coworkers based their research on novel synthetic methods starting from substituted fulvenes. A fulvenes is an organic molecule with the chemical formula C_6H_6, which consists of a five-membered ring system and can be used to easily introduce functional groups into titanocenes (Figure 8.23) [11].

Reaction of substituted fulvenes with titanium dihalides via a reductive dimerisation process leads to the so-called *ansa*-titanocenes, which are normally used as catalysts. These *ansa*-titanocenes are characterised by a carbon–carbon bridge between the cyclopentadienyl (Cp) rings, which restricts the geometry of the Cp

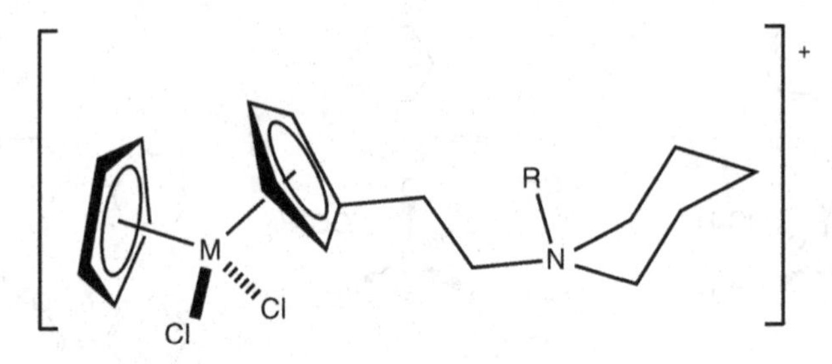

Figure 8.22 Examples of substituted titanocenes

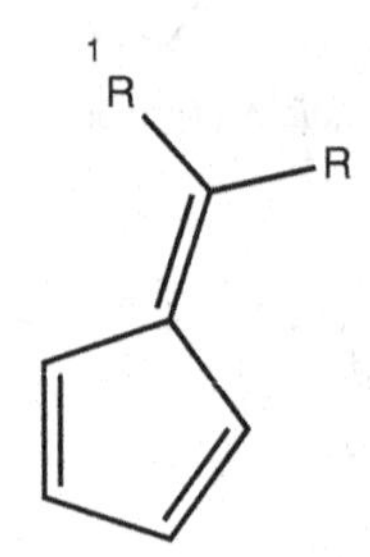

Figure 8.23 Chemical structure of fulvene

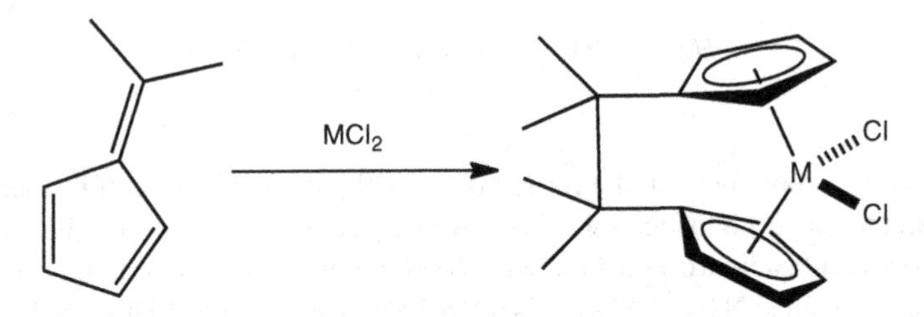

Figure 8.24 Synthesis of ansa-titanocenes using reductive dimerisation of fulvenes

rings. *In vitro* testing showed that they exhibit a moderate cytotoxic activity, which is similar or better than Cp_2TiCl_2 when tested against a model of renal cell cancer (Figure 8.24) [11].

The reaction of substituted fulvenes with superhydride ($LiBEt_3H$) in a so-called hydridolithiation reaction or the reaction with an organolithium compound in a carbolithiation reaction followed by transmetallation with titanium tetrachloride allows access to a variety of substituted titanocenes. Hydridolithiation allows access to benzyl-bridged titanocenes, whilst titanocenes obtained via carbolithiation typically contain more functional groups. This can be of advantage, for example, for the introduction of groups that can be easily ionised in order to improve the water solubility of these titanocenes, but can also be a disadvantage as additional stereocentres are potentially introduced (Figures 8.25 and 8.26).

These titanocenes have been intensively studied with regard to their potential anticancer activity. It has been shown that some of these compounds are active against a variety of human cancer types. The so-called titanocene Y has been the most intensively studied titanocene of this series. It has shown good activity against

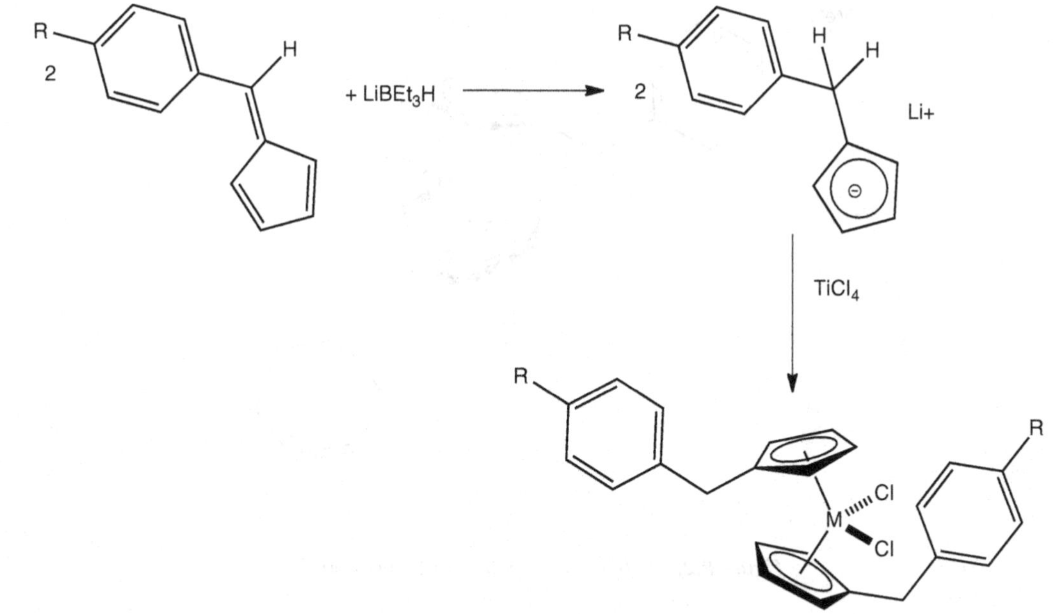

Figure 8.25 Synthesis of benzyl-substituted titanocenes using hydridolithiation (Adapted from [11].)

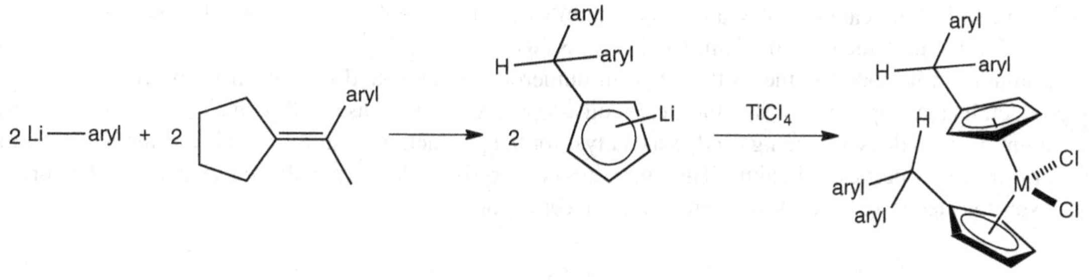

Figure 8.26 Synthesis of substituted titanocenes using carbolithiation

a model of renal cell cancer as well as other human cancer cell lines in a variety of *in vitro* experiments (Figure 8.27).

In general, titanocenes obtained via these methods reached IC_{50} values in the low micromolar range when tested against a renal cell cancer model. This represents an up to 2000-fold improvement compared to Cp_2TiCl_2. Nevertheless, the main problems are the potential presence of stereocentres and still a poor aqueous solubility. Further research has been carried out with the aim to replace the chloride ligands by other groups. This includes the incorporation of chelating ligands similar to the research undertaken for cisplatin. Nevertheless, this research did not result in any major improvement with regard to the cytotoxic activity. Also, some initial formulation studies have been carried out with only limited success. The mode of action of these titanocenes is so far not clear. It is believed that they coordinate to DNA via its phosphate backbone. It has also been shown that they can use transferrin as a transporter molecule into the cancer cell [11].

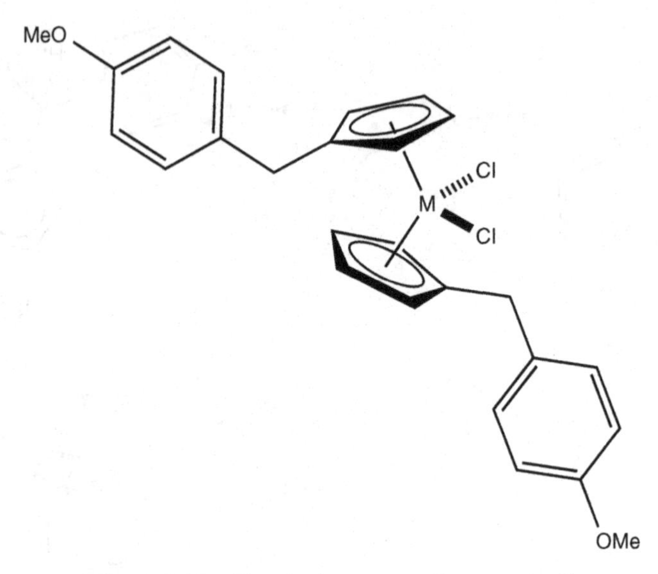

Figure 8.27 Chemical structure of titanocene Y

8.5 Vanadocenes

Vanadium, with chemical symbol V and atomic number 23, is a member of the d-block metals and belongs to group 5 of the periodic table of elements (Figure 8.28).

Vanadium can be found in the earth's crust in numerous minerals and is isolated from ores mostly as a by-product. Its main application is in the steel industry, where it is used as an alloy in combination with iron. Vanadium pentaoxide is also being used as a catalyst for the production of sulfuric acid. The metal vanadium has very similar properties to titanium. Therefore, it is not surprising that its metallocene, vanadium dichloride, was also subjected to research as a potential anticancer agent.

H																	He
Li	Be											B	C	N	O	F	Ne
Na	Mg											Al	Si	P	S	Cl	Ar
K	Ca	Sc	Ti	**V**	Cr	Mn	Fe	Co	Ni	Cu	Zn	Ga	Ge	As	Se	Br	Kr
Rb	Sr	Y	Zr	Nb	Mo	Tc	Ru	Rh	Pd	Ag	Cd	In	Sn	Sb	Te	I	Xe
Cs	Ba	La-Lu	Hf	Ta	W	Re	Os	Ir	Pt	Au	Hg	Tl	Pb	Bi	Po	At	Rn
Fr	Ra	Ac-Lr	Rf	Db	Sg	Bh	Hs	Mt	Ds	Rg	Uub						

Figure 8.28 Periodic table of elements; the element vanadium is highlighted

$[VO_4]^{3-} \rightarrow [VO_3 \times OH]^{2-} \rightarrow \cdots \rightarrow [V_3O_9]^{3-} \rightarrow V_2O_5 \times (H_2O)_n \rightarrow [V_{10}O_{28}]^{6-} \rightarrow [VO_2]^+$

pH 12 pH 9 pH 6.5 pH 2.2 pH < 1

Figure 8.29 *Vanadium oxide formation depending on pH [12]*

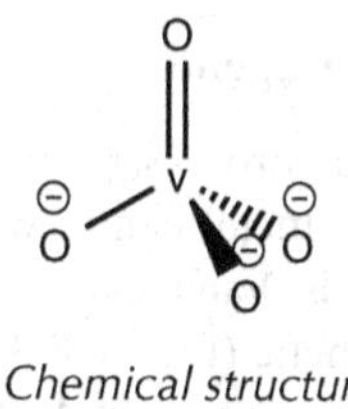

Figure 8.30 *Chemical structure of vanadate*

Vanadium is easily passivated by an oxide film, and the metal is insoluble in nonoxidising acids. Typical oxidation states are +II, +III, +IV and +V, whilst the biologically active oxidation states are +IV and +V. Vanadium reacts to vanadium halide by reacting the metal with the corresponding halogen under heating, whilst it also reacts with oxygen with the formation of V_2O_5. Vanadium (+V) oxides are amphoteric and, as a result, vanadates (VO_4^{3-}) and dioxovanadium ions (VO_2^+) are formed in aqueous solutions depending on the pH (Figure 8.29) [4].

Vanadium is an essential trace metal in the human body, but still very little is known about its biological function. Vanadium is mainly found in its ionic state bound to proteins. As mentioned, the metal mostly occupies oxidation states +V and +IV in biological systems, resulting in electron configurations of [Ar]$3d^0$ for V^{+5} and [Ar]$3d^1$ for V^{+4}. The chemical formula for the tetrahedral ion vanadate is written as VO_4^{3+}; whereas the diatomic oxovanadium(+IV) ion, also called *vanadyl*, has the chemical formula VO^{2+} (Figure 8.30).

Vanadium compounds are well known for their toxicity. The most famous example is the poisonous mushroom toadstool, *Amanita muscaria. A. muscaria* contains the toxic compound amavadin, which is a toxic octahedral vanadium complex (see Figure 8.31) [13].

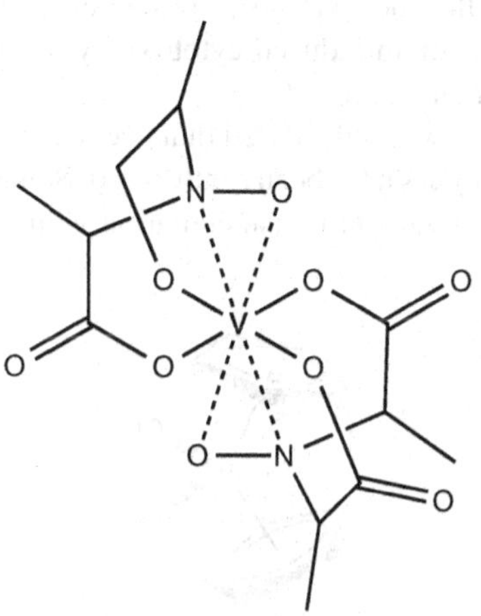

Figure 8.31 *Chemical structure of amavadin*

Vanadate and vanadyl are known to cause adverse effects in mammals, including loss of body weight, gastrointestinal problems, reproductive toxicity and morbidity. However, their toxicity depends on a variety of factors such as the chemical form, oxidation state, route of administration and duration of exposure. Nevertheless, toxic effects of vanadate or vanadyl are observed only at dose levels significantly greater than usual uptake through diet [14]. Nevertheless, it is important to improve the understanding of the adverse and toxic effects of vanadium compounds before any compound can be successfully developed for clinical use.

8.5.1 Vanadocene dichloride as anticancer agents

Vanadocene dichloride [$(\eta^5\text{-}C_5H_5)_2VCl_2$, dichloro bis($\eta^5$-cyclopentadienyl)vanadium(IV)] is structurally very similar to Cp_2TiCl_2. It also consists of a metal centre with an oxidation number of +IV, in this case vanadium, and two Cp^- and two chloride ligands. Vanadocene dichloride is a 17-electron complex containing an unpaired electron and is therefore paramagnetic (Figure 8.32).

Vanadocene dichloride has found application as a catalyst for polymerisation reactions, but was also intensively studied as an anticancer agent in parallel to Cp_2TiCl_2 because of their structural similarities. Vanadocene dichloride has proven to be even more effective than its titanium analogue as an antiproliferative agent against both animal and human cell lines in preclinical testing. The main problems are the difficult characterisation of the active vanadium compounds and their fast hydrolysis. Because of their paramagnetic character, it is difficult to apply standard classical analysis techniques such as NMR (nuclear magnetic resonance) to identify the antiproliferative vanadium species. Furthermore, vanadocene dichloride undergoes fast hydrolytic processes and is even more prone to hydrolysis than titanocene dichloride. This poses even more challenges for its potential clinical application [15].

In recent years, researchers have shown renewed interest in the use of substituted vanadocene dichlorides as potential anticancer agents. A selection of substituted vanadocenes have been synthesised and tested for their cytotoxic activity against testicular cancer. Examples of these compounds include vanadocenes containing substituted cyclopentadienyl ligands and/or replacement groups for the chloride ligands – similar to the research being undertaken for cisplatin analogues. Results of *in vitro* studies show that these compounds exhibit good but variable cytotoxic activity depending on the substitution pattern and induce apoptosis (cell-induced cell death). Interestingly, only organometallic vanadium(+IV) complexes showed cytotoxic activity against testicular cancer. When the purely inorganic compound vanadyl(IV) sulfate was tested in the same study, no cytotoxic effect was observed at the same concentrations. It is also important to note that titanocene dichloride and other metallocenes had no cytotoxic effect against testicular cancer. It was concluded that the mode of action of vanadium-induced cytotoxicity must be different from that of titanocene dichloride and other metallocenes (Figure 8.33) [16].

In parallel to the research undertaken with substituted titanocene dichlorides as potential chemotherapeutic agents, some of their vanadocene analogues have been synthesised. Some examples include the hydrolithiation of fulvenes (see Section 8.4.2) and subsequent transmetallation with vanadium tetrachloride. The resulting

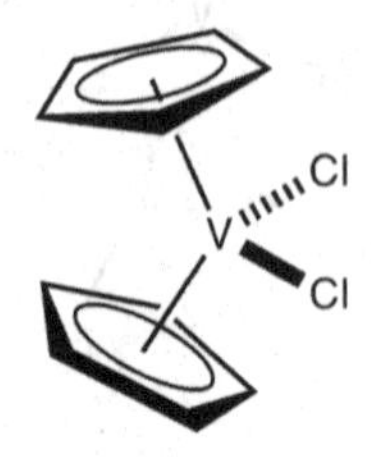

Figure 8.32 *Chemical structure of vanadocene dichloride*

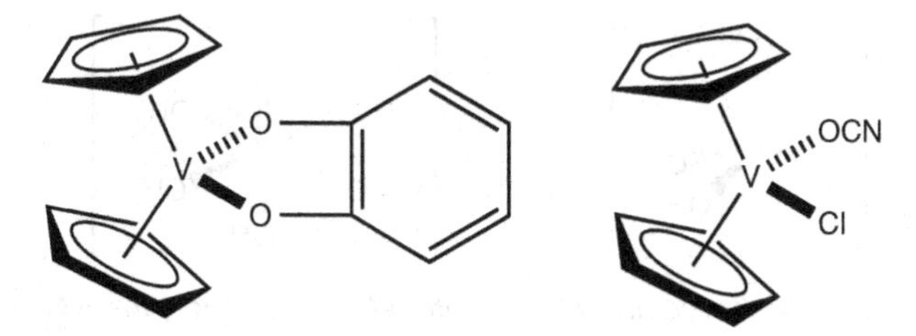

Figure 8.33 Examples of substituted vanadocenes dichlorides

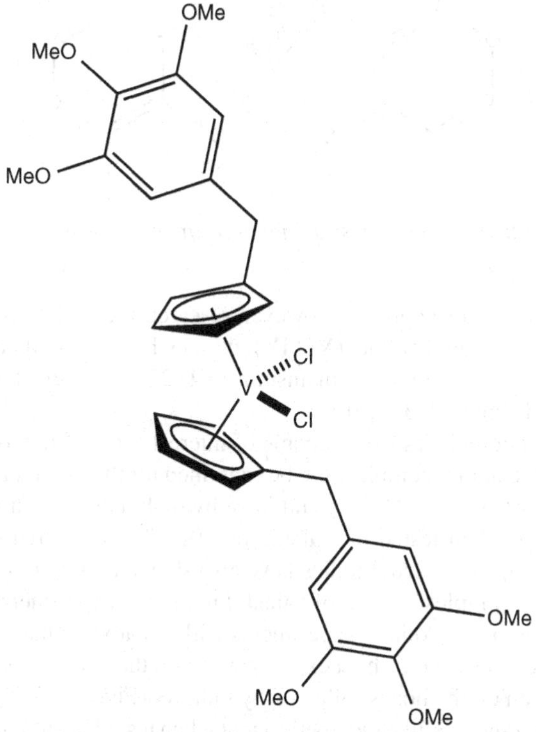

Figure 8.34 Example of a substituted vanadocenes synthesised via transmetallation

substituted vanadocene dichlorides were found to be highly toxic compounds when tested *in vitro* against a model of renal cell cancer and more potent than the corresponding titanocene. Further preclinical studies are still needed (Figure 8.34) [15].

8.5.2 Further vanadium-based drugs: insulin mimetics

Towards the end of the nineteenth century, inorganic vanadium compounds were under evaluation as potential treatment options for Diabetes Mellitus (DM) as so-called insulin mimetics. Sodium vanadate ($Na_3V(+V)O_4$) was tested for its ability to lower glucose levels in the blood of candidates with and without DM. The inorganic vanadium compound showed mild effects in some of the patients suffering from DM, whilst no severe

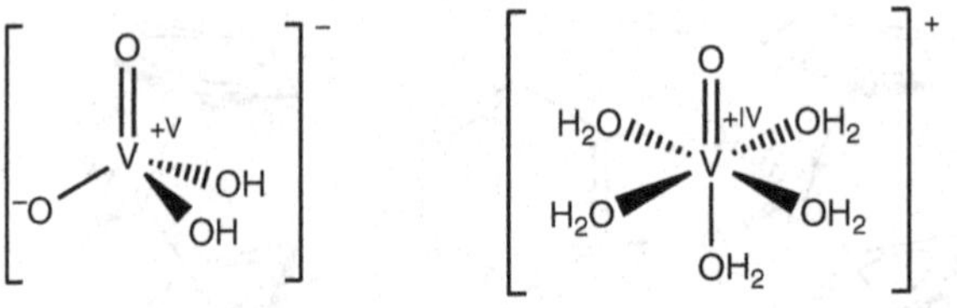

Figure 8.35 Chemical structures of vanadate, VO_4^{3-}, (a) and vanadyl oxycation (b)

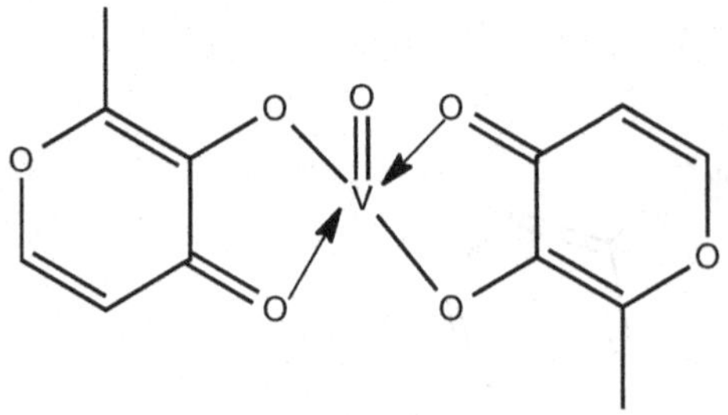

Figure 8.36 Chemical structure of bis(maltolato)oxovanadium

side effects were reported for the dose applied. However, research focused more on the less toxic inorganic vanadium compounds, such as vanadyl sulfate ($V(+IV)OSO_4$) which is significantly less toxic than sodium vanadate. Nevertheless, with the development of insulin in 1922, the interest in vanadium compounds as antidiabetic drugs diminished (Figure 8.35) [9].

In more recent years, metal complexes have become of interest for a variety of clinical applications. This also renewed the interest for vanadium complexes to be examined for the treatment of diabetes. The vanadium complexes bis(maltolato)oxovanadium (BMOV) and bis(ethylmaltolato)oxovanadium (BEOV) have shown to be unique insulin mimetics when tested in diabetic rats [9, 17]. An increase in uptake and tolerability compared to the inorganic form was noted. Studies have also shown that there is a difference in distribution between the inorganic and the complexed form of vanadyl in *in vivo* experiments, which might relate to the differences in uptake and tolerability. Animal experiments with vanadyl sulfate have shown accumulation of vanadium mainly in the kidneys and liver, whilst experiments with the vanadyl complexes BMOV and BEOV resulted in a high accumulation on the bones followed by kidneys (Figure 8.36) [14, 18].

BMOV has proven itself as a successful antidiabetic agent when tested in animal models. Nevertheless, only very little is known about its mode of action. It is believed that BMOV acts as a competitive and reversible inhibitor of the enzyme protein tyrosine phosphatase (PTP). Other vanadium complexes are also known to inhibit PTP, but mostly inhibiting it irreversibly [17a].

PTPs belong a family of enzymes that remove phosphate groups from phosphorylated tyrosine residues on proteins. Its member protein tyrosine phosphatase 1B (PTP1B), which is located in the cytosol, has been identified as a negative regulator of insulin signal transduction. Resistance to insulin can be observed in different tissues such as muscles, liver and fat, which are all crucial for the homeostasis of glucose levels in the human body. In the healthy human body, the transport of glucose into the cell occurs through the activation of the insulin receptor including the phosphorylation of the tyrosine residue. As a result, the so-called insulin receptor substrate (IRS) is recruited, followed by the activation of several enzymes. Finally, the glucose transporter GLUT4 is translocated, which mediates the transport of glucose into the cell.

PTP1B seems to be a key regulator for the activity of the insulin receptor, including all downstream signalling processes [19]. It works by the dephosphorylation of the phosphotyrosine residues at the activated insulin receptor kinase and therefore ultimately hinders the uptake of glucose. PTP1B has been identified as a promising target for new drugs treating DM Type 2. Blocking the PTP1B-mediated dephosphorylation of insulin receptor kinase by an inhibitor of PTB1B is believed to lead to an increase in insulin sensitivity.

As previously mentioned, BMOV is believed to be a potent and competitive inhibitor of the PTP1B activity and additionally seem to support the autophosphorylation of the insulin receptor leading to an increased sensitivity towards insulin. Research has shown that varying the organic ligand has an influence on the effectiveness and bioavailability of the resulting vanadium compound. It is believed that factors such as absorption, tissue uptake and distribution are affected most. Interestingly enough, X-ray crystal data of PTP1B soaked with BMOV showed only vanadate $[V(+V)O_4^{3-}]$ at the active site. This would emphasise that the organic ligands are only carriers of the active compound and play no role in the enzyme inhibition itself. Furthermore, in aqueous solution, V(IV) is rapidly and reversibly oxidised to V(V), supporting the possible formation of vanadate [17a].

BEOV entered clinical trials and successfully finished phase IIa trials for the treatment of DM Type 2. In phase I trials, doses of 10–90 mg were given to healthy nondiabetic volunteers and no adverse side effects were seen. In the phase IIa clinical trial, seven diabetic patients were treated with 20 mg/day of BEOV and showed a reduction of around 15% of their blood glucose levels. Two patients were treated with a placebo, and no reduction in blood glucose levels was observed. It was also interesting to note that the glucose level reduction lasted for 1 week after finishing the treatment [9].

8.6 Exercises

8.6.1 Write the electronic configuration for the following elements or ions:

(a) Ti^{4+}
(b) Fe^{2+}
(c) Fe^{3+}
(d) Ru^{2+}

8.6.2 What is the oxidation state of the central metal atom in the following complexes?

(a) $[Fe(C_5H_5)_2]^+$
(b) $TiCl_4$
(c) $Ti(C_5H_5)_2Cl_2$

8.6.3 Draw the energy diagrams displaying the d-orbital splitting for the low and high-spin complexes of the following examples assuming an octahedral complex.

(a) Fe^{3+}
(b) Fe^{2+}
(c) Ti^{4+}

8.6.4 Predict the geometry of the following complexes:

(a) Platinum tetrachloride
(b) Vanadium hexacarbonyl

8.7 Case study – titanium dioxide

Integrated sun protection in cosmetics is becoming increasingly important. Titanium dioxide nanoparticles are a commonly used as an inorganic UV filter. New formulation techniques allow its integration without the previously known whitening effect. Sunscreens contain typically between 5% and 20% w/v nanosized titanium dioxide.

There are a variety of different methods to analyse the quantity of titanium dioxide in cosmetic formulations. One method is based on the reduction of Ti(+IV) and subsequent re-oxidation with a ferric solution. Titanium dioxide is typically dissolved in hot sulfuric acid and reduced by adding metallic aluminium. The resulting Ti(+III) is then titrated against a standard solution of ammonium iron(III) sulfate in the presence of potassium thiocyanate as indicator.

(a) Research the type of titration described.
(b) Describe the chemical structure and mode of action of the indicator.
(c) Formulate all relevant reaction equations.
(d) The package states that the sunscreen contains 10% w/v titanium dioxide. For the analysis, a volume containing the theoretical value of 0.5 g of titanium dioxide is dissolved in sulfuric acid and reacted with metallic aluminium. The resulting solution is titrated against a 0.5 M solution of ammonium iron(III) sulfate using potassium thiocyanate as indicator.

For each titration, the following volume of ammonium iron(III) sulfate has been used:

12.55 ml	12.50 ml	12.60 ml

Calculate the real amount of titanium dioxide present in your sample. Express your answer in grams and moles.

(e) How many millilitres of sunscreen have been used for the analysis?
(f) Discuss the result in relation to the typically accepted error values.

References

1. F. A. Cotton, G. Wilkinson, C. A. Murillo, M. Bochmann, *Advanced inorganic chemistry*, 6th ed., Wiley, New York; Chichester, **1999**.
2. G. Gasser, I. Ott, N. Metzler-Nolte, *J. Med. Chem.* **2011**, *54*, 3–25.
3. J. D. Newman, A. P. F. Turner, *Biosens. Bioelectron.* **2005**, *20*, 2435–2453.
4. C. E. Housecroft, A. G. Sharpe, *Inorganic chemistry*, 3rd ed., Pearson Prentice Hall, Harlow, **2008**.
5. A. Vessieres, S. Top, W. Beck, E. Hillard, G. Jaouen, *Dalton Trans.* **2006**, *4*, 529–541.
6. K. M. Buettner, A. M. Valentine, *Chem. Rev.* **2012**, *112*, 1863–1881.
7. E. R. Tiekink, M. Gielen, *Metallotherapeutic drugs and metal-based diagnostic agents: the use of metals in medicine*, Wiley, Chichester, **2005**.
8. E. Alessio, *Bioinorganic medicinal chemistry*, Wiley-VCH, Weinheim, **2011**.
9. J. C. Dabrowiak, *Metals in medicine*, Wiley-Blackwell, Oxford, **2009**.
10. O. R. Allen, L. Croll, A. L. Gott, R. J. Knox, P. C. McGowan, *Organometallics* **2004**, *23*, 288–292.
11. K. Strohfeldt, M. Tacke, *Chem. Soc. Rev.* **2008**, *37*, 1174–1187.
12. J. D. Lee, *Concise inorganic chemistry*, 5th ed., Chapman & Hall, London, **1996**.
13. R. E. Berry, E. M. Armstrong, R. L. Beddoes, D. Collison, S. N. Ertok, M. Helliwell, C. D. Garner, *Angew. Chem. Int. Ed.* **1999**, *38*, 795–797.
14. H. Sakurai, *Chem. Rec.* **2002**, *2*, 237–248.
15. B. Gleeson, J. Claffey, M. Hogan, H. Muller-Bunz, D. Wallis, M. Tacke, *J. Organomet. Chem.* **2009**, *694*, 1369–1374.
16. P. Ghosh, O. J. D'Cruz, R. K. Narla, F. M. Uckun, *Clin. Cancer Res.* **2000**, *6*, 1536–1545.
17. (a) K. G. Peters, M. G. Davis, B. W. Howard, M. Pokross, V. Rastogi, C. Diven, K. D. Greis, E. Eby-Wilkens, M. Maier, A. Evdokimov, S. Soper, F. Genbauffe, *J. Inorg. Biochem.* **2003**, *96*, 321–330; (b) B. Kasibhatla, C. Winter, D. D. Buchanan, L. Fei, S. J. Samuelsson, J. S. Lange, K. G. Peters, *FASEB J.* **2003**, *17*, A594-A594.
18. L. Zhang, Y. Zhang, Q. Xia, X. M. Zhao, H. X. Cai, D. W. Li, X. D. Yang, K. Wang, Z. L. Xia, *Food Chem. Toxicol.* **2008**, *46*, 2996–3002.
19. T. O. Johnson, J. Ermolieff, M. R. Jirousek, Protein tyrosine phosphatase 1B inhibitors for diabetes. *Nature Reviews Drug Discovery* **2002**, *1*, 696–709.

Further Reading

W. Kaim, B. Schwederski, *Bioinorganic chemistry: inorganic elements in the chemistry of life: an introduction and guide*, Wiley, Chichester, **1994**.

H.-B. Kraatz, N. Metzler-Nolte, *Concepts and models in bioinorganic chemistry*, Wiley-VCH [Chichester: John Wiley, distributor], Weinheim, **2006**.

G. A. McKay, M. R. Walters, J. L. Reid, *Lecture notes. Clinical pharmacology and therapeutics*, 8th ed., Wiley-Blackwell, Chichester, **2010**.

R. M. Roat-Malone, *Bioinorganic chemistry: a short course*, Wiley, Hoboken, N.J. [Great Britain], **2002**.

G. J. Tortora, B. Derrickson, *Principles of anatomy and physiology*, 12th ed., international student/Gerard J. Tortora, Bryan Derrickson. ed., Wiley [Chichester: John Wiley, distributor], Hoboken, N.J., **2009**.

9
The Clinical Use of Lanthanoids

The lanthanoid series consists of 14 elements (cerium to lutetium) and they are also called *f-block metals*, as the valence shell of lanthanoids contains 4f orbitals. Lanthanoids are named after the element lanthanum (La), which itself is a d-block metal. Nevertheless, because of its very similar chemical behaviour, lanthanum is often also classified as a lanthanoid. The term *rare earth metals* describes the lanthanoid series together with lanthanum (La), scandium (Sc) and yttrium (Y).

> *The **lanthanoid series** is often given the symbol Ln and referred to the elements La–Lu.*

The trivalent cation is the most stable oxidation state, as the lanthanoids tend to lose three electrons ($6s^2$ and $5d^1$) because of their electronic configuration. The resulting trivalent cations contain a xenon (Xe) core with regard to their electronic configuration with the addition of varying numbers of 4f electrons. The 4f electrons can be found closer to the nucleus, whilst the $5s^2$ and $5p^6$, which are full sub-shells, shield them. Trivalent lanthanoids are often referred to as *triple positive charged noble gases* (Table 9.1).

9.1 Biology and toxicology of lanthanoids

Mostly, lanthanoids are used in the production of batteries, lasers and other technological devices. Some lanthanoids salts, such as the salts of lanthanum, cerium and gadolinium (highlighted in Figure 9.1), are increasingly used in a clinical setting, for example, as a phosphate binder in the treatment of renal osteodystrophy or as MRI (magnetic resonance imaging) contrast agents (CAs).

Lanthanoids (Ln) show a biological behaviour very similar to that of Ca^{2+}, as they have similar ionic radii. Lanthanoids are mostly trivalent and therefore possess a higher charge than Ca^{2+}. Lanthanoids display a high binding affinity to calcium-binding sites in biological molecules and to water molecules. The coordination number for lanthanoids varies from 6 to 12. Mostly, eight or nine water molecules are coordinated to the lanthanoid ion. This is a significantly lower coordination number compared to that of calcium, which is 6.

Essentials of Inorganic Chemistry: For Students of Pharmacy, Pharmaceutical Sciences and Medicinal Chemistry,
First Edition. Katja A. Strohfeldt.

Companion website: www.wiley.com/go/strohfeldt/essentials

Table 9.1 *Lanthanoid series*

Element name	Symbol	Ground-state electronic configuration
Lanthanum	La	$[Xe]6s^2 5d^1$
Cerium	Ce	$[Xe]4f^1 6s^2 5d^1$
Praseodymium	Pr	$[Xe]4f^3 6s^2$
Neodymium	Nd	$[Xe]4f^4 6s^2$
Promethium	Pm	$[Xe]4f^5 6s^2$
Samarium	Sm	$[Xe]4f^6 6s^2$
Europium	Eu	$[Xe]4f^7 6s^2$
Gadolinium	Gd	$[Xe]4f^1 6s^2 5d^1$
Terbium	Tb	$[Xe]4f^9 6s^2$
Dysprosium	Dy	$[Xe]4f^{10} 6s^2$
Holmium	Ho	$[Xe]4f^{11} 6s^2$
Erbium	Er	$[Xe]4f^{12} 6s^2$
Thulium	Tm	$[Xe]4f^{13} 6s^2$
Ytterbium	Yb	$[Xe]4f^{14} 6s^2$
Lutetium	Lu	$[Xe]4f^{14} 6s^2 5d^1$

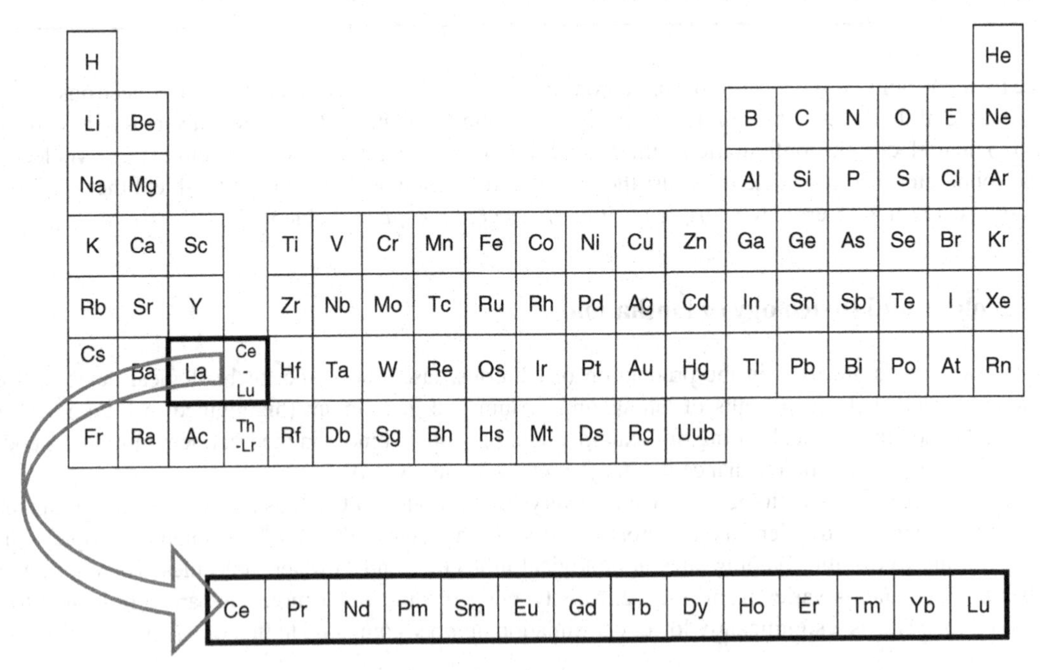

Figure 9.1 *Periodic table of elements; lanthanoids clinically used are highlighted*

Within the human body, lanthanoid ions are known to block the receptor-operated calcium channels. Lanthanoids cannot cross the cell membrane, but they still block the Na^+/Ca^{2+} synaptic plasma membrane exchange and therefore inhibit muscle contraction (e.g. in skeletal muscle or cardiac muscle). Lanthanoids can also displace calcium in proteins and enzymes, which can either lead to inhibition or activation of their catalytic activity. In general, lanthanoids mimic the biological behaviour of calcium ions and as a result lanthanoids can be used to study the mode of action of calcium ions in a variety of biological applications.

It is very interesting to look at the toxicity of lanthanoids. In general, lanthanoids are not regarded as toxic, as they cannot cross the cell membrane and therefore are not absorbed if administered orally. Lanthanoids are toxic if they are administered intravenously, as they can then interact with a variety of biological targets. A sudden decrease in blood pressure and sudden cardiovascular complications are signs of acute toxicity. Chronic toxicity manifests itself in liver damage and oedema. After intravenous administration, lanthanoids are often quickly distributed to the liver and the bones [1].

9.2 The clinical use of lanthanum carbonate

The chemical element lanthanum has the symbol La and atomic number 57. It is a silvery white metal and represents the start of the lanthanoid series (Ln).

Lanthanum has two oxidation states, +II and +III, and the latter is the more stable one. The electronic configuration of the resulting La^{3+} ion is $[Xe]4f^0$. Lanthanum burns in the presence of air and forms lanthanum(III) oxide:

$$4La + 3O_2 \rightarrow 2La_2O_3$$

Lanthanum reacts with water with the formation of lanthanum hydroxide owing to the electropositive nature of the metal. It also reacts with halogens and forms the respective lanthanum halide salt.

Lanthanum carbonate $La_2(CO_3)_3$ see Figure 9.2 is the only lanthanum salt approved for clinical use. It is used in the managementof hyperphosphataemia, which is defined as high levels of phosphate in the serum blood (Figure 9.3).

$$2La + 6H_2O \rightarrow 2La(OH)_3 + 3H_2$$

$$2La + 3Cl_2 \rightarrow 2LaCl_3$$

Figure 9.2

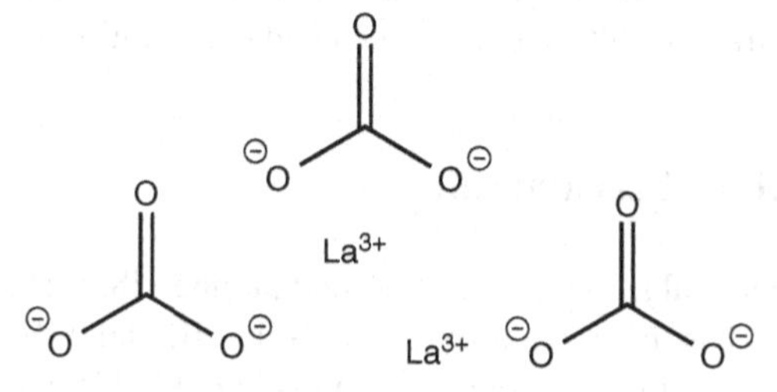

La3+
La3+

Figure 9.3 *Chemical structure of lanthanum carbonate*

Patients with end-stage renal failure (ESRF) often present high levels of phosphate in the serum as a result of the failure of the kidneys. The serum phosphate level of a healthy human is in the range 0.8–1.5 mmol/l [2], whereas this level is significantly increased in patients with kidney failure. High phosphate levels are linked to a decrease of calcium serum levels and the release of the so-called parathyroid hormone (PTH). This can manifest itself in renal osteodystrophy, which can have severe pathological consequences such as bone malfunction (see Section 3.4.4). Also, increased levels of PTH are observed, which can lead to secondary calcification of muscles and vascular tissue. Nearly half of the deaths of dialysis patients with ESRF are due to cardiac events.

The average phosphate intake ranges from 1000 to 1500 mg/day. In a healthy human, phosphate is absorbed in the gastrointestinal (GI) tract and excreted via the kidneys. In patients with ESRF, phosphate excretion via the kidneys is reduced and therefore accumulates in the serum. A common treatment option includes the binding of phosphate already in the GI tract before it can enter the blood stream. The ideal phosphate binder should bind phosphate with high affinity, should not be absorbed in the GI tract and should be excreted via the faeces. Aluminium salts were used until the early 1980s as phosphate binders. Aluminium phosphate is readily formed but not absorbed. Unfortunately, the aluminium salt (aluminium hydroxide) itself is absorbed in the GI tract and has been found to be toxic. Specific toxicity to the central nervous system (CNS) was observed.

As an alternative treatment option, calcium salts were and still are used as phosphate binders. Calcium salts, such as calcium acetate and calcium carbonate, are very successful treatment options especially in patients undergoing dialysis. The main issue is that calcium ions can be absorbed and this can lead to hypercalcaemia, which is defined as high levels of serum calcium ions. This can further increase the risk of tissue calcification and cardiac events.

Further research has led to a variety of drugs, with sevelamer hydrochloride being one of the most successful ones. Sevelamer is a hydrogel containing cross-linked polyallylamine chains. The negatively charged phosphate can be bound to the positively charged amine groups of the hydrogel in the intestines and removed via the faeces.

Furthermore, lanthanum carbonate has been successfully studied as a phosphate binding agent. Lanthanum carbonate fulfils the criteria for a good phosphate binder as stipulated above: nontoxic, rapid binding of phosphate, not absorbed in the GI tract and easily excreted. A comparative study of a variety of lanthanum salts showed that $La_2CO_3{\cdot}4H_2O$ has the best phosphate binding properties at a variety of pHs. This means that phosphate can be bound in the stomach (very low pH) and the complex remains intact whilst travelling through the GI tract where the pH is higher. Pharmacological studies have shown that lanthanum carbonate is poorly absorbed when administered orally and that more than 90% is excreted via the faeces. No specific toxicity has been observed either. Lanthanum carbonate hydrate is marketed under the name Fosrenol and has received approval in Europe and the United States for its clinical use in patients with chronic renal failure [1].

Lanthanum carbonate hydrate is usually given with an initial dose of 0.75–2.25 g of elemental lanthanum in divided doses with meal. The dose needs to be reviewed every 2–3 weeks until a maintenance dose (1.5–3 g) is achieved. Patients should be advised to chew the tablets before swallowing. Most common side effects include disturbances of the GI tract, resulting in diarrhoea and constipation [3].

9.3 The clinical application of cerium salts

The element cerium has the chemical symbol Ce and atomic number 58. It is a silvery and soft metal, which easily is oxidised in air. Cerium has three oxidation states, +II, +III and +IV, the last being the more stable one. The electronic configuration of the resulting Ce^{4+} ion is $[Xe]4f^0$. Ce^{+2} is the rarest oxidation state and the resulting electronic configuration is $[Xe]4f^2$.

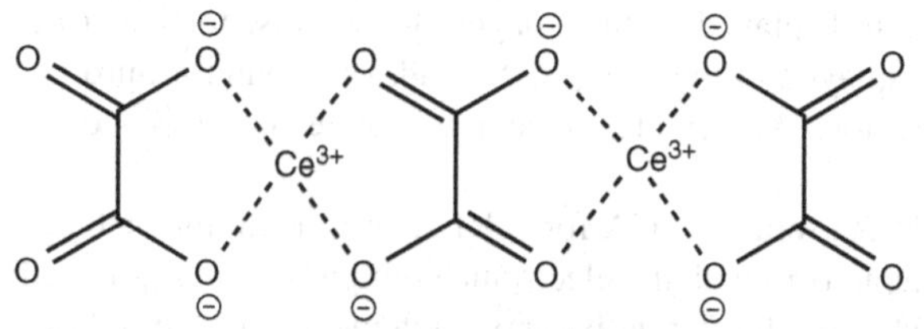

Figure 9.4 Chemical structure of cerium(III) oxalate

The clinical use of cerium dates back to the mid-nineteenth century, when cerium(III) oxalate was used as an antiemetic. The exact mode of action is unknown, but it is believed to be a local effect limited to the GI tract. This is based on the fact that lanthanoids are not easily absorbed after oral administration, as previously discussed, and that cerium oxalate has a low aqueous solubility. Nevertheless, the clinical use of cerium salts as antiemetic was eventually replaced by antihistamines (Figure 9.4) [1].

At the end of the nineteenth century, several Ce^{3+} salts were under investigation for their antibacterial activity in burn wounds. Especially, cerium nitrate $[Ce(NO_3)_3]$ showed broad activity against a variety of pathogens and was subsequently used in combination with silver sulfadiazine. Initial studies were very successful and an estimated reduction of 50% of the mortality rate was suggested. It was believed that this result was due to the synergistic antimicrobial effect of both reagents [1].

Recent studies have shown that cerium nitrate has no significant effect on pathogens from burn wounds. Furthermore, research has shown that the suppression of the immune system in patients with serious burn wounds is a main factor for mortality. It has also been shown in animal models that cerium nitrate is a modulator of the burn-associated immune response. Nowadays, this is believed to be the main role of cerium(III) nitrate when used as part of a combination treatment of burn victims. Currently, the cerium salt is used in combination with silver sulfadiazine in individual cases for the treatment of life-threatening burn wounds. Reports suggest that wound healing improves, mortality rates drop and graft rejection rates are also significantly lower [1].

9.4 The use of gadolinium salts as MRI contrast agents

The chemical element gadolinium has the chemical symbol Gd and atomic number 64. It is a silvery white metal, which is malleable and ductile. The dominant oxidation state of gadolinium salts is +III, with the resulting electronic configuration $[Xe]4f^7$.

Gadolinium is a relatively stable metal upon exposure to dry air. Nevertheless, it tends to oxidise once it is exposed to moisture, as it slowly reacts with water. Gadolinium is used in microwave applications and in the manufacturing of compact discs and computer memory. Gadolinium is often used in alloys. With as little as 1% gadolinium, the properties and workability of iron and chromium improve.

Solutions containing gadolinium salts are used as CAs for MRI as a clinical application. Using this method, it is possible to detect and observe pathological and physiological alteration to living tissue. MRI used in medicine uses the so-called relaxation properties of excited hydrogen nuclei found in water and lipids in the human tissue. Relaxation refers to an effect known in physics and chemistry, where there is a delay between the application of an external stress to the system and its response. Within a strong magnetic environment, which is produced in an MRI scanner, excited hydrogen nuclei show different behaviours depending on their environment.

MRI can be carried out without any CAs, but the use of a suitable CA can enhance the imaging properties. The basic idea is that the water relaxation rates are altered in the presence of a CA, which leads to additional and/or enhanced information displayed in the images. In contrast to their role in the X-ray imaging, the CAs themselves are not displayed in the images. Nowadays, around a third of MRI scans are undertaken using CAs, and organ perfusions, kidney clearance and changes in the blood–brain barrier can be detected (Figure 9.5).

The trivalent cation Gd^{3+} is useful as a CA for MRI, as it is a paramagnetic compound (see Section 7.1.2 for the definition of paramagnetism) with the electronic configuration [Xe]$4f^7$. Gd^{3+} has seven unpaired electrons in its valence shell, which endows it with the paramagnetic properties. Gd^{3+}-containing CAs affect the image quality of an MRI in two ways: (i) Paramagnetic Gd^{3+} complexes can coordinate water molecules and exchange them for water molecules in their environment, which are coordinated to the metal centre. A typical Gd^{3+} complex is an eight-coordinated metal complex in which the ninth binding site is available for the coordination of water. As previously discussed, the typical coordination number for lanthanoid complexes

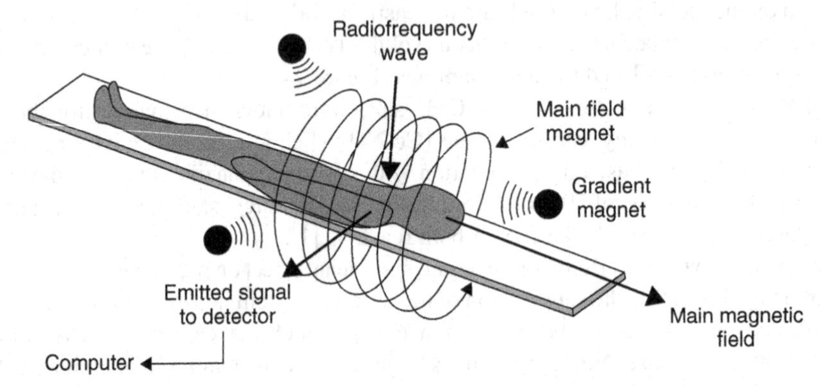

Figure 9.5 Schematic diagram of a patient in a magnetic resonance imaging (MRI) instrument. The main magnetic field is created by an electromagnet, and the field in many specific planes within the tissue of the patient is modified by gradient electromagnets. After analysis using a computer program, a 3D image of the relaxation lifetimes of water molecules in the tissue of the patient is obtained [4] (Reproduced with permission from [2]. Copyright © 2009, John Wiley & Sons, Ltd.)

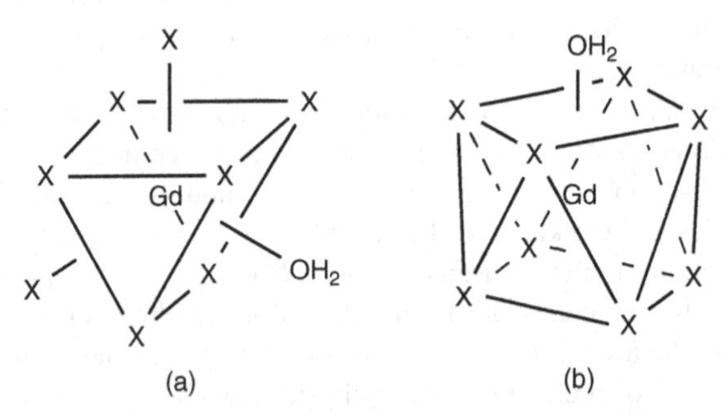

Figure 9.6 Coordination geometries of Gd^{3+} MRI CAs: (a) tricapped trigonal prism and (b) monocapped square antiprism. Depending on the structure of the octadentate ligand, which contributes eight donor groups to Gd^{3+}, denoted by X, and the location of the water molecule, each geometry can exist in a number of isomeric forms [4] (Reproduced with permission from [2]. Copyright © 2009, John Wiley & Sons, Ltd.)

is 9. (ii) In addition, the paramagnetic Gd^{3+} creates a small local magnetic field, which influences the water molecules that are not coordinated to the metal centre (Figure 9.6).

Solutions of paramagnetic organic Gd^{3+} compounds are administered intravenously, and they enhance the images obtained by MRI. The trivalent ion itself has no physiological function in the human body and is highly toxic. Therefore, Gd^{3+} is administered as stable chelate in which the lanthanoid is chelated by an

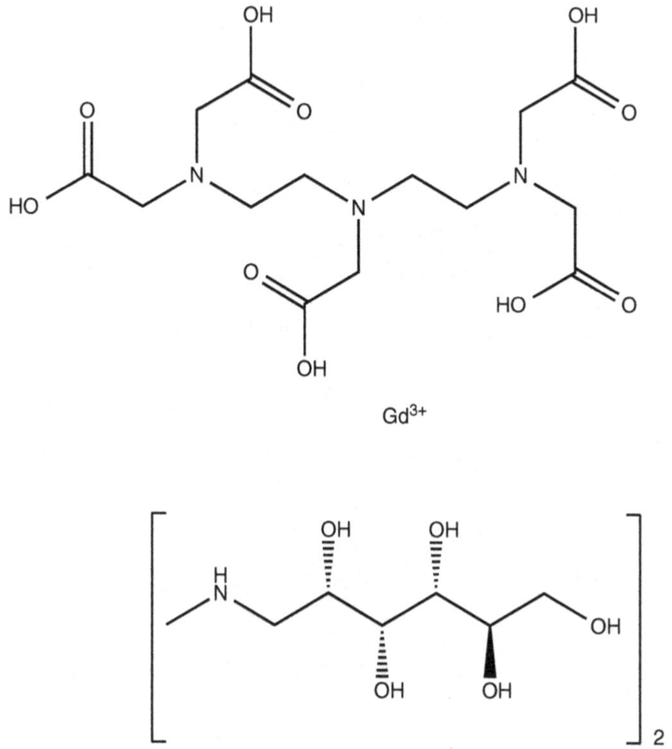

Figure 9.7 *Chemical structure of Magnevist*

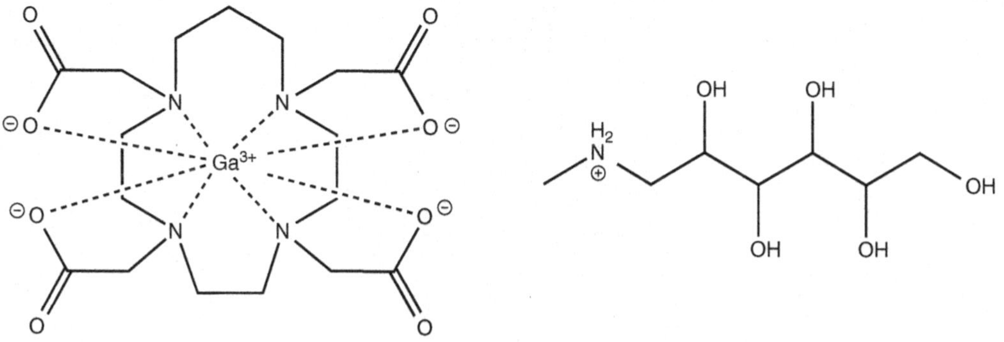

Figure 9.8 *Chemical structure of Dotarem*

organic ligand and as a result exhibits a significantly lower toxicity. The most widespread clinical examples are Magnevist™ and Dotarem™ (see Figures 9.7 and 9.8), both approved by the FDA for their use as CAs in MRI. Gadolinium-containing chelates are around 50 times less toxic than the 'free' Gd^{3+} in salts such as $GdCl_3$. Gadolinium-containing chelates are typically cleared from the blood plasma within a few hours (half-life <2 h) and excreted via the urine [1]. In contrast, free Gd^{3+} ions remain in blood serum for significantly longer; only 2% is excreted after 7 days [1].

9.5 Exercises

9.5.1 Write the electronic configuration for the following:

(a) La
(b) La^{2+}
(c) Eu
(d) Eu^{3+}
(e) Er
(f) Ce^{3+}

9.5.2 Draw the Lewis structures for the following compounds:

(a) Lanthanum nitrate
(b) Lanthanum sulfate
(c) Lanthanum acetate

9.5.3 A typical solution of Dotarem for injection is labelled with gadoteric acid, 0.5 mmol/ml/100 ml. Calculate

(a) the amount of gadoteric acid in gram;
(b) the corresponding amount of DOTA;
(c) the corresponding amount of gadolinium oxide.

9.6 Case study: lanthanum carbonate tablets

Research the properties and clinical use of lanthanum carbonate tablets. Summarise your findings in a drug monograph, which could be published in a Pharmacopoeia. Your drug monograph should contain the following aspects:

1. Chemical structure
2. Action and use
   (a) Preparation
3. Chemical formula
4. Definition
5. Content
6. Characteristics
   (a) Appearance
   (b) Solubility
7. Identification (two different forms of identification)
   (a) Identification A
   (b) Identification B
8. Assay to quantify the API (active pharmaceutical ingredient)
9. Impurities.

References

1. S. P. Fricker, *Chem. Soc. Rev.* **2006**, *35*, 524–533.
2. J. Wright, A. H. Gray, V. Goodey, *Clinical pharmacy: pocket companion*, Pharmaceutical Press, London, **2006**.
3. S. C. Sweetman, *Martindale: the complete drug reference*, 36th ed., Pharmaceutical Press, London; Chicago, I.L., **2009**.
4. J. C. Dabrowiak, *Metals in medicine*, Wiley-Blackwell, Oxford, **2009**.

Further Reading

E. Alessio, *Bioinorganic medicinal chemistry*, Wiley-VCH, Weinheim, **2011**.

F. A. Cotton, G. Wilkinson, C. A. Murillo, M. Bochmann, *Advanced inorganic chemistry*, 6th ed., Wiley, New York; Chichester, **1999**.

C. E. Housecroft, A. G. Sharpe, *Inorganic chemistry*, 3rd ed., Pearson Prentice Hall, Harlow, **2008**.

W. Kaim, B. Schwederski, *Bioinorganic chemistry: inorganic elements in the chemistry of life: an introduction and guide*, Wiley, Chichester, **1994**.

H.-B. Kraatz, N. Metzler-Nolte, *Concepts and models in bioinorganic chemistry*, Wiley-VCH [Chichester, John Wiley, distributor], Weinheim, **2006**.

J. D. Lee, *Concise inorganic chemistry*, 5th ed., Chapman & Hall, London, **1996**.

R. M. Roat-Malone, *Bioinorganic chemistry: a short course*, Wiley, Hoboken, N.J. [Great Britain], **2002**.

E. R. Tiekink, M. Gielen, *Metallotherapeutic drugs and metal-based diagnostic agents: the use of metals in medicine*, Wiley, Chichester, **2005**.

10

Radioactive Compounds and Their Clinical Application

The preparation, handling and dispensing of radioactive materials that are being used for the diagnosis and/or treatment of diseases form a specialised area of pharmacy, the so-called nuclear pharmacy.

> ***Nuclear pharmacy*** *is defined as the area of pharmacy dealing with the compounding and dispensing of radioactive material to be used in nuclear medicine.*

In order to understand what a pharmacist's obligations are when handling nuclear material, it is important to understand what radioactivity is and how it is used in a clinical setting.

10.1 What is radioactivity?

10.1.1 The atomic structure

It is fundamental to look at the structure of an atom in order to understand what radioactivity is. As discussed in Chapter 1 and briefly summarised here, an atom consists of a positively charged nucleus formed from the so-called nucleons, which is surrounded by negatively charged electrons, which may occupy different energy levels. Protons and neutrons form the nucleus. Protons are nucleons with a positive charge and a mass of 1.6726×10^{-24} g. The atomic number (Z) expresses the number of protons. A neutron is a nucleon without a charge and a mass similar to that of protons (1.6749×10^{-24} g). The so-called neutron number (N) describes the total number of neutrons. Neutrons and protons are held together by nuclear binding forces and therefore form the nucleus. The letter A stands for the number of nucleons, which is the sum of number of protons (Z) and neutrons (N). Electrons have a mass of 9.1094×10^{-28} g and they move in energy levels around the nucleus. Lower orbitals, which are defined as the orbitals closer to the nucleus, possess a higher kinetic energy. If an electron moves to an orbital closer to the nucleus, energy is released, whilst the energy is required to move it

Essentials of Inorganic Chemistry: For Students of Pharmacy, Pharmaceutical Sciences and Medicinal Chemistry,
First Edition. Katja A. Strohfeldt.

Companion website: www.wiley.com/go/strohfeldt/essentials

$$^{A}_{Z}E$$

E = element symbol

A = number of protons + number of neutrons = mass number

Z = number of protons = number of electrons = atomic number

Figure 10.1

away from the nucleus. The number of electrons should be equal to the number of protons in order to have an element without any charge. Typically, the number of neutrons equals the number of protons (Figure 10.1).

10.1.2 Radioactive processes

Radioactive decay, also known as *radioactivity*, describes the process by which an unstable nucleus spontaneously loses energy in order to form a stable nucleus. This energy loss is achieved by emitting particles of ionising radiation. An element or material that spontaneously emits energy in this form is considered as radioactive. Radiation can take place in the form of α, β^-, β^+, X-rays and γ-rays.

> ***Radioactivity** is defined as the process whereby an unstable nucleus spontaneously loses energy in order to form a more stable nucleus.*

The terms *nuclide* and *radionuclide* describe identifiable atomic species (nonradioactive or radioactive), which are characterised by their exact number of protons (Z) and neutrons (N). In contrast, an element is only defined by its number of protons (Z). The number of neutrons can vary, which leads to different isotopes of the same element as described in Chapter 1. An element can have a number of isotopes, some of which are stable and some are not stable and therefore are classified as radioactive.

10.1.3 Radioactive decay

It is important to understand the different forms of radioactive decay, as not all forms of radiation are useful for clinical applications. It is also crucial to understand the reason for the occurrence of radioactivity in order to provide the patient with the best and safest possible treatment option.

There are two reasons for the occurrence of radioactive decay: one is that the ratio of neutrons to protons is greater or less than 1. The second reason is that radioactive decay takes places as a result of an energy imbalance within the atom, which means that the atom needs to get rid of energy in order to reach a stable form. The different forms of radiation can be summarised as α, β, γ-decay and X-ray emission.

Alpha-(α-)radiation is defined as the emission of helium particles, precisely $^{4}_{2}He^{2+}$. α-Decay occurs in elements with a so-called heavy nucleus, principally in elements with a higher atomic mass (typically in elements with $Z > 82$). α-Particles are fairly heavy particles and follow a straight path when penetrating through a material. They only display a short-range activity and the radiation can be easily shielded off with a piece

α–particle = $^{4}_{2}He^{2+}$

$^{A}_{Z}X_{N}$	→	$^{4}_{2}He^{2+}$	+	$^{A-4}_{Z-2}Y_{N-2}$
Parent isotope		α–particle		daughter isotope

Figure 10.2 *α-Particle*

Equation for β^--decay: $n \rightarrow p + \beta^- + \nu$

$^{A}_{Z}X_{N}$	→	$^{A}_{Z+1}Y_{N-1}$	+	$\beta^- + \nu$
Parent isotope		daughter isotope		anti-neutrino β^-–particle

Figure 10.3 *β^--Radiation*

of paper. Their clinical application is very limited and includes only a few therapeutic examples. Current research includes the use of monoclonal antibodies as radiopharmaceuticals. The idea is to deliver radioisotopes directly to the tumour cells, minimising the exposure of healthy cells to radiation.

In general, interaction of α-radiation with neighbouring matter can occur in two ways – through ionisation or excitation. Excitation means that an α-particle can, upon collision, promote an electron to a higher energy level (higher outer shell). Once the electron falls back to its original energy level, energy is emitted. The more important interaction is the ionisation of an atom. This occurs when an α-particle collides with its target and 'strips' away an electron, leaving behind a positively charged molecule (Figure 10.2).

Beta-(β-)decay occurs when basically an electron is ejected from the nucleus. This occurs when the 'neutron to proton ratio' is >1. In order for this to happen, within the nucleus a neutron is converted into a proton and a negatively charged β-particle (negatron, β^-). Additionally a so-called antineutrino (ν) is 'produced', carrying away any excess binding energy from the nucleus. These processes result in an increase in the proton number (Z) to $Z+1$ and a decrease in the neutron number (N) to $N-1$. Negatively charged β-particles have the appearance of electrons, but they originate from the nucleus and carry energy. In contrast, electrons that are present in the orbit outside the nucleus have no energy and obviously their origin differs (Figure 10.3).

β-Particles differ significantly from α-particles as they are considered as extremely light and fast particles. As a consequence, they travel much further and their clinical applications include imaging methods as well as therapeutic ones, with the emphasis being on the latter.

Positron emission is the ejection of positively charged β-particles from a proton-rich nucleus. This occurs when the neutron to proton ratio is <1. For this to happen, a proton is converted into a neutron, a positron (β^+) and a neutrino (ν). A neutrino is the opposite of an antineutrino, a small particle carrying no mass or charge. These processes result in a decrease in the proton number (Z) to $Z-1$ and an increase in the neutron number (N) to $N+1$. The differences between a positron and a neutron manifest in a lower energy and range for positrons. Positrons and positron-emitting elements are used mainly for imaging purposes in the so-called positron emission tomography (PET, see Chapter 10) (Figure 10.4).

Gamma-(γ-)emission is the elimination of excess energy by the emission of photons. A γ-photon has no charge or mass and occurs as electromagnetic radiation. The radiation occurs at short wavelengths and is therefore of high energy. It has the longest range of all nuclear emissions discussed. The nucleus that emits the γ-photon does not undergo any change of the neutron number; mostly isomers are formed. Isomers are

Equation for β^+-decay: $p \rightarrow n + \beta^+ + \nu$

$$^{A}_{Z}X_{N} \rightarrow {}^{A}_{Z-1}Y_{N-1} + \beta^+ + \nu$$

Parent isotope — daughter isotope — neutrino positron

Figure 10.4 Positron emission

^{99m}Tc

Figure 10.5 Element symbol technetium

$^{0}_{0}\gamma$

$$^{A(m)}{}_{Z}X \rightarrow {}^{A}_{Z}Y + \gamma$$

Figure 10.6 γ-Emission

defined as nuclides of the same atomic mass (*A*) and number (*Z*) with the only difference being that one isomer is in an excited (metastable (m)) state. This is marked with ^{Am}E as the atomic number (Figures 10.5 and 10.6).

γ-Emission normally occurs following another nuclear decay. It has a high penetrating power of several metres as the γ-photons are not charged. They typically interact with matter through direct collision with nuclei and electrons of the orbital. γ-Emission finds its clinical application mainly as part of radiopharmaceutical imaging processes.

Electron capture and X-ray emission constitutes another form of positron emission and occurs in unstable atoms where the 'neutron to proton' ratio is <1. In order to convert to a stable atom, a proton from the nucleus catches an electron from the orbital and transforms into a neutron and a neutrino, with the neutrino carrying any excess energy. Typically, a cascade reaction follows, in which electrons from the outer (higher energy) orbitals move closer to the nucleus and fill the vacated orbitals. Orbitals closer to the nucleus are of lower energy, and the energy difference is given off as X-rays, a form of electromagnetic radiation. In contrast to γ-decay, which originates predominantly from the nucleus, X-rays stem from outside the nucleus. Additionally, X-rays have a wavelength longer than γ-photons. The result of electron capture and X-rays are the so-called isobars, similar to β-decay.

***Isobars** are nuclides with the same mass number (A) but different atomic number (Z) and different neutron number (N).*

Figure 10.7 shows the effect of charge on different forms of radiation. A radioactive sample is placed in a container, where radiation is released in only one direction. This directed radiation is then exposed to a negatively and a positively charged electrode. As a result, the α-particles (positive charged particles) are directed towards the negative charge, whereas the gamma rays are not affected by the charge at all. Consequently, the negatively charged β-particles are directed towards the positive charge.

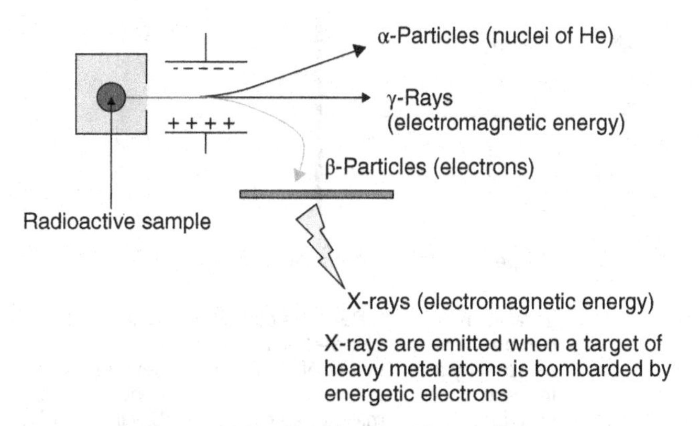

Figure 10.7 *Types of ionising radiation*

10.1.4 Penetration potential

It is important to understand the different penetration potentials of the various forms of radioactive decay in order to evaluate their clinical potential and the safety concerns of radiopharmaceutical compounds. α-Particles have the shortest range and can be stopped by paper. Skin is typically thick enough to provide sufficient protection. It is dangerous to ingest α-particles as it can cause serious damage to the affected areas. Unfortunately, it is difficult to monitor α-radiation. α-Particles have typically a long half-life and cause cell death, and therefore can sometimes be used in cancer treatment.

β-Particles cause ionisation and excitation (similar to α-particles) when they interact with matter. Ingestion is also a serious problem, as many β-radioisotopes are isotopes of carbon, hydrogen, sulfur, phosphorus and other essential elements and can easily be incorporated into biological material in the human body. This can lead to extensive damage of DNA and tissue. Additionally, many β-particles have relatively long half-lives. Fortunately, they can be shielded by a few millimetres of aluminium or plastic and are easy to monitor.

γ-Photons also cause ionisation and excitation, but not as successfully as α-particles or β-particles. The energy of γ-photons is usually larger than that of chemical bonds and can destroy biological structures in the human body. Their penetration potential is the highest amongst the types of radiation discussed here. γ-Photons can penetrate through skin and tissue easily. Shielding requires a few inches of lead or a few feet of concrete.

10.1.5 Quantification of radioactivity

Radioactivity can be quantified, and there are several units being used in order to describe the energy, exposure and the dose of radiation. It is crucial for a nuclear pharmacist to understand these units in order to dispense and handle radioactive material correctly.

10.1.5.1 Units of radioactivity

The activity of a radioactive source is defined as the number of transformations per unit time. The old traditional unit of radiation is curie (Ci). One curie is defined as 3.7×10^{10} disintegrations per second (dps). Nevertheless, the SI unit is becquerel (Bq), which is equal to 1 dps and is a metric unit (Figure 10.8, Tables 10.1 and 10.2).

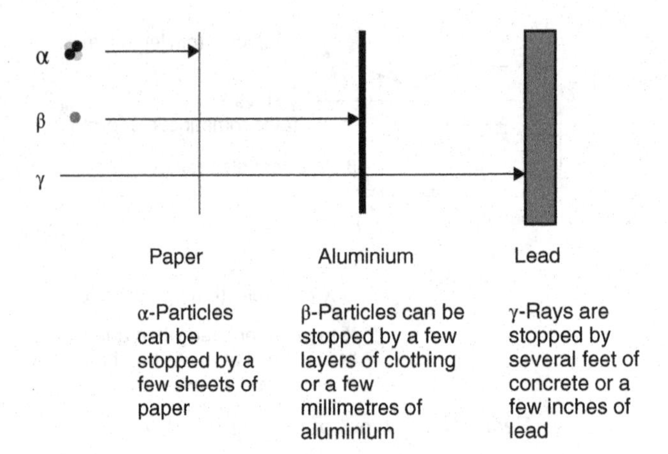
α
β
γ
Paper
Aluminium
Lead
α-Particles can be stopped by a few sheets of paper
β-Particles can be stopped by a few layers of clothing or a few millimetres of aluminium
γ-Rays are stopped by several feet of concrete or a few inches of lead

Figure 10.8 *Penetration potential of different forms of radioactivity*

Table 10.1 *Conversion between curie and SI units*

1 Bq (becquerel)	1 dps	2.7×10^{-11} Ci
1 kBq (kilobecquerel)	10^3 dps	2.7×10^{-6} Ci
1 MBq (megabecquerel)	10^6 dps	2.7×10^{-4} Ci
1 GBq (gigabecquerel)	10^9 dps	2.7×10^{-2} Ci
1 TBq (terabecquerel)	10^{12} dps	2.7 Ci

Table 10.2 *Conversion between curie and SI units*

1 Ci (curie)	3.7×10^{10} Bq	37 GBq
1 mCi (millicurie)	3.7×10^{7} Bq	37 MBq
1 mCi (microcurie)	3.7×10^{4} Bq	37 kBq
1 nCi (nanocurie)	3.7×10 Bq	37 Bq

For a pharmacist, it is crucial to know the specific activity of a radioactive preparation. This is the activity of a particular radionuclide per unit mass of this element, usually expressed in grams. The radioactive concentration of a solution is defined as the activity of a particular radionuclide per unit volume. The absorbed dose is the energy deposited per unit mass of the material. The unit 1 gray (Gy) is equal to 1 J of energy absorbed in 1 kg of material. As a comparison, ~25 Gy is need to kill bacteria when sterilising. The 'dose equivalent' takes also into account variations in the biological effectiveness of different radiation. The unit is sievert (1 Sv) (Figure 10.9).

10.1.5.2 Half-life ($t_{1/2}$)

For nuclear pharmacists, it is also important to understand the term *half-life* ($t_{1/2}$), as this gives information on how fast the radioactive decay takes place. The shorter the half-life, the faster the radioisotope decays. The

The **activity** of a radioactive source is the number of nuclear transformations per unit time

- 1 becquerel (Bq) is one nuclear transformation per second

1 megabecquerel (1MBq) =	10^6 Bq
1 kilobecquerel (1kBq) =	10^3 Bq

- 1 curie (Ci) –3.700 × 10^{10} nuclear transformations per second or disintegrations per second (dps)

1 millicurie (1 mCi) =	10^{-3} curie
1 microcurie (1 μCi) =	10^{-6} curie
1 nanocurie (1 nCi) =	10^{-9} curie

Figure 10.9 *Radiation units and definitions*

Dose equivalent = Absorbed dose × Quality factor

The quality factor for β- and γ-radiation is 1 and therefore 1 Sv = 1 Gy

Figure 10.10 *Dose equivalent*

half-life of radioactive elements can vary from several years to less than a second. Typical examples include ^{14}C (5730 years), ^{24}Na (15 h) and ^{18}Kr (13 s). Radiopharmaceuticals are typically divided into products with long (>12 h) and short (<12 h) half-lives.

Half-life *($t_{1/2}$) is defined as the time it takes for the activity (or the amount of radioactivity) to reduce by 50%. The shorter the half-life, the faster the isotope decays and the more unstable it is. The half-life is unique for any given radioisotope.*

Radioactive decay follows an exponential curve and it is therefore possible to determine the half-life by plotting a graph of activity versus time (see Figure 10.10). The first half-life is the point on the time axis at which only 50% of the initial activity remains. Subsequently, the second half-life is the time point at which the activity has halved again, that is, 25% of the original activity is left (Figure 10.11).

The so-called decay constant λ is related to half-life $t_{1/2}$ and can be calculated by using the following formula:

$$\lambda = \frac{\ln 2}{t_{1/2}} = \frac{0.693}{t_{1/2}}$$

Example

The decay constant for ^{99m}Tc is $0.1153\,h^{-1}$. Calculate the half-life ($t_{1/2}$) of this radioisotope (Table 10.3).

$$\lambda = \frac{\ln 2}{t_{1/2}} = \frac{0.693}{t_{1/2}}$$

$$t_{1/2} = \frac{0.693}{\lambda} = \frac{0.693}{0.1153\,h^{-1}} = 6.01\,\text{hours}$$

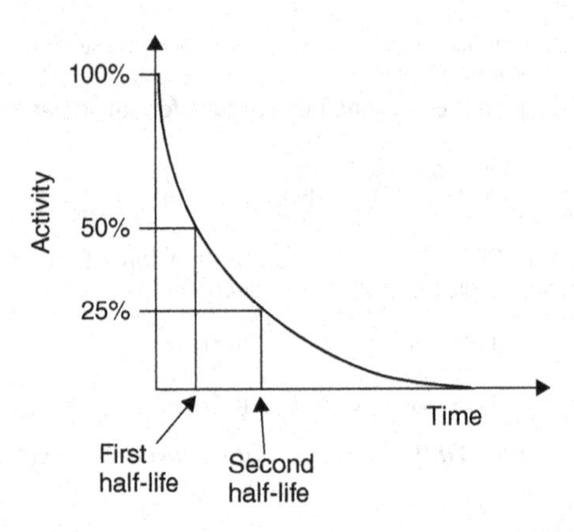

Figure 10.11 Exponential graph depicting half-life

Table 10.3 Half-life of commonly used radioisotopes [1]

Radioisotope	Half-life
Tritium (^{3}H)	12.33 yr
^{11}C	20.385 min
^{15}O	122.24 s
^{32}P	14.26 d
^{33}P	25.34 d
^{66}Ga	9.49 h
^{67}Ga	3.26 d
^{68}Ga	67.63 min
^{89}Sr	50.53 d
^{90}Sr	28.74 yr
^{99}Mo	65.94 h
^{99m}Tc	6.01 h
^{123}I	13.27 h
^{125}I	59.402 h
^{126}I	13.11 d
^{131}I	8.02 d

10.1.5.3 Calculation of radioactive decay

It is important for the healthcare professional to predict the activity of the radioactive material at any point in time before or after the assay being undertaken, as it is crucial to know the exact activity at administration to the patient. The radioactive decay can be described as the average number of radioactive isotopes (N) disintegrating per unit time (=disintegration rate). The disintegration rate is defined as $-\mathrm{d}N/\mathrm{d}t$. The disintegration rate is proportional to the number of undisposed radioisotopes, and can be also expressed as the activity (A).

$$-\frac{\mathrm{d}N}{\mathrm{d}t} = \lambda \times N = A$$

$$A = A_0 \cdot e^{-\lambda \cdot t}$$

A – specific activity at time t
A_0 – initial activity
$\lambda = \ln 2/t_{1/2} = 0.693/t_{1/2}$ – decay constant
t – elapsed time

Figure 10.12 *Radionuclide decay equation*

Upon integration, the radioactive decay of any radioactive sample can be calculated by applying the so-called radionuclide decay equation (see Figure 10.12). In order to calculate the radioactivity at a specific time point t, it is important to know the initial activity A_0, the elapsed time t and the decay constant λ. Half-life is the time that passes by until the activity has halved.

Example

A radioactive sample has a half-life of 8.05 days and contains 150 mCi radioactivity. Calculate the radioactivity left after 20 days.

$$A_0 = 150\,\text{mCi} \quad t_{1/2} = 8.05\,\text{days} \quad t = 20\,\text{days}$$

$$\lambda = \frac{\ln 2}{t_{1/2}} = \frac{0.693}{t_{1/2}} = \frac{0.693}{8.05\,\text{days}} = 0.0861\,\text{days}^{-1}$$

$$A_t = A_0 \times e^{-\lambda t} = 150\,\text{mCi} \times e^{-0.0861\times 20} = 26.8\,\text{mCi}$$

10.1.5.4 Dispensing of radioisotopes: percentage activity and activity concentration

For the dispensing pharmacist, it is crucial to know the activity of a radioactive sample at any given time. The factor $e^{-\lambda t}$ from the decay equation $A = A_0 e^{-\lambda t}$ is called the *decay factor* and can be used to calculate the percentage activity remaining after given time t. The percentage activity can be calculated using the following equation:

$$\text{Percentage activity} = 100 \times e^{-\lambda t}$$

Example

^{131}I has a half-life of 8.05 days. Calculate the percentage of ^{131}I remaining after 4.5 days.

$$\lambda = \frac{\ln 2}{t_{1/2}} = \frac{0.693}{t_{1/2}} = \frac{0.693}{8.05\,\text{days}} = 0.0861\,\text{days}^{-1}$$

$$\text{Percentage activity after 4.5 days} = 100 \times e^{-\lambda t} = 100 \times e^{-0.0861\times 4.5} = 67.88\%$$

Radioactive decay charts, that is, charts showing the percentage activity at certain time points, are often used in hospitals for quick reference. A representative table illustrating the percentage activity for the radioisotope ^{99m}Tc is shown in Table 10.4.

Radioisotopes are often dispensed according to their specific activity and activity concentration. The specific concentration is defined as the amount of radioactivity, usually expressed in curie, per unit mass. For example, a sample of containing 50 mg of ^{99m}Tc-albumin has an activity of 100 mCi. Therefore, the specific activity can be determined as 100 mCi/50 mg = 2 mCi/mg.

Table 10.4 Radioactive decay chart for ^{99m}Tc

Time (h)	% Activity remaining
2	79.4
4	63.1
6	50.1
8	39.8
10	31.6
12	25.1
14	19.9

The decay constant (λ) is calculated as (0.693/6.01 h^{-1}). The decay factor is $e^{-\lambda t}$. The percentage activity remaining (in percentage) is calculated as $100 \times e^{-\lambda t}$.

Most radioactive products are dispensed as solutions, and therefore the term *activity concentration* is important to understand. The activity concentration is defined as the radioactivity expressed in curie, per unit volume. A 5 ml solution of ^{99m}Tc-albumin has a radioactivity of 100 mCi. Therefore, the activity concentration is 100 mCi/5 ml = 20 mCi/ml. Once the activity concentration is known, the right dose (volume) containing the correct radioactivity can be dispensed.

$$\text{Quantity(ml)} = \frac{\text{Activity required}}{\text{Activity concentration}}$$

Example

The current activity concentration of a radiopharmaceutical is 20 mCi/ml. A solution with a dose of 15 mCi has been ordered. What quantity needs to be dispensed in order to provide a sample with the correct radioactivity?

$$\text{activity concentration} : 20\,\text{mCi/ml} \quad \text{activity requested} : 15\,\text{mCi}$$

$$\text{Quantity needed} = 15\,\text{mCi}/20\,\text{mCi/ml} = 0.75\,\text{ml} = \text{amount to be dispensed}$$

10.2 Radiopharmacy: dispensing and protection

Radiopharmacy deals with the manufacture and dispensing of radioactive materials that are used as radioactive medicines (or better known as *radiopharmaceuticals*). Radiopharmaceuticals can be used as diagnostic or therapeutic tools. Radionuclides that are used for a diagnosis should have as little an impact as possible on the health of the patient. Therefore, radioactive elements with a short-half-life and ones that only emit γ-radiation are seen as ideal. The radionuclide ^{99m}Tc in combination with a gamma camera is often used for imaging purposes, as the former has a half-life of 6 h and only emits γ-rays. Radionuclides that emit β-particles are more suitable for a therapeutic use. ^{131}I with its β-radiation is used for the treatment of hyperthyroidism (overactive thyroid) and metastatic diseases of the thyroid gland. ^{131}I also emits gamma radiation, which can be used to diagnose renal function and determine exactly the glomerular filtration rate (see Section 10.3.1).

Radiation can cause harmful effects in humans, which include nausea, skin burns, cancer, sterility, hair loss and even death. Nevertheless, all of these side effects depend on the type of radiation and its energy, the penetration power and the time scale of exposure. If radiation is used correctly, it can offer a range of useful applications. These include the treatment of cancer, sterilisation of medical instruments and, away from clinical applications, the generation of energy and dating of archaeological remains.

The correct protection from radiation is crucial for the safe handling of radioactive material. Radiation protection can be achieved by shielding; plastic and aluminium can shield from β rays, whereas lead or tungsten is needed to effectively shield from gamma rays. Furthermore, distance and time scale of exposure are important factors for the effective protection from radiation. The radiation dose is inversely proportional to the square of the distance from the radiation source. Also, minimising the time of exposure helps to reduce the risk of side effects from radiation.

The role of a specialised pharmacist, amongst other things, focusses on the correct dispensing of the radiopharmaceutical, which is more complicated than the dispensing of a nonradioactive item. The pharmacist is responsible to ensure that the proper prescribed dose is prepared and dispensed. This is not as simple as it sounds, as radioactive material undergoes continuous decay. Therefore, it is important to state when the activity was measured and what the half-life of this radionuclide is. Radiopharmaceuticals are typically dispensed in doses of units of activity (mainly kilobecquerel or megabecquerel).

10.3 Therapeutic use of radiopharmaceuticals

Radiopharmaceuticals that are used therapeutically are molecules with radiolabelling. This means that certain atoms in this molecule have been exchanged by their radioactive isotopes. These radiolabelled molecules are designed to deliver therapeutic doses of ionising radiation (mostly β-radiation) to specific disease sites around the body. The more specific the targeting is, the fewer the side effects expected. For any design of a treatment regime including radiopharmaceuticals, it is important to consider what the decay properties of the radionuclide are and what the clearance route and rate from nontarget radiosensitive tissue is.

10.3.1 ^{131}Iodine: therapy for hyperthyroidism

Iodine has the chemical symbol I and atomic number 53. It is a member of the halogens (group 17 of the periodic table of elements) (Figure 10.13).

Elemental iodine is characterised by the purple colour of its vapour. Free iodine typically exists (like the other halogens) as the diatomic molecule I_2. Iodide (I^-) is the highly water-soluble anion, which is mainly found in the oceans. Iodine and its compounds are mainly used in nutrition. It has relatively low toxicity and is easy to include into organic compounds, which has led to its application as part of many X-ray contrast agents. Iodine is required by humans to synthesise the thyroid hormones, and therefore iodine will accumulate in the thyroid gland. Iodine has only one stable isotope ($^{127}{}_{53}I$), but it has several radioactive isotopes. Some of these are used for medicinal purposes including diagnostic tests and treatment. Radioisotopes of iodine will accumulate in the thyroid gland and therefore can be used clinically. The radioactive isotope ^{129}I has a half-life of 15.7 million years, ^{125}I has 59 days and ^{123}I has 13 h. The last one is used in nuclear medicine as an imaging agent because of its gamma radiation and its short half-life. Using a gamma camera, images of the human body can be made showing areas of accumulation of the radioisotope.

^{131}I is the product of nuclear fission (as experienced during the Chernobyl disaster) and is a β-emitting radioisotope which will be transported to the thyroid gland if inhaled. Fortunately, it can be replaced by treatment with potassium iodide (nonradioactive), which will replace the radioisotope. Nevertheless, ^{131}I can be used as a therapeutic agent against thyroid cancer when applied in high doses. Paradoxically, the β-emitting radioisotope causes cancer when it is applied in low doses, but it will destroy its surrounding tissue if the dose is high enough. Therefore, preparations containing $^{131}I^-$ are often used to treat hyperthyroidism. These preparations are normally administered orally either as capsules or solution.

H																	He
Li	Be											B	C	N	O	F	Ne
Na	Mg											Al	Si	P	S	Cl	Ar
K	Ca	Sc	Ti	V	Cr	Mn	Fe	Co	Ni	Cu	Zn	Ga	Ge	As	Se	Br	Kr
Rb	Sr	Y	Zr	Nb	Mo	Tc	Ru	Rh	Pd	Ag	Cd	In	Sn	Sb	Te	I	Xe
Cs	Ba	La-Lu	Hf	Ta	W	Re	Os	Ir	Pt	Au	Hg	Tl	Pb	Bi	Po	At	Rn
Fr	Ra	Ac-Lr	Rf	Db	Sg	Bh	Hs	Mt	Ds	Rg	Uub						

Figure 10.13 *Periodic table of elements showing the element iodine (highlighted)*

H																	He
Li	Be											B	C	N	O	F	Ne
Na	Mg											Al	Si	P	S	Cl	Ar
K	Ca	Sc	Ti	V	Cr	Mn	Fe	Co	Ni	Cu	Zn	Ga	Ge	As	Se	Br	Kr
Rb	Sr	Y	Zr	Nb	Mo	Tc	Ru	Rh	Pd	Ag	Cd	In	Sn	Sb	Te	I	Xe
Cs	Ba	La-Lu	Hf	Ta	W	Re	Os	Ir	Pt	Au	Hg	Tl	Pb	Bi	Po	At	Rn
Fr	Ra	Ac-Lr	Rf	Db	Sg	Bh	Hs	Mt	Ds	Rg	Uub						

Figure 10.14 *Periodic table of elements showing the element strontium*

10.3.2 ^{89}Strontium

Strontium is an alkaline-earth element with the atomic number 38, is a member of group 2 in the periodic table of elements and has the chemical symbol Sr (Figure 10.14).

Strontium is a soft grey metal and is more reactive with water than calcium. On contact with water, it produces strontium hydroxide and hydrogen gas. In order to protect the element, strontium metal is usually kept under mineral oil to prevent oxidation. Natural strontium is formed of a mixture of four stable isotopes – ^{84}Sr, ^{86}Sr, ^{87}Sr and ^{88}Sr, with the last one being the predominant one.

^{89}Sr is an artificial radioisotope and is a β-emitter with a half-life of 50.5 days. It is a product of the neutron activation of ^{88}Sr and decays to the stable ^{89}yttrium. Metastron is a product containing ^{89}Sr and is licensed by the FDA. It comes in a ready-to-use vial and expires within 28 days. It is supplied with a calibration vial, so that the pharmacist will be able to ensure that the patients get the accurate dose prescribed [2].

Because of the similarity of strontium and calcium (neighbouring elements in the periodic table of elements), strontium is believed to be metabolised in the human body in a similar way and accumulates, for example, in the bones. This has led to its application as a treatment option for pain caused by bone metastasis. It is known that >50% of patients with prostate, breast or lung cancer will develop painful bone metastasis. The exact mechanism of relief from bone pain is not known. $^{89}SrCl_2$ is administered intravenously and, as its distribution in the human body is similar to that of calcium, it is quickly cleared from the blood and deposited in the bone mineral. Strontium can be found in the hydroxyapatite cells of the bones rather in bone marrow cells. The radioisotope ^{89}Sr delivers localised β-radiation, inducing a pain relieving effect. A majority of the administered $SrCl_2$ is actively distributed to the metastases. Any free $SrCl_2$ is excreted renally or along with the faeces [2].

Low platelet count is the most likely side effect occurring in patients being treated with $^{89}SrCl_2$. Platelet counts should return to preadministration levels after 6 months once treatment is finished. Treatment with $^{89}SrCl_2$ is not recommended in patients with an already low platelet or white blood cell count, and for patients receiving this treatment the blood parameters have to be regularly checked even after the treatment is completed.

10.3.3 Boron neutron capture therapy (BNCT)

Boron has two stable isotopes, ^{10}B and ^{11}B, and 14 radioisotopes with very short half-lives. ^{11}B is the most abundant isotope and represents 80% of natural boron, whilst ^{10}B (~20%) finds a significant clinical application in the so-called boron neutron capture therapy (BNCT).

BNCT is a noninvasive treatment option for malignant tumours, especially brain tumours and head and neck cancers, and is currently under clinical trials. The patient is injected with a nonradioactive ^{10}B-containing compound that acts as a neutron-capturing agent and shows high selectivity to cancer tissues. Once the compound has reached the tumour, the patient is exposed to a beam of low-energy neutrons, the so-called epithermal neutrons. These neutrons lose their energy once they penetrate the skin, but they can still interact with the neutron-capturing agent and initiate a nuclear reaction. This reaction of ^{10}B with a neutron results in the conversion to the nonradioactive isotope ^{7}Li and low-energy gamma radiation together with the emission of α-radiation ($^{4}{}_{2}He^{2+}$ particles). α-Radiation is a radiation with a short range and bombards the local tumour tissue from within the tumour cells. The linear energy transfer (LET) of these α-particles ranges approximately one cell diameter, which means there is minimum exposure to healthy tissue (Figure 10.15).

$$^{10}B + n_{th} \rightarrow [^{11}B] \rightarrow {}^{4}He + {}^{7}Li + 2.31\,MeV$$

A variety of carrier molecules for ^{10}B have been investigated, including carbohydrates, antibodies, liposomes and amino acids. There are currently only two boron compounds as BNCT delivery agents used in clinical trials. Sodium mercaptoundecahydro-*closo*-dodecaborate ($Na_2B_{12}H_{11}SH$), known as *borocaptate* (BSH), was mainly used in clinical trials in Japan, whereas the boron-based amino acid (L)-4-dihydroxy-borylphenylalanine (BPA, boronophenylalanine) is used in clinical trials in Europe and the United states (Figure 10.16) [3, 4].

10.4 Radiopharmaceuticals for imaging

Radiopharmaceuticals are typically administered intravenously and then distributed to a particular organ. The molecule itself, and not the radiolabelling, will determine to which organ the radioactive molecule is transported. γ-Radiation is detected externally by using a special scintillation detector, also known as a *gamma*

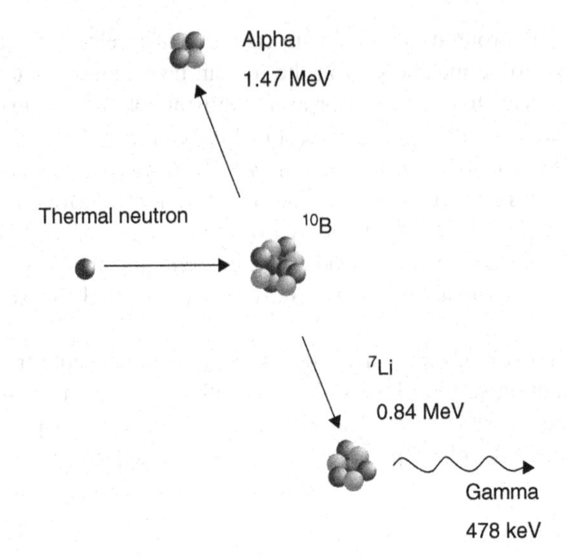

Figure 10.15 *The mechanism of BNCT [3] (Reproduced with permission from [3]. Copyright © 2005, John Wiley & Sons, Ltd.)*

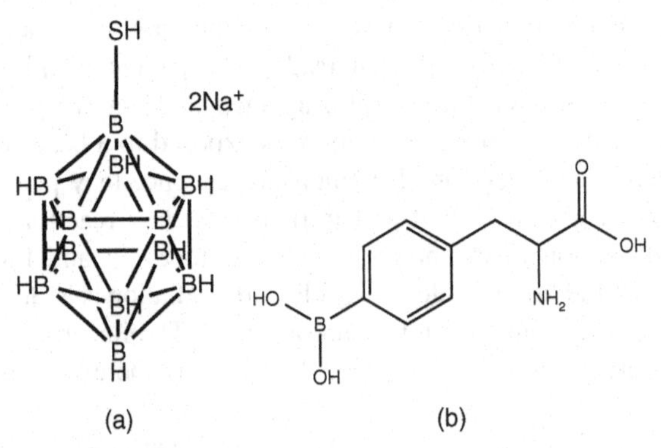

Figure 10.16 *Chemical structure of sodium mercaptoundecahydro-*closo*-dodecaborate (BSA) (a) and (L)-4-dihydroxy-borylphenylalanine (BPA) (b)*

camera. The camera captures the emitted radiation and forms a two-dimensional image. This diagnostic test is also called *scintigraphy*.

In contrast, PET produces a three-dimensional image of the functional processes in the human body. The method is based on the use of positron-emitting radionuclides and their indirectly emitted gamma rays. Radionuclides, the so-called tracers, are introduced to the body as parts of biologically active molecules. PET also uses gamma cameras to detect the internally applied radiation, but in modern scanners, three-dimensional images are often achieved with the aid of a CT X-ray scan performed at the same time as part of the same machine.

Diagnostic X-ray uses external radiation, which is sent through the body to produce a two-dimensional image, whereas scintography is based on the internal accumulation of radionuclides.

H																	He
Li	Be											B	C	N	O	F	Ne
Na	Mg											Al	Si	P	S	Cl	Ar
K	Ca	Sc	Ti	V	Cr	Mn	Fe	Co	Ni	Cu	Zn	Ga	Ge	As	Se	Br	Kr
Rb	Sr	Y	Zr	Nb	Mo	**Tc**	Ru	Rh	Pd	Ag	Cd	In	Sn	Sb	Te	I	Xe
Cs	Ba	La-Lu	Hf	Ta	W	Re	Os	Ir	Pt	Au	Hg	Tl	Pb	Bi	Po	At	Rn
Fr	Ra	Ac-Lr	Rf	Db	Sg	Bh	Hs	Mt	Ds	Rg	Uub						

Figure 10.17 *Periodic table of elements showing the element technetium (highlighted)*

10.4.1 ^{99m}Technetium

Technetium has the chemical symbol Tc and atomic number 43. It is the lightest element that has no stable isotope. It is a silvery-grey transition metal (Figure 10.17).

^{99m}Tc (also referred to as *technetium-99m*) is the metastable isomer of ^{99}Tc, which is a gamma-emitting nuclide routinely used in diagnostic medicine. It has a short half-life of around 6 h, which is ideal for diagnostic applications (but not for therapeutic applications) as it helps to keep the radiation exposure to the patient low. The use of a gamma camera allows detection of the radioactive tracer in the body and creates images of the area in question (Figure 10.18).

One challenge of using a radioactive material is to safely manufacture the products and deliver them to the clinical setting. Radionuclides with long half-lives are usually prepared commercially using a nuclear reactor and supplied as the finished product. Products containing radionuclides with a short half-life cannot be delivered as the finished product because of their rapid decay. Therefore, they are delivered to the clinical setting as radionuclides with a long half-life and the desired radionuclide is then generated and formulated at the moment of use. ^{99m}Tc and its compounds are generated *in situ* for use as an imaging agent using a so-called ^{99m}Tc generator. The generator is loaded with molybdenum-99 (^{99}Mo), which is often referred to as the commercially available transportable source of ^{99m}Tc. The general idea is that the generator contains a long-lasting ‘parent’ compound, which decays and produces the ‘daughter’ radionuclide. In the case of the ^{99m}Tc generator, it contains $^{99m}MoO_4^{2-}$ absorbed on an alumina column. $^{99m}MoO_4^{2-}$ decays to $^{99m}TcO_4^-$, which can be removed as $Na^{99m}TcO_4$ when the column is washed with a NaCl solution. Hospitals tend to buy these generators on a regular basis to provide a continuous supply of $^{99m}TcO_4^-$ (Figure 10.19).

Compounds containing ^{99m}Tc can be used for imaging a variety of functions and structures in the human body. The use of different molecules containing ^{99m}Tc determines to which part of the body the radionuclide is transported and which structure can be imaged. There are a variety of different molecules, but, for example, ^{99m}Tc-aerosol can be used for the imaging of lung ventilation, whereas ^{99m}Tc-albumin is generally used for judging cardiac function. ^{99m}Tc-albumin is an injectable solution prepared by combining sodium pertechnetate ($NaTcO_4$) and human albumin in the presence of a reducing agent such as a tin salt [1].

^{99m}Tc-medronate is used for skeletal imaging, and the succimer analogue is used for preparing images of the kidney. ^{99m}Te succimer injection is prepared by reacting sodium pertechnetate ($NaTcO_4$) with *meso*-2,3-dimercaptosuccinic acid in the presence of a reducing agent such as a stannous salt (Figure 10.20) [1].

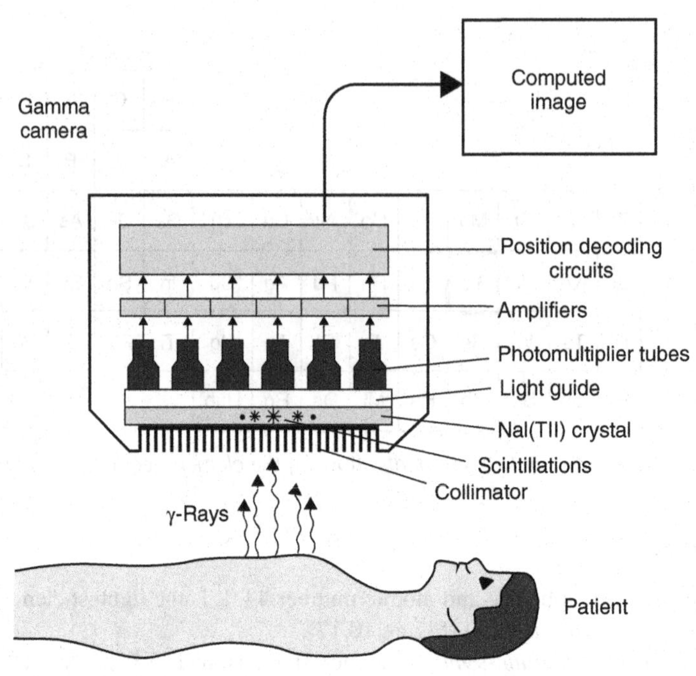

Figure 10.18 *Scheme showing the use of a gamma camera on a patient treated with a* ^{99m}*Tc imaging agent [5] (Reprinted with permission from the Federation of American Scientists. http://www.fas.org/irp/imint/docs/rst/Intro/img003.gif.)*

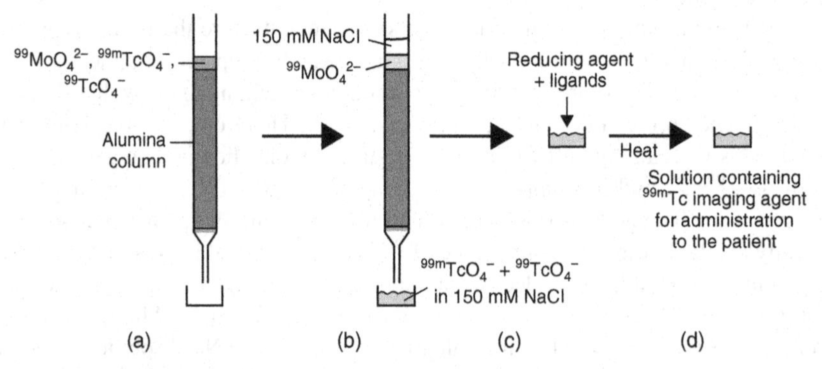

Figure 10.19 *(a–d) Illustration of a* ^{99m}*Tc generator [5] (Reproduced with permission from [5]. Copyright © 2009, John Wiley & Sons, Ltd.)*

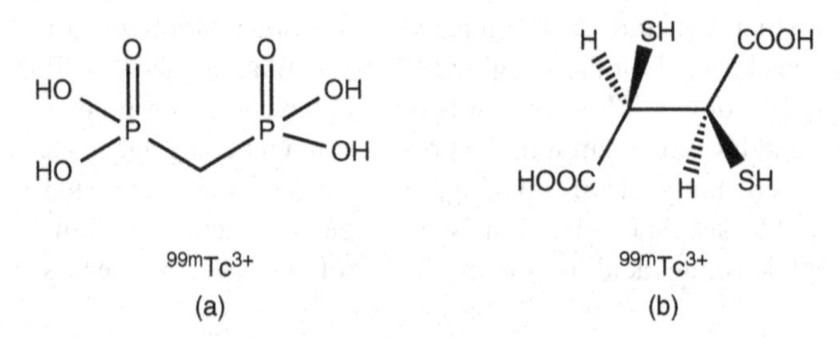

Figure 10.20 *Chemical structures of* ^{99m}*Tc-medronate (a) and* ^{99m}*Tc-succinate (b)*

Cardiolite is an organometallic compound based on ^{99m}Tc, which has become one of the most used nuclear imaging agents to visualise the heart muscle and abnormalities of the parathyroid. Cardiolite is the trade name of ^{99m}Tc-sestamibi, which is a coordination complex of ^{99m}Tc with six so-called MIBI ligands. MIBI stands for methoxyisobutylisonitrile. The full chemical name is (OC-6-11)-hexakis[1-(isocyano-κC)-2-methoxy-2-methylpropane][^{99m}Tc]technetium(I) chloride. A typical solution for injection is prepared by heating a solution tetrakis[(2-methoxy-2-methylpropyl-1-isocyanide)copper(I)] tetrafluoroborate, which is a weak chelating agent, and sodium pertechnetate ($NaTcO_4$) in the presence of a stannous salt (Figure 10.21) [1].

^{99m}Tc-exametazime is a ^{99m}Tc preparation that can be used to visualise damage to the brain, for example, in the evaluation and localisation of stroke damage, head trauma, dementia and cerebral function impairment (Figure 10.22 and Table 10.5) [2].

Each of these ^{99m}Tc-containing compounds is freshly prepared by the radiopharmacist strictly following a standard protocol issued by the supplier. Usually, all ingredients are supplied in closed vials, mostly

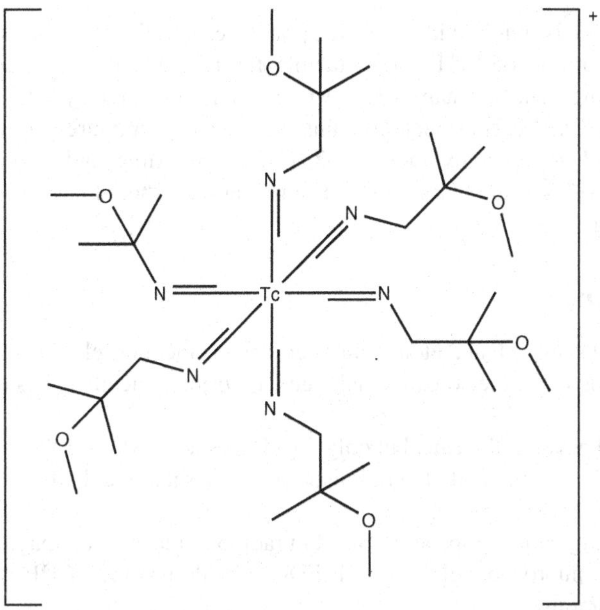

Figure 10.21 Chemical structure of Cardiolite™

Table 10.5 Some examples of common ^{99m}Tc radiopharmaceuticals for imaging procedures

Radiopharmaceutical	For imaging of
^{99m}Tc-aerosol	Lung ventilation
^{99m}Tc-sestamibi	Heart
^{99m}Tc-albumin	Cardiac function
^{99m}Tc-exametazime	Brain
^{99m}Tc-medronate	Bones
^{99m}Tc-succimer	Kidney

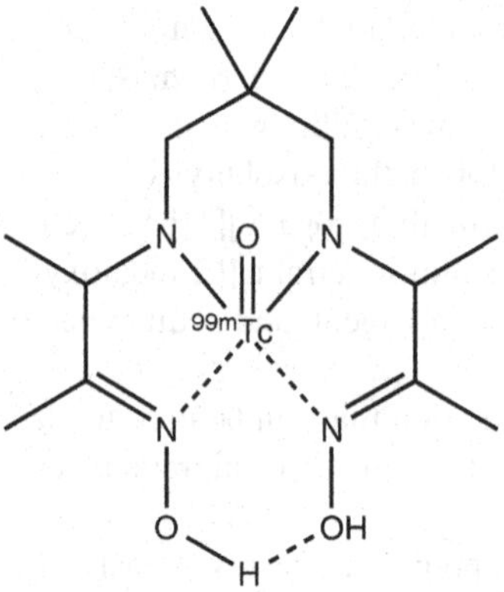

Figure 10.22 Chemical structure of ^{99m}Tc-exametazime

characterised as reagent vials, buffer vials and, if applicable, a vial containing stabiliser. For illustration purposes, only the preparation of ^{99m}Tc-exametazime for injection (as supplied by GE Healthcare) is explained in the following. The nonstabilised formulation is prepared by adding 54 mCi of $^{99m}TcO_4^-$ to a 5 ml reagent vial. The reagent vial contains the racemic mixtures of the ligand exametazime [(3*RS*,9*RS*)-4,8-diaza-3,6,6,9-tetramethylundecane-2,10-dione bisoxime] and stannous chloride dehydrate as reducing agent together with sodium chloride [1]. The preparation should have a pH of 9.0–9.8 and should be used within 30 min [2].

10.4.2 ^{18}Fluoride: PET scan

Fluorine has the chemical symbol F and atomic number 9 and is the most electronegative element. It belongs to group 17 of the periodic table, the so-called halogens. Fluorine typically exists as a diatomic molecule at room temperature.

There are 18 isotopes known of fluorine, but only 1 (^{19}F) is stable. Most of the radioactive isotopes have a very short half-life, mostly <1 min. Only the radioisotope ^{18}F has a longer half-life of around 110 min and is clinically used (Figure 10.23).

^{18}F is a positron-emitting radioisotope and is used in radiopharmaceutical imaging such as PET scanning. Two compounds, namely fluorodeoxyglucose (^{18}F-FDG) and derivatives of ^{18}F choline, are under intense clinical investigation and/or use.

^{18}F-FDG is a glucose derivative that contains a radiolabel (^{18}F) at the 2′ position replacing the hydroxyl group. ^{18}F-FDG is administered intravenously and is used as an assessment of problems with glucose metabolism, especially in the brain, often associated with epilepsy and in cancer. Areas where an increased absorption of ^{18}F-FDG are visible correlate to areas where an increased glucose metabolism is present. ^{18}F-FDG is distributed around the body similar to glucose and is cleared renally. There are no known contraindications known to ^{18}F-FDG (Figure 10.24).

^{18}F-FDG is the main radioimaging agent used in PET scanning. Examples include studies of heart, where it is used to differentiate between dead and live tissue in order to assess the myocardium. In neurology, it can be used to diagnose dementia, seizure disorders or tumours of the brain. ^{18}F-FDG is generally used to assess the extent of the tumour in a cancer patient. Cancerous tissue is characterised by increased cell proliferation, which requires energy, and therefore an increased amount of glucose. This leads to an accumulation of ^{18}F-FDG in

H																	He
Li	Be											B	C	N	O	**F**	Ne
Na	Mg											Al	Si	P	S	Cl	Ar
K	Ca	Sc	Ti	V	Cr	Mn	Fe	Co	Ni	Cu	Zn	Ga	Ge	As	Se	Br	Kr
Rb	Sr	Y	Zr	Nb	Mo	Tc	Ru	Rh	Pd	Ag	Cd	In	Sn	Sb	Te	I	Xe
Cs	Ba	La-Lu	Hf	Ta	W	Re	Os	Ir	Pt	Au	Hg	Tl	Pb	Bi	Po	At	Rn
Fr	Ra	Ac-Lr	Rf	Db	Sg	Bh	Hs	Mt	Ds	Rg	Uub						

Figure 10.23 *Periodic table of elements showing the element fluorine (highlighted)*

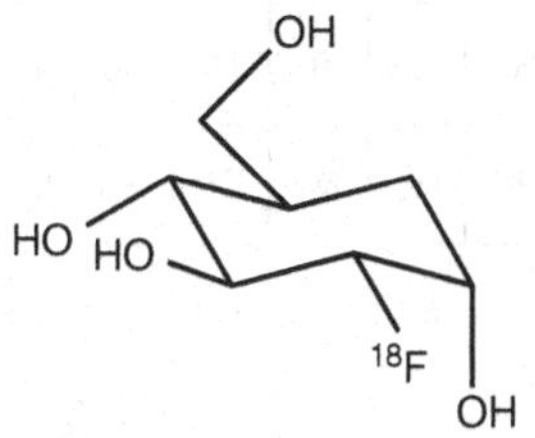

Figure 10.24 *Chemical structure of ^{18}F-FDG*

malignant tumours and allows judging the degree of metastasis formed. This information is important for any surgical procedure and also for the initial assessment of the cancer stage.

Unfortunately, there are limitations to the use of ^{18}F-FDG, as its uptake is not very specific. As a result, other conditions can also cause an accumulation of ^{18}F-FDG and can lead to misdiagnosis. These conditions include inflammation and healing of wounds, which also show increased glucose metabolism.

Therefore, a variety of other ^{18}F-labelled compounds are under intense scrutiny as alternative PET scanning agents, mainly compounds with a more specific biological pathway. This includes ^{18}F-choline. Choline is a compound incorporated into the cell membrane and therefore cells dividing at a fast rate have an increased need for this substance. Studies for a range of tumours were undertaken, but most studies focussed on prostate cancer. In comparison to ^{18}F-FDG, ^{18}F-choline showed less activity in the bladder and a prolonged elimination via the kidneys. Additionally, biological processes other than cancer also include rapid division of cells and can lead to misdiagnosis (Figure 10.25).

10.4.3 ^{67}Gallium: PET

As previously mentioned (see Chapter 4), gallium consist of two stable isotopes (^{69}Ga and ^{71}Ga) and there are two radioisotopes (^{67}Ga and ^{68}Ga) that are commercially available. ^{67}Ga has a half-life of 3.3 days, whereas ^{68}Ga has an even shorter half-life of 68 min (Figure 10.26).

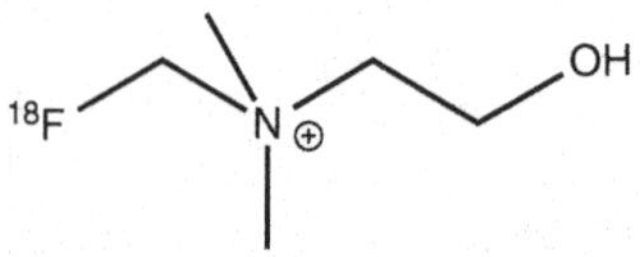

^{18}F — N⊕ — OH

Figure 10.25 Chemical structure of ^{18}F-choline

H																	He
Li	Be											B	C	N	O	F	Ne
Na	Mg											Al	Si	P	S	Cl	Ar
K	Ca	Sc	Ti	V	Cr	Mn	Fe	Co	Ni	Cu	Zn	**Ga**	Ge	As	Se	Br	Kr
Rb	Sr	Y	Zr	Nb	Mo	Tc	Ru	Rh	Pd	Ag	Cd	In	Sn	Sb	Te	I	Xe
Cs	Ba	La-Lu	Hf	Ta	W	Re	Os	Ir	Pt	Au	Hg	Tl	Pb	Bi	Po	At	Rn
Fr	Ra	Ac-Lr	Rf	Db	Sg	Bh	Hs	Mt	Ds	Rg	Uub						

Figure 10.26 Periodic table of elements showing the element gallium (highlighted)

^{67}Ga decays via electron capture and subsequently emits γ-rays, which can be detected with a gamma camera. ^{68}Ga is a positron-emitting isotope and is used for PET. Because of its short half-life, fresh ^{68}Ga for clinical applications is obtained through generators. The generator is equipped with the parent compound ^{68}Ge, which has a half-life of 271 days and decays via electron capture to form the 'daughter' ^{68}Ga.

It has been reported that radioactive gallium-67 citrate accumulates in malignant cells when injected into animals that are infected with tumours. This has led to the development of ^{67}Ga scans, which have been used over the past two decades mostly for the detection of residual cancer cells in patients with Hodgkin's and non-Hodgkin's lymphomas after chemo or radiotherapy. The level of ^{67}Ga present in lymphoma cells correlates with their metabolic activity and directly with their proliferation rate. Therefore, a positive ^{67}Ga scan (mostly undertaken after chemotherapy) indicates the survival of malignant cells and the need for further treatment (Figure 10.27).

As previously mentioned (see Chapter 4), Ga^{3+} is mainly transported by transferrin. *In vitro* studies have shown that the uptake of the radioactive gallium into the cancer cells was significantly increased when transferrin was added to the medium [6].

10.4.4 ^{201}Thallium

The element thallium belongs to the boron group, and has the chemical symbol Tl and atomic number 81. Thallium is a soft grey metal, which cannot be found as the free metal in nature (Figure 10.28).

The common oxidation states for thallium are +3, which resembles the oxidation states of other group members, and +1, which is actually the far more dominant oxidation state for thallium ions. Thallium ions

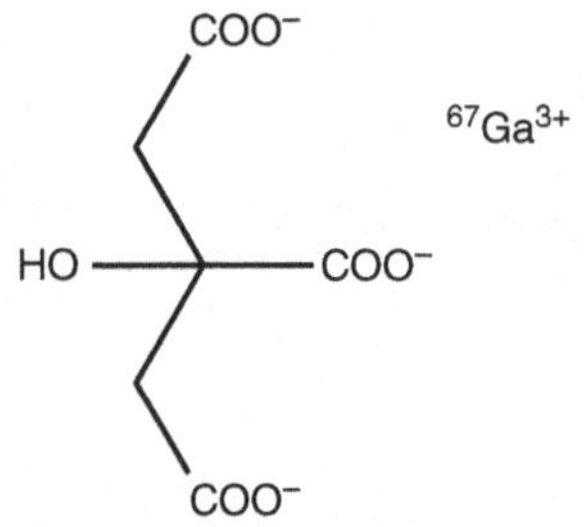

COO⁻ / HO — C — COO⁻ / COO⁻ $^{67}Ga^{3+}$

Figure 10.27 Chemical structure of gallium citrate

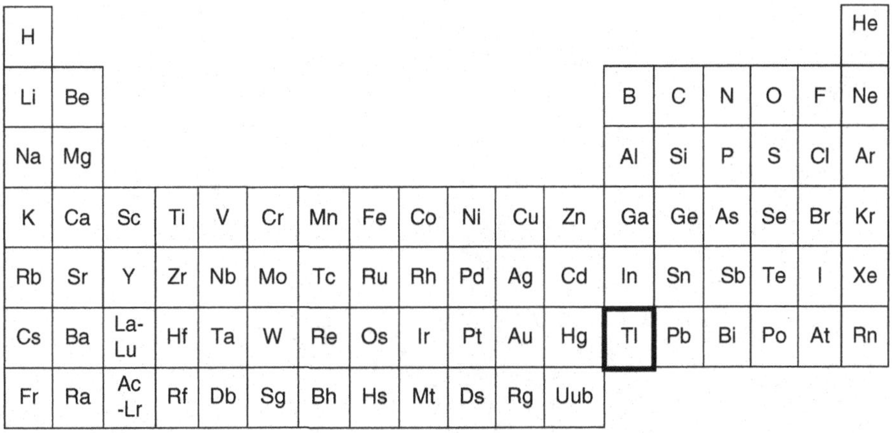

H																	He
Li	Be											B	C	N	O	F	Ne
Na	Mg											Al	Si	P	S	Cl	Ar
K	Ca	Sc	Ti	V	Cr	Mn	Fe	Co	Ni	Cu	Zn	Ga	Ge	As	Se	Br	Kr
Rb	Sr	Y	Zr	Nb	Mo	Tc	Ru	Rh	Pd	Ag	Cd	In	Sn	Sb	Te	I	Xe
Cs	Ba	La-Lu	Hf	Ta	W	Re	Os	Ir	Pt	Au	Hg	**Tl**	Pb	Bi	Po	At	Rn
Fr	Ra	Ac-Lr	Rf	Db	Sg	Bh	Hs	Mt	Ds	Rg	Uub						

Figure 10.28 Periodic table of elements showing the element thallium

with the oxidation state +1 follow alkali metals in their chemical behaviour and are handled in biological systems similar to potassium (K^+) ions.

Thallium and many of its compounds are toxic. In particular, the Tl^+ cation displays good aqueous solubility and it can enter the body via the potassium-based uptake processes as its behaviour is similar to that of K^+. Unfortunately, there are differences in the chemistry of both ions that affect, for example, their binding to sulfur-containing molecules and lead to the toxicity of thallium ions. Thallium-based compounds were used as rat poison, but their use is nowadays discontinued as their toxic properties are not very specific. Signs of thallium poisoning include hair loss, nerve damage and, ultimately, at high enough doses, sudden death.

The radioactive thallium isotope ^{201}Tl was the main substance used for nuclear imaging in cardiology. It was used for the so-called thallium nuclear cardiac stress test, where a radiotracer such as $^{201}TlCl$ (thallous chloride-201) is injected into a patient during exercise. After a short waiting period (in order to ensure good distribution of the radioactive substance), images of the heart are taken with a gamma camera and the blood flow within the heart muscle is evaluated. Nowadays, the radio isotope has been mostly replaced by ^{99m}Tc imaging. The radio isotope ^{201}Tl has a half-life of 73 h and can be generated using a transportable thallium-201 generator. This generator uses ^{201}Pb (lead-201) as the 'parent', which decays via electron capture to the 'daughter' ^{201}Tl. ^{201}Tl decays by electron capture and has good imaging characteristics [7].

10.5 Exercises

10.5.1 Write the elemental formula for the following radioisotopes:

(a) Technetium containing 48 neutrons
(b) Radon containing 136
(c) Francium containing 136 neutrons
(d) Radium containing 138 neutrons

10.5.2 Write the equation for the radioactive decay of the following elements:

(a) ^{210}Po (α-emitter)
(b) ^{226}Ra (α-emitter)
(c) ^{91}Tc (β-emitter – positron)
(d) ^{227}Ac (β-emitter – negatron)

10.5.3 A ^{201}Tl chloride injection has a labelled activity of 450 μCi. Express this answer in megabecquerel.

10.5.4 A radioactive material has an activity of 12.25 mCi. How many disintegrations per second are represented by this?

10.5.5 A radioactive material has been labelled with an activity of 112 MBq. Convert this activity into curie.

10.5.6 If a radioactive element has a half-life of 2 h, what percentage of material is left after

(a) 2 h
(b) 6 h
(c) 8 h

10.5.7 The disintegration constant of ^{24}Na is 0.0462 year^{-1}. Determine the half-life of this radioisotope.

10.6 Case studies

10.6.1 A sample containing ^{99m}Tc was found to have a radioactivity of 15 mCi at 8 a.m. when the sample was tested.

(a) Research the half-life of ^{99m}Tc.
(b) Calculate its activity at 5 a.m. on the same day, when it was prepared.
(c) Calculate its activity at 3 p.m. on the same day, when it was administered to the patient.

State your answer in curies and SI units.

10.6.2 A typical intravenous dose of ^{99m}Tc-albumin used for lung imaging contains a radioactivity of 4 mCi

(a) Convert the dose to SI unit.
(b) What radioactive dose is left after 12 h, when the technetium is cleared from the body?
(c) The pharmacist prepared the sample actually 2 h before the administration. What activity did this sample have at that point of preparation?
(d) The pharmacist prepared a 2 ml solution for injection. What is the activity concentration at the point of administration?

10.6.3 Develop a quick-reference radioactive decay chart for ^{131}I

Research the half-life of ^{131}I and calculate the percentage activity remaining for 20 days using 1-day intervals. Create your own quick-reference radioactive decay chart.

Day	Percentage activity remaining (%)
1	
2	
3	
4	
5	
6	
⋮	

References

1. *British pharmacopoeia*, Published for the General Medical Council by Constable & Co, London, **2014**.
2. B. T. Smith, *Nuclear pharmacy: concepts and applications*, Pharmaceutical Press, London, **2010**.
3. E. R. Tiekink, M. Gielen, *Metallotherapeutic drugs and metal-based diagnostic agents: the use of metals in medicine*, Wiley, Chichester, **2005**.
4. (a) J. W. Hopewell, G. M. Morris, A. Schwint, J. A. Coderre, *Appl. Radiat. Isot.* **2011**, *69*, 1756–1759; (b) R. F. Barth, *Appl. Radiat. Isot.* **2009**, *67*, S3–S6.
5. J. C. Dabrowiak, *Metals in medicine*, Wiley-Blackwell, Oxford, **2009**.
6. C. R. Chitambar, *Int. J. Environ. Res. Publ. Health* **2010**, *7*, 2337–2361.
7. C. Rosendorff, *Essential cardiology: principles and practice*, 2nd ed., Humana Press, Totowa, N.J., **2005**.

Further Reading

E. Alessio, *Bioinorganic medicinal chemistry*, Wiley-VCH, Weinheim, **2011**.

F. A. Cotton, G. Wilkinson, C. A. Murillo, M. Bochmann, *Advanced inorganic chemistry*, 6th ed., Wiley, New York; Chichester, **1999**.

J. D. Lee, *Concise inorganic chemistry*, 5th ed., Chapman & Hall, London, **1996**.

G. A. McKay, M. R. Walters, J. L. Reid, *Lecture notes. Clinical pharmacology and therapeutics*, 8th ed., Wiley-Blackwell, Chichester, **2010**.

G. J. Tortora, B. Derrickson, *Principles of anatomy and physiology*, 12th ed., international student/Gerard J. Tortora, Bryan Derrickson. ed., Wiley [Chichester: John Wiley, distributor], Hoboken, N.J., **2009**.

11

Chelation Therapy

Chelation therapy is considered as one treatment option for heavy-metal poisoning. It is also used as a potential therapy for Wilson disease, which is caused by a build-up of copper ions in the body (see Chapter 7). The basic principle behind chelation therapy is to capture the metal ions and remove them from the body.

11.1 What is heavy-metal poisoning?

Heavy-metal poisoning is defined as the accumulation of toxic metal in the body, mainly in the soft tissue. There is an on-going debate on how to define heavy metals. Mostly, it is defined as an element that has more than five times the density of water. With regard to danger to human health, all metals that cause harm to the body should be included, such as lead, mercury, arsenic, thallium and cadmium. Some heavy metals, such as zinc, copper, chromium, iron and manganese, are required by the body in small amounts but are toxic in larger quantities (see Chapter 7) (Figure 11.1).

Heavy metals can be taken up by ingestion (food or drink), through air (inhalation) or also by absorption through the skin. Heavy metals are then mostly stored in the soft tissue. Places of exposure can often be traced back to the work place (industrial work, pharmaceutical industry or agriculture). Contaminated sand and soil on playgrounds are known to have been responsible for heavy-metal poisoning in children. Within the human body, toxic heavy metals typically compete with essential metals, such as magnesium, zinc, iron, calcium and others for their receptors. This can lead to irreversible organ damage.

Depending on the type and quantity of heavy metal absorbed, the patient will display varying symptoms. These may include vomiting, nausea, diarrhoea, sweating, headache and a metallic taste in the mouth. In severe cases, heavy-metal poisoning can lead to impairment of cognitive, motor and language skills in the patients. The famous expression 'mad as a hatter' originated in seventeenth-century France, where hat makers used a toxic mercury salt (mercuric nitrate) to soak animal hides (Figure 11.2).

Heavy-metal poisoning is diagnosed using blood and urine tests, or by hair, tissue and X-ray analysis. Upper concentration limits in blood depend on the individual metal. Whilst very high concentration levels lead to serious health concerns, it has been shown that, especially in children, even lower levels can lead to chronic health problems. Additionally, the length of exposure can be crucial for developing serious health problems. Exposure to a heavy metal of a long period can be highly toxic even at low levels. Diagnosing the concentration

Essentials of Inorganic Chemistry: For Students of Pharmacy, Pharmaceutical Sciences and Medicinal Chemistry,
First Edition. Katja A. Strohfeldt.

Companion website: www.wiley.com/go/strohfeldt/essentials

H																	He
Li	Be											B	C	N	O	F	Ne
Na	Mg											Al	Si	P	S	Cl	Ar
K	Ca	Sc	Ti	V	Cr	Mn	Fe	Co	Ni	Cu	Zn	Ga	Ge	As	Se	Br	Kr
Rb	Sr	Y	Zr	Nb	Mo	Tc	Ru	Rh	Pd	Ag	Cd	In	Sn	Sb	Te	I	Xe
Cs	Ba	La-Lu	Hf	Ta	W	Re	Os	Ir	Pt	Au	Hg	Tl	Pb	Bi	Po	At	Rn
Fr	Ra	Ac-Lr	Rf	Db	Sg	Bh	Hs	Mt	Ds	Rg	Uub						

Figure 11.1 *Periodic table showing examples of metals that are harmful to the human body (black), and metals that are essential in small amounts and toxic in larger quantities (light grey)*

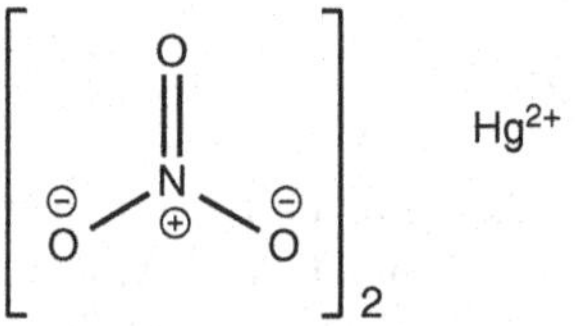

$$[NO_3^-]_2 \quad Hg^{2+}$$

Figure 11.2 *Chemical structure of mercuric nitrate*

of heavy metal, which rapidly clears from the blood, can be fairly difficult. Arsenic clears rapidly from the blood stream, and it might be possible to detect arsenic poisoning in the urine up to ~48–72 h. Acute arsenic poisoning especially after ingestion of arsenic may be visualised by using X-ray diagnosis, as arsenic is opaque to X-rays. Furthermore, arsenic can be detected in hair and nail for a month following the exposure.

The main treatment option for heavy-metal poisoning is the so-called chelation therapy. Chelation therapy is based on the principle that the toxic heavy metal is bound to a chelating agent, which reduces its toxicity and removes it from the human body.

11.2 What is chelation?

Chelation is the binding of molecules to a metal ion and is one of the most important concepts within the area of transition-metal complexes. The molecules or ligands in question are organic compounds and can form more than one bond to the metal ion. These ligands are classified as polydentate ligands, which means that this single ligand has more than one atom that can bind to the central atom in a coordination complex. The ligands are called *chelators* or *chelating ligands* (Figure 11.3).

> ***Chelation*** *is defined as the formation of two or more coordinate bonds between a polydentate ligand and a single atom.*

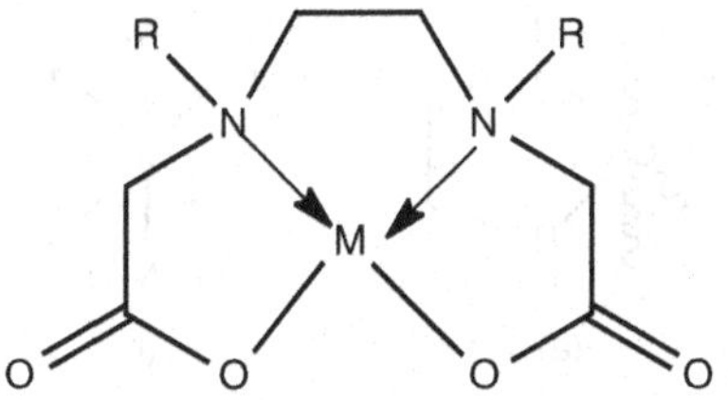

Figure 11.3 Chemical structure of an example for a chelation complex

In terms of the nomenclature, the denticity of a ligand is denoted by the Greek letter κ (kappa). If the denticity equals 1, it is a monodentate ligand and only one bond from the ligand to the metal is formed. For bidentate ligands, the denticity equals 2, which means that two coordination bonds are formed from the ligand to the metal. One of the most known examples is EDTA (ethylenediaminetetraacetic acid), which is a hexadentate ligand. EDTA can coordinate to a metal ion via six atoms and therefore is denoted as κ^6-EDTA. It is important not to confuse denticity with hapticity (see Chapter 8), which only refers to bonds that are contiguous (Figure 11.4).

The term *chelate effect* is used to describe the preferred binding of a chelating ligand to a metal ion in comparison to the corresponding amount of monodentate (nonchelating) ligand under the same reaction conditions. The most widely known example is the reaction of a transition metal (e.g. Cu^{2+}) with 1 equiv of the bidentate ligand ethylenediamine (en) or with 2 equiv of the monodentate ligand amine (NH_3) (Figure 11.5).

The reaction of the bidentate ligand ethylenediamine results in the formation of a chelate complex in which the ligand forms two bonds to the metal centre and a five-membered ring is formed (see Figure 11.6). The same reaction takes places with 2 equiv of the monodentate ligand amine. Nevertheless, under the same reaction conditions, the concentration of the five-membered coordination complex (product A) will be significantly higher than product B. This is called the *chelate effect*.

The chelate effect increases with the number of chelate rings formed. This explains why EDTA is a very good chelating agent, as it can form up to six coordinating bonds to the metal centre. In general, the chelate

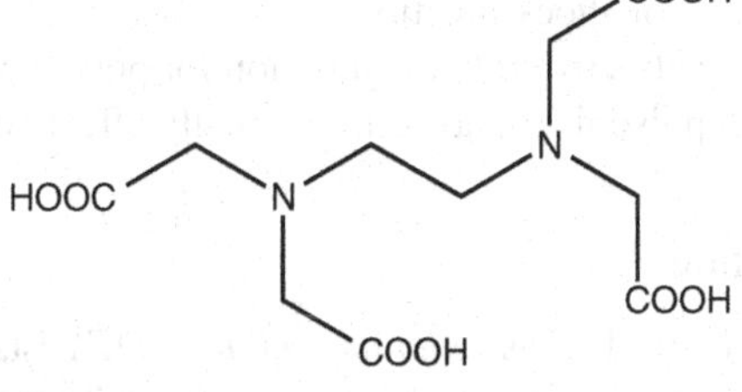

Figure 11.4 Chemical structure of EDTA

$$Cu^{2+} + 3en \rightarrow Cu(en)_3{}^{2+}(A)$$

$$Cu^{2+} + 6NH_3 \rightarrow Cu(NH_3)_6{}^{2+}(B)$$

Figure 11.5 Complexation of Cu^{2+} with ethylenediamine (en) or amine

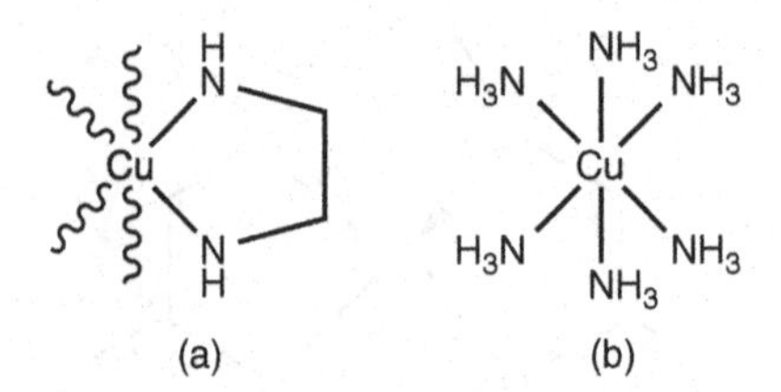

Figure 11.6 *Coordination of Cu^{2+} with ethylenediamine (a) and NH_3 (b)*

effect can be explained by using a thermodynamic approach, which considers the equilibrium constants for the reaction. Nevertheless, this will not be further discussed here, as it is beyond the scope of this book.

11.3 Chelation therapy

As previously mentioned, chelation therapy is one treatment option for most heavy-metal poisoning. A chelating agent (antidote), which is specific to the metal involved, is administered to the patient orally, intramuscularly or intravenously. Dimercaprol (BAL, British anti-Lewisite), calcium disodium edetate and penicillamine are the three most common chelating agents used in the treatment of heavy-metal poisoning. The unlicensed drug Succimer (DMSA, *meso*-2,3-dimercaptosuccinic acid) may be valuable in the treatment of most forms of heavy-metal poisoning including lead, arsenic and mercury. These and other chelating agents such as unithiol (DMPS, 2,3-dimercapto-1-propanesulfonic acid) and α-lipoic acid (ALA) are also used in alternative medicine, which has led to much criticism and discussion. So far, no medical study has proven the effectiveness of chelation therapy for any clinical application other than heavy-metal poisoning.

The mode of action is based on the ability of the chelating agent to bind to the metal in the body's tissues and form a chelate. This complex is then released from the tissue and travels in the bloodstream. Finally, the complex is filtered by the kidneys and excreted in the urine. Unfortunately, this process requires admission to the hospital because it may be painful and it is important to stabilise the vital functions of the patient. The patient additionally may require treatment for complications associated with heavy-metal poisoning, including anaemia and kidney failure or shock reactions.

Chelation therapy is an especially effective treatment option for poisoning with lead, mercury and arsenic. It is very difficult to treat cadmium poisoning, and so far no really effective therapy has been found.

11.3.1 Calcium disodium edetate

Calcium disodium edetate, also referred to as *calcium sodium EDTA*, stands for the chemical compound 2,2′,2″,2‴-(ethane-1,2-diyldinitrilo)tetraacetic acid, which is a synthetic amino acid. Edetate refers to the calcium disodium salt of the chelating agent with the formula $(HO_2CCH_2)_2NCH_2CH_2N(CH_2CO_2H)_2$. In the United States, it is found under the name calcium disodium versenate. Edetate is mainly used to complex di and trivalent metal ions. Edetate can bind to metals via the four carboxylate and two amine groups, and it forms specially strong complexes with Co(III), Cu(II), Mn(II) and Fe(III) (Figure 11.7).

Edetate is indicated for the treatment of lead poisoning, which was especially a big problem for navy personnel after World War II. Staff repainting the navy ships with lead-based paint were exposed to the heavy metal and suffered from symptoms of lead poisoning. Around this time, edetate was introduced as a medicinal chelating agent. The side effects include nausea, diarrhoea and abdominal pain. It can lead to

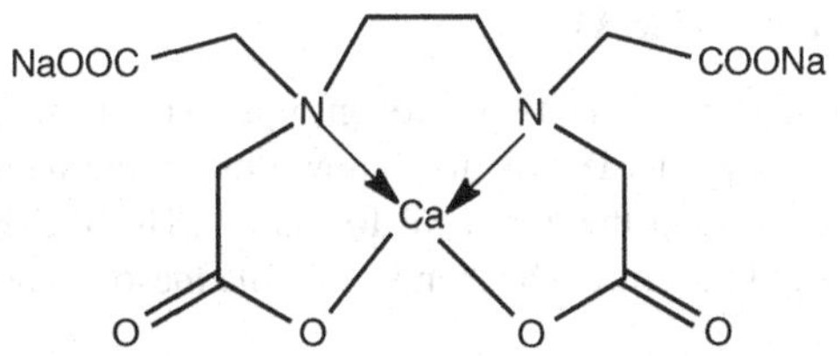

Figure 11.7 Chemical structure of edetate

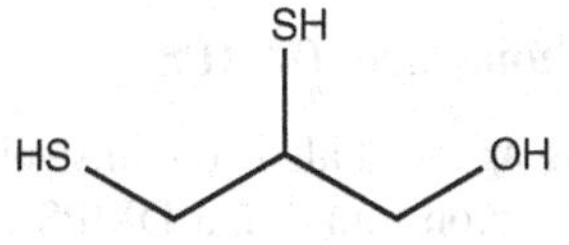

Figure 11.8 Chemical structure of dimercaprol

renal damage if given as over dosage [1]. Nevertheless, these side effects are less serious/invasive than those of most other chelating agents being used. Edetate is typically administered by intravenous infusion for up to 5 days [1].

Other clinical applications include the application of chromium-EDTA, which can be used to evaluate kidney function. It is administered intravenously and its filtration into the urine is monitored. Chromium-EDTA exits the human body only via glomerular filtration as it is not secreted or metabolised in any other way. EDTA and its salts can act as an anticoagulant for blood samples and is therefore often found as additives in blood sampling bottles.

11.3.2 Dimercaprol (BAL)

Dimercaprol (BAL) is a chelating agent used as an antidote for arsenic, antimony, bismuth, gold and mercury poisoning [1]. It has the chemical name 2,3-dimercapto-1-propanol and is a clear, colourless or slightly yellow liquid (Figure 11.8).

British scientists at the University of Oxford also developed BAL during World War II as an antidote to Lewisite. Lewisite is an arsenic-based chemical warfare agent used in form of a blister gas. Further research showed that it can be used as an antidote against a variety of toxic metals. Additionally, it was used in the treatment of Wilson disease, which is a chronic disease in which the body retains excess amounts of copper (see Chapter 7).

Heavy-metal poisoning often results from the coordination of the metal to sulfhydryl groups of enzymes, which means that these enzymes are blocked for their activity. BAL also contains sulfhydryl groups and basically competes with the enzymes for the coordination of the metal. The chelated complex is then excreted in the urine. Whilst BAL removes a range of heavy metals, it also seems to increase the concentration of some metals in the human body and therefore limits its use. It is not indicated as an antidote for cadmium (increased levels are found in the kidneys after treatment), selenium or iron poisoning [1].

Unfortunately, BAL itself is very toxic and has only a narrow therapeutic window. Its multiple side effects include, amongst others, hypertension, malaise, tachycardia, nausea, diarrhoea, burning sensation and muscle pain. The administration is by intramuscular injection and is fairly painful. Because of its instability in water, it is formulated with peanut oil as solvent.

11.3.3 Dimercaptosuccinic acid (DMSA)

DMSA is a modification of BAL containing two thiol groups, which are responsible for the unpleasant smell, and two carboxylic acid groups. DMSA is also known under the name Succimer. It chemical name is *meso*-2,3-dimercaptosuccinic acid and the chemical formula is $HO_2CCH(SH)CH(SH)CO_2H$. There are two diastereomeric forms, meso and the chiral DL forms, with the meso isomer being used as chelating agent (Figure 11.9).

DMSA was developed in the 1960s and replaced BAL and edetate in some countries for the treatment of lead, arsenic and mercury poisoning. Furthermore, the dimethylester modification of DMSA has been successfully used for the treatment of heavy-metal poisoning.

11.3.4 2,3-Dimercapto-1-propanesulfonic acid (DMPS)

DMPS is also a thiol-containing chelating agent. It also contains sulfhydryl groups and an additional sulfate group. Researchers in the former Soviet Union found that DMPS is a useful chelating agent and has some effect as an antidote to mercury (Figure 11.10).

11.3.5 Lipoic acid (ALA)

Lipoic acid, also known as α-*lipoic acid* (*alpha-lipoic acid*) or *thioctic acid*, has the formula $C_8H_{14}S_2O_2$ and systematic name 5-(1,2-dithiolan-3-yl)pentanoic acid. It contains a disulfide group, which can be transformed in the body to a dithiol group (Figure 11.11).

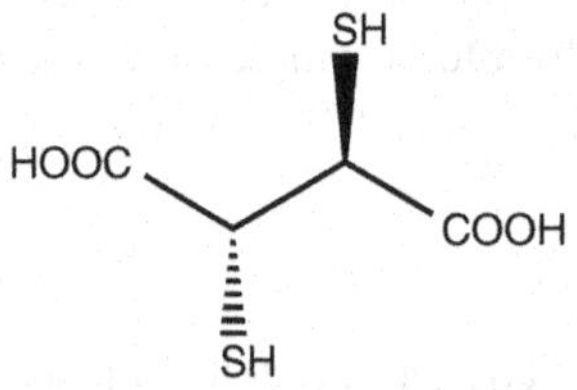

SH
HOOC
COOH
SH

Figure 11.9 Chemical structure of DMSA

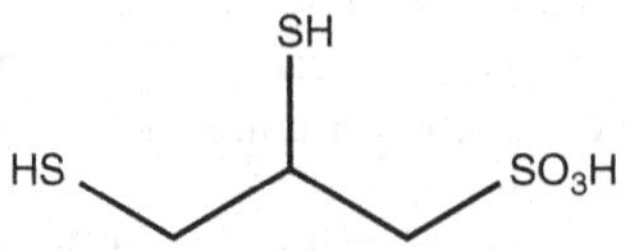

SH
HS
SO_3H

Figure 11.10 Chemical structure of DMPS

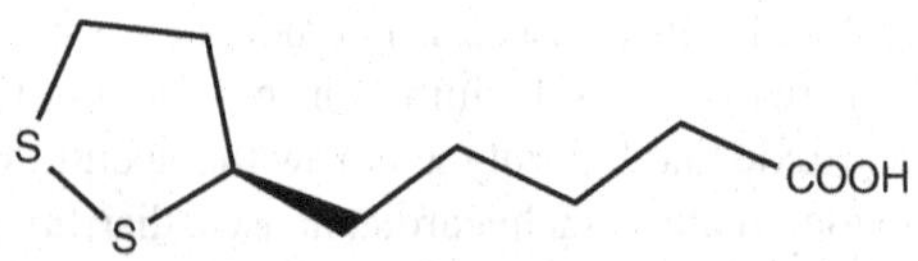

S
S
COOH

Figure 11.11 Chemical structure of ALA

ALA has been on the market since the 1950s as a dietary supplement. It is a natural antioxidant usually made by the body. The advantage of ALA over other antioxidants such as vitamin C and E is that it is soluble both in water and in fat [2]. Researchers in the former Soviet Union found that ALA can chelate mercury once it is transformed into the dithiol-containing compound. ALA can penetrate both the blood–brain barrier and the cell membrane and therefore would be a very interesting chelating agent. Nevertheless, there is much debate about its mode of action, side effects and effectiveness. Other antidotes, such as BAL and DMSA, are more efficient in the removal of heavy metals. ALA has not received FDA approval as a chelating agent, but it is still sold as a food supplement.

11.4 Exercises

11.4.1 Sodium calcium edetate is usually administered by intravenous infusion. A 5 ml ampule contains 200 mg/ml sodium calcium edetate.

(a) What is the number of moles of calcium sodium edetate present in one ampule? Express your answer in moles.

(b) What is the concentration of calcium sodium edetate in one ampule? Express you answer in moles per litre.

11.4.2 Dimercaprol is usually administered by intramuscular injection. A 2 ml ampule typically contains 50 mg/ml dimercaprol.

(a) What is the number of moles of dimercaprol present in one ampule? Express your answer in moles.

(b) What is the concentration of dimercaprol in one ampule? Express you answer in moles per litre.

11.4.3 Draw the structure of all enantiomers of DMSA and discuss which one is used clinically.

11.4.4 Draw the chemical structure of the cobalt(II)–EDTA complex ion.

11.4.5 Draw the chemical structure of Hg^{2+} coordinated by dimercaprol.

11.5 Case studies

11.5.1 Disodium edetate

Your pharmaceutical analysis company has been contacted by an important client and asked to analyse a batch of solutions for infusion containing disodium edetate. According to your brief, you are supposed to analyse the active pharmaceutical ingredient (API), following standard quality assurance guidelines.

Typical analysis methods used for quality purposes are based on titration reactions. A certain amount of the solution is diluted in water and hexamethylenetetramine together with HCl is added. The resulting solution is titrated against a solution of lead nitrate using xylenol orange triturate as indicator [3].

(a) Research the type of titration described. Describe the chemical structure and mode of action of the indicator.
(b) Formulate the relevant chemical equations.
Typically, disodium edetate is dispensed as 5 ml ampules containing 200 mg/ml of the API. For the analysis, an amount of solution containing theoretically 0.3 g disodium edetate $\times 2H_2O$ is dissolved in water. Hexamethylenetetramine (2 g) and dilute HCl are added to this solution, which is then titrated against 0.1 M lead nitrate solution in the presence of xylenol orange triturate as the indicator [3].
(c) How many milligrams of the API are there in one ampule?
(d) For each titration, the following volume of lead nitrate has been used:

9.0 ml	8.8 ml	8.6 ml

Calculate the real amount of disodium edetate $\times 2H_2O$ present in your sample. Express your answer in grams and moles.
(e) Critically discuss your result in context with the stated value for the API.
(f) Research the typically accepted error margins.

11.5.2 Dimercaprol

Dimercaprol injections can be used as for the treatment of heavy-metal poisoning. Your pharmaceutical analysis company has been contacted by an important client and asked to analyse a batch of solutions for injections containing dimercaprol. The chelating agent is typically dispensed in 2 ml ampules containing 50 mg/ml of dimercaprol. According to your brief, you are supposed to analyse the API, following standard quality assurance guidelines.

Typical analysis methods used for quality purposes are based on titration reactions. A certain amount of injection is dissolved in methanol. Diluted HCl and an iodine solution are added and the resulting solution is titrated against sodium thiosulfate [3].

(a) Research the type of titration described and how the endpoint is detected.
(b) Formulate the relevant chemical equations.
According to its label, a 2 ml ampule of dimercaprol contains 50 mg/ml. An amount of solution theoretically containing 0.10 g dimercaprol is dissolved in methanol. Dilute hydrochloric acid (20 ml, 0.1 M) is added together with 50.0 ml of a 0.05 M iodine solution. After 10 min, the solution is titrated against a 0.1 M sodium thiosulfate solution [3].
(c) How many milligram of the API are there in one ampule?

(d) For each titration, the following volume of sodium thiosulfate has been used:

16.95 ml	17.10 ml	17.00 ml

Calculate the real amount of dimercaprol present in your sample. Express your answer in grams and moles.

(e) Critically discuss your result in context with the stated value for the API.

(f) Research the typically accepted error margins.

References

1. *British national formulary*, British Medical Association and Pharmaceutical Society of Great Britain, London.
2. H. Moini, L. Packer, N. E. L. Saris, *Toxicol. Appl. Pharmacol.* **2002**, *182*, 84–90.
3. *British pharmacopoeia*, Published for the General Medical Council by Constable & Co, London.

Further Reading

E. Alessio, *Bioinorganic medicinal chemistry*, Wiley-VCH Verlag GmbH, Weinheim, **2011**.

F.A. Cotton, G. Wilkinson, C.A. Murillo, M. Bochmann, *Advanced inorganic chemistry*, 6th ed., Wiley, New York; Chichester, **1999**.

W. Kaim, B. Schwederski, *Bioinorganic chemistry: inorganic elements in the chemistry of life: an introduction and guide*, Wiley, Chichester, **1994**.

H.-B. Kraatz, N. Metzler-Nolte, *Concepts and models in bioinorganic chemistry*, Wiley-VCH [Chichester: John Wiley, distributor], Weinheim, **2006**.

J. D. Lee, *Concise inorganic chemistry*, 5th ed., Chapman & Hall, London, **1996**.

G. A. McKay, M. R. Walters, J. L. Reid, *Lecture notes. Clinical pharmacology and therapeutics*, 8th ed., Wiley-Blackwell, Chichester, **2010**.

R. M. Roat-Malone, *Bioinorganic chemistry: a short course*, Wiley, Hoboken, N.J. [Great Britain], **2002**.

E. R. Tiekink, M. Gielen, *Metallotherapeutic drugs and metal-based diagnostic agents: the use of metals in medicine*, Wiley, Chichester, **2005**.

G. J. Tortora, B. Derrickson, *Principles of anatomy and physiology*, 12th ed., international student/Gerard J. Tortora, Bryan Derrickson. ed., Wiley [Chichester: John Wiley, distributor], Hoboken, N.J., **2009**.

Index

Essentials of Inorganic Chemistry: For Students of Pharmacy, Pharmaceutical Sciences and Medicinal Chemistry,
First Edition. Katja A. Strohfeldt.

Companion website: www.wiley.com/go/strohfeldt/essentials

Printed in the United States
By Bookmasters